NO-CODE AI

Concepts and Applications in Machine Learning, Visualization, and Cloud Platforms

NO-CODE AI

Concepts and Applications in Machine Learning, Visualization, and Cloud Platforms

Min Soo Kang
Eulji University, South Korea

Sung Yul Park
LG U+, South Korea

Myung-Ae Chung
Eulji University, South Korea

Dong-hun Han
Eulji Medical Intelligence Information Center, South Korea

NEW JERSEY • LONDON • SINGAPORE • BEIJING • SHANGHAI • HONG KONG • TAIPEI • CHENNAI • TOKYO

Published by

World Scientific Publishing Co. Pte. Ltd.
5 Toh Tuck Link, Singapore 596224
USA office: 27 Warren Street, Suite 401-402, Hackensack, NJ 07601
UK office: 57 Shelton Street, Covent Garden, London WC2H 9HE

Library of Congress Cataloging-in-Publication Data
Names: Kang, Min Soo, author.
Title: No-code AI : concepts and applications in machine learning, visualization, and cloud platforms / Min Soo Kang, Eulji University, South Korea, Sung Yul Park, LG U+, South Korea, Myung-Ae Chung, Eulji University, South Korea, Dong-hun Han, Eulji Medical Intelligence Information Center, South Korea.
Description: Singapore ; Hackensack, NJ : World Scientific, [2024] | Includes bibliographical references and index.
Identifiers: LCCN 2024018829 | ISBN 9789811293887 (hardcover) | ISBN 9789811293917 (paperback) | ISBN 9789811293894 (ebook for institutions) | ISBN 9789811293900 (ebook for individuals)
Subjects: LCSH: Artificial intelligence--Computer programs--Popular works. | Artificial intelligence--Mathematics--Popular works. | Deep learning--Popular works. | Algorithms--Popular works.
Classification: LCC Q336 .K36 2024 | DDC 006.3--dc23/eng/20240617
LC record available at https://lccn.loc.gov/2024018829

British Library Cataloguing-in-Publication Data
A catalogue record for this book is available from the British Library.

For any available supplementary material, please visit
https://www.worldscientific.com/worldscibooks/10.1142/13858#t=suppl

Desk Editors: Sanjay Varadharajan/Amanda Yun

Typeset by Stallion Press
Email: enquiries@stallionpress.com

Preface

Artificial intelligence (AI) refers to the implementation of human intellectual capabilities, either partially or fully, within computers. It encompasses the development of Information and Communications Technologies (ICT) for the acquisition and utilization of high-level judgment and analysis outcomes through various information processing functions such as cognition, learning, reasoning, and understanding. Artificial intelligence technology and big data are being recognized as key drivers of the Fourth Industrial Revolution, which serve as major sources of national industrial competitiveness.

Although the term "Artificial Intelligence" was coined in 1956, AI has recently become more mainstream due to the increase in data volume, advanced algorithms, and improvements in computing power and storage. Early AI research in the 1950s began with exploring themes like problem-solving and symbolic methods, before evolving into decision support and smart search systems that have paved the way for the automation and formal reasoning we see in computers today; thereby supplementing and enhancing human capabilities.

Recently, AI has been used in various breakthroughs, from Deep Blue's victory in chess in 1997, AlphaGo's win in Go in 2016, the successful trial runs of autonomous vehicles, to telecom companies' AI speakers with voice recognition, and AI support in medical diagnostics. Furthermore, not just in the scientific and technological realms but also in Hollywood movies and science fiction novels, AI has been depicted and developed to offer numerous benefits across all industries.

Consequently, there's a growing trend of researchers in the field of AI, though unfamiliar terms and vague definitions for concepts such as machine learning (ML) and deep learning (DL) make it challenging from the start.

For those just beginning with AI, it might seem necessary to understand the history of AI, its definitions, logic, mathematics, and algorithms from an academic perspective. However, non-experts might find that a rapid application of coding and algorithms in their field is more engaging. Thus, the authors introduce ways for those already familiar with AI technologies to easily use programs in the cloud, and for those without expert knowledge to apply AI technologies, prompting the publication of this book.

This book is written for anyone interested in AI, explaining what AI is and how one can use their data to apply AI, even without prior knowledge of programming or mathematics. However, one must follow the instructions as provided in the book. Open-source users should read the book carefully to decide whether or not to choose open source. While open source is advantageous due to being free, applying necessary libraries or algorithms has limitations, often requiring more time to find or modify the desired source than investing in the actual research.

Comparing Windows/iOS and Linux, users can easily clarify this: Windows/iOS users purchase their operating system (OS) but receive various supports, and programs are developed based on that OS. In contrast, Linux users, though not purchasing, may find convenience and support limited. This description of OS is meant to aid understanding and should not be misconstrued.

Therefore, cloud-based AI programs from Microsoft Azure Machine Learning and Amazon Web Services (AWS) have been utilized. The book is structured into three parts:

Part 1 briefly introduces the history of AI, ML, DL, and algorithms, including how data can be utilized, i.e., data mining and big data processing for data science.

Part 2 covers Sage Maker from AWS and Azure Machine Learning from Microsoft, explaining their use and providing necessary examples. Sage Maker allows building, training, and deploying ML models on a large scale within a single Integrated Development Environment (IDE), offering tools such as notebooks, debuggers, profilers, pipelines, Machine Learning Operations (MLOps), and even methods to apply AI technologies without coding. Azure Machine Learning enables ML experts, data scientists, and engineers to use it in their workflows to train and deploy models and manage MLOps, facilitating the use of algorithms for supervised and unsupervised learning through an easy selection menu. The book illustrates each step, starting with Sage Maker and Azure Machine Learning, to completing example programs, including using real public and medical data. Thus, even beginners to machine learning can utilize cloud-based programs like Sage Maker and Azure Machine Learning if they read this book thoroughly.

Part 3 explains visualization types of Business Intelligence (BI), using programs such as AWS's Quicksight and Microsoft's Power BI to support better data-informed decisions by analyzing business data and converting it into actionable insights. It allows data preparers or administrators to ask questions as if speaking, receive relevant visualization materials using an ML-based engine, discover hidden insights from data, perform accurate predictions and what-if analyses, or add easy-to-understand natural language sentences to dashboards. Whether on the move or needing to create rich interactive reports, these tools swiftly connect, shape, visualize, and share data information, maintaining reports and connectivity to data anytime, anywhere.

Therefore, this book enables all readers — whether individuals starting with AI, those unfamiliar with AI, or those knowledgeable about AI — to develop better technologies and learn about visualization all at once, by simply applying AI.

About the Authors

Minsoo Kang is a professor in the Department of Big Data Medical Convergence at Eulji University, Korea, and he also serves as the Director of the Integrated Computing Center at Eulji University Medical Center. He participated in AI-related projects and research in the medical field in collaboration with the cardiology, Rheumatology, and Laboratory medicine departments at Eulji University Hospital from 2019 to 2021. Recently, he has been conducting AI-related research with the otorhinolaryngology department. Professor Kang obtained his Ph.D. degree in Control and Instrumentation Engineering from Kwangwoon University, Korea, and is currently conducting research on the Internet of Things and Artificial Intelligence.

He previously served as an adjunct professor at Hanyang University, Korea, and as a professor in the Department of Information and Communication Engineering at Hanyang Cyber University, Korea. He also served as the Director of the Radio Frequency Identification/Ubiquitous Sensor Network (RFID/USN) Program under the Ministry of Knowledge Economy in South Korea, overseeing the planning and evaluation of knowledge economy R&D projects. He currently also holds the position of President of the Korea Artificial Intelligence Association.

Prof. Kang has authored publications such as *Machine Learning: Concepts, Tools and Data Visualization* (English version), *Getting Started with Machine Learning using Microsoft Azure Machine Learning*, *RFID Basics* (a textbook approved by the Superintendent of Education in Seoul), *Digital Circuit Design Using VHDL*, and *RFID GL Qualification Test* (National qualification exam).

Park Sung Yul is a vice president of LG U+, Korea, and leads the B2B Service/Product Development Group at LG U+, overseeing all B2B business domains such as AICC, Data Center, Wired/Wireless Solutions, and Connected Car, catering to clients in the enterprise, Government, small and medium enterprises (SME), and small office/home office (SOHO) sectors. Under his leadership, the division manages an annual revenue of approximately ₩2.5 trillion.

Recognizing the transformative potential of artificial intelligence (AI), Park spearheaded the company's entry into the new AICC market in 2023, leveraging AI technology to disrupt the traditional IPCC (IP Contact Center) landscape. This strategic move resulted in approximately ₩30 billion in orders in 2023. This year, Park aims to expand the AICC market to over ₩500 billion by shifting away from price competition and towards differentiation through value-added AI services, especially for major financial services clients like insurance, banking and credit companies.

Park Sung Yul's strategic vision extends to expanding LG U+'s presence in the SME market, where its current market share stands at 40%, and the SOHO market, in which it currently holds a 10% market share. He has already introduced AI services, including the SOHO Store AI launched in October 2023 and is currently focused on enhancing the customer experience through additional value-added services.

Looking ahead, Park envisions applying AI technologies like large language models (LLMs) and small Large Language Model (sLLM) to develop chatbots, drive process innovation, and enhance legacy services across all B2B industry verticals. By integrating AI and prioritizing value-added services, he aims to deliver practical benefits to enterprises, governments, and all B2B clients.

Myeong-Ae Chung graduated from the Department of Chemistry at Ewha Womans University, Korea, and obtained her master's degree at the graduate school of the same university. After earning her Ph.D. in Polymer Physical Chemistry from the Clausthal University of Technology, Germany, she worked as a post-Doc at the same university.

She then served as a researcher at the Max Planck Institute for Polymer Research in Mainz, Germany, before returning to Korea to work as a researcher at the Korea Research Institute of Standards and Science. She moved to the Electronics and Telecommunications Research Institute in Korea, where she conducted research and developed biosensors in the medical field, worked on biometric recognition (iris, fingerprint recognition, etc.), and carried out projects on new functional medical devices. She has held positions such as the head of the Future Technology Research Laboratory.

Later, she served as the chairperson of Korea's Presidential Advisory Council on Science and Technology and participated as a member of various national science and technology committees, including the National Information and Communication Promotion Committee.

She is currently a professor in the Department of Big Data Medical Convergence at Eulji University, conducting research on artificial intelligence-based cardiovascular disease prediction and monitoring technologies.

Dong-hun Han is a researcher specializing in data analysis and machine learning. He is currently working at the Medical Intelligence Information Center, affiliated with Eulji University, Korea.

His projects have included the automation of manual inspection processes for organic light emitting diodes (OLEDs) through the use of artificial intelligence (AI), leading to increased yields, and conducting pseudo-3D cell simulations based on the unique Automatic Affine Generator (A^2G) algorithm, in collaboration with the Electronics and Telecommunications Research Institute (ETRI), Korea. He is actively involved in writing various academic papers and is currently working with the Institute for Information & Communication Technology Planning & Evaluation (IITP), Korea, on developing AI prediction models based on big data for various cardiovascular diseases.

Additionally, he is participating in a project in collaboration with the Department of Otorhinolaryngology at a university hospital, focusing on the self-diagnosis of sleep apnea. This project utilizes the STOP-BANG questionnaire and AI technology to create an enhanced self-diagnosis questionnaire.

Book Reviews

Prof. Dr. Andreas Offenhäusser, Forschungszentrum Jülich GmbH
Modern chemical research is no longer confined to traditional laboratory work; data analysis and modeling play a crucial role. In particular, the understanding and prediction of complex chemical reactions, such as material synthesis, are increasingly reliant on artificial intelligence (AI) technologies. This book opens up avenues for researchers without programming experience to access and utilize these technologies, making it an essential guide for those looking to apply AI in chemical research. Designed to be accessible to researchers with limited technical background knowledge, this book can significantly contribute to advancing research and innovation in the field of chemistry.

Prof. Dr. rer. nat. habil. Jörg Adams Institut für Physikalische Chemie
In modern society, as AI leads innovation in numerous fields, many individuals aspire to integrate it into their daily lives and work routines. AI is being utilized across all fields, making it desirable for non-experts to apply it within their own domains. This book doesn't require knowledge of programming. It is illustrated and provides step-by-step instructions, enabling readers to see the results of applying AI technology. Additionally, it allows for the construction of simple reports and dashboards using data. Therefore, I wholeheartedly recommend this book.

Prof. Dr. Kyung Hak Lee, Kwangwoon University
The book offers a unique opportunity for students and professionals in electronic engineering to explore AI without the barrier of needing advanced mathematical or programming skills. It encourages a multidisciplinary approach, showing that the future of technological innovation lies in the intersection of different fields of study. By providing a clear and accessible introduction to AI, along with practical examples of its application, this book is poised to become an essential guide for

those in electronic engineering and related disciplines looking to expand their skill set and explore new research horizons in the era of AI.

Ki Bong Yoo, Assistant Professor, Yonsei University
This book was written to help people who do not have math or programming skills to pick up machine learning. There is a part about visualization that makes it easier to understand the results of data analysis.

This book is considered highly beneficial for studying AI because it covers topics such as AI programs, visualization, and cloud computing.

Elle Lee, Nurse Practitioner, Optum Inc., USA
Sections discussing the role of AI in hospitals and its applications in areas as diverse as image and text data processing resonate with the needs of the health insurance industry today. These insights are invaluable for professionals at the intersection of healthcare delivery and insurance to anticipate and adapt to the evolving landscape of healthcare delivery and policy planning.

The book also provides a solid foundation for understanding how AI can transform data analysis and interpretation in insurance and healthcare by taking a systematic approach from AI fundamentals to advanced applications using BI tools. Mastering these tools, as the authors of this book have done, means not only improving operational efficiency but also enabling data-driven strategies to improve upon older statistical methods.

Marie Pradel, RN, Jefferson Hospital, USA
Sections discussing the role of AI in hospitals and its applications in various areas such as image and text data processing align with the current needs of the health insurance industry. These insights are crucial for healthcare practitioners at the crossroads of healthcare delivery and insurance, enabling them to foresee and adapt to the changing landscape of healthcare services and policy formulation. The book lays a solid groundwork for comprehending how AI can revolutionize data analysis and interpretation in the insurance and healthcare sectors. It adopts a structured approach, moving from AI fundamentals to sophisticated uses of BI tools. For healthcare practitioners, mastering these tools, as the authors of this book have achieved, signifies not merely enhancing operational efficiency but also fostering data-driven strategies to surpass traditional statistical methods.

Shin Eui Cho, Vice President, Opasnet Co. ltd., Korea (KOSDAQ)
The history of machine learning is short but well-organized and explains Microsoft's approach to the Microsoft Azure Machine Learning Studio program. Even people who don't know a programming language can easily program if they are interested in machine learning. In Part 2, an example program is written step by step with Microsoft Azure Machine Learning Studio. This example can be

followed by the reader in order to get results. In addition, it is expected that the use of real data will be helpful to researchers in data science and analytics.

Chun Oh Park, CEO, PNPSECURE
This book has the basic concept of AI, and even people without an engineering background can run AI programs. An AI program is possible, statistics can be analyzed with the R program, and visualization is also made so that the entire cycle from a collection of AI data to visualization can be easily progressed.

Geon Bok Lee, Executive Managing Director, Microsoft Korea IoT Solution Architect
While many books already explain AI, it was hard to find a book that is appropriate for developers in the age of AI's democratization. This book provides a systematic approach to making AI easy to use for developers who do not have a deep knowledge of AI statistics. Reading this book is the best way to gain the vivid experience and know-how of a professional who has worked with Microsoft for years.

Yeonghoon Jeong, Vice President Unit, LGU+, AI/DX Business
To learn AI technology, one acquires skills through online courses and tutorials, projects and practice, as well as books and resources. Based on these learning methods, this book is recommended as it covers AI theory, AI programs (including programs without coding), and business intelligence (BI), allowing readers to learn AI technology comprehensively in one go.

Byoungjoon Kim, AWS, Account Executive/Higher Education Sales Team
This book provides easy-to-understand explanations of AI technology for those who are not proficient in math or coding. It also explains SageMaker in detail, a platform where even those without basic knowledge of AI can expect results from applying AI technology. Additionally, it covers QuickSight for visualization, offering detailed explanations and visualizations. By also explaining global companies' AI and visualization content, it helps readers understand the advantages of the programs. Furthermore, it includes explanations about the cloud, making it a useful book for non-specialists in AI to study easily.

Min Sun Kim, CEO, Korea Institute of Business Analysis & Development (KIBA)
AI is being applied in all fields of the industry, and especially in the security field. The AI described in this book is well organized so that people who don't know programming or math can understand it. By the end of this book, even those who have not studied AI in the field of engineering will be able to utilize AI.

Contents

Part 1

Introduction to Artificial Intelligence

Chapter 1

Overview of Artificial Intelligence

1.1 Definition of Artificial Intelligence

Artificial intelligence (AI) has several definitions, including "intelligence made from machines" and "artificial embodiment of some or all of the intellectual abilities possessed by humans." The various definitions of AI can be summarized as "to enable human intelligence to be equipped with machines by understanding human judgment, behavior, and cognition." Intelligence is said to be "the ability to apply prior knowledge and experience to achieve challenging new tasks." This can be said to refer to human intellectual ability. This ability can be used to respond flexibly to a variety of situations and problems and is also related to learning ability. The dictionary definition of the term "learning ability" is the ability of an individual to acquire and understand new knowledge, skills, experiences, etc. This is associated with various cognitive processes such as intellectual ability, memory, problem-solving skills, and more. Learning ability plays a crucial role in acquiring, understanding, applying, and utilizing new information when necessary. In the field of AI, learning ability is defined as the adaptability of a system to efficiently and effectively solve a problem, based on modifying and enhancing its knowledge derived from the inference process after performing a task. This involves incorporating experiences gained during the inference process into the system' s knowledge for improved performance in solving similar problems or the same problem in the future. In other words, it refers to the ability to learn, and having excellent learning ability means being able to grasp and understand content that others may find difficult to learn. It also implies the capability to learn either the same content more quickly or to learn it more broadly in terms of practical application. Therefore, intelligence is generally defined as the ability to solve problems. We need advanced intelligence to solve various problems. For example, when you play chess, you will use information learned through many documents, many chess games, and the Internet to make predictions and then take action. Such actions require high human intelligence, and showing the results learned based on this intelligence stems from thinking. If we were to scientifically explain thinking, it could be broadly referred to as consciousness at a higher level and in a broader sense, it is akin to "experience" related to a subject. Experience is also a process of knowledge, and a person who has had considerable experience in a particular field may gain a reputation as an expert. Thus, intelligence is the recognition, analysis, and understanding of the thoughts and experiences that human beings may have, and the AI that makes it artificial is AI. To create a machine that thinks like humans, it is necessary to study human thoughts and behaviors, including listening, speaking, seeing, and acting. It seems very easy, but having the ability to listen, speak and act on what a computer has heard, and seen is a very difficult task, so it would be the fastest to emulate what humans think. Therefore, as a scientific approach, we need human-like thoughts and actions. To put this into practice, research should investigate whether repetitive learning, akin to a young child's experiential learning that leads to thinking and acting on their own, can enable self-thinking and self-action. Prof. Kathleen McKeown has four main categories of human thought and behavior.

1.1.1 *Systems that think like humans*

Cognitive modeling goes beyond theoretical concepts and represents the process of autonomously interacting with the surrounding environment. This is utilized in expert systems, natural language processing, neural networks, robotics, and the field of AI. This is a cognitive modeling approach with a system that can make similar human-like thinking and decision-making as defined by Hodgeland in 1985 and Bellman in 1978. The goal is to develop computer models that mimic the cognitive processes of human thinking. This is associated with AI and involves modeling human thought and decision-making abilities to enable computer systems to perform tasks and make decisions in a manner similar to humans. Therefore, to realize systems that think like humans, various AI technologies such as machine learning, deep learning, natural language processing, computer vision, and others need to be researched and developed.

1.1.2 *Systems that think rationally*

This is a law-of-thought approach to thinking through a system with mental abilities such as perception, reasoning, and behavior through a calculation model,[1] as

[1] Rules of Thought approach: An approach in which AI systems model human thought processes for problem solving and decision making.

defined by Charniak and McDermott in 1985 and Winston in 1992. There are two main obstacles to this approach.

First, it is not easy to take informal knowledge and state it in the formal terms required by logical notation, particularly when the knowledge is less than 100% certain. Second, there is a big difference between being able to solve a problem "in principle" and doing so in practice. This is because it is one of the approaches aiming to model human logical reasoning processes. These systems are designed to solve specific problems and make optimal decisions based on given information and rules.

1.1.3 *Systems that act like humans*

This is a Turing test approach,[2] defined by Kurzweil in 1990 and Rich and Knight in 1991, with a system that allows the machine to mimic any action that requires human intelligence. Alan Turing's proposed test in 1950 is the ability to determine if a machine has intelligence based on how similar a machine can communicate with a human, that is, to achieve the same level of performance as a human. In other words, it involves placing humans and machines in separate spaces, and the questioner engages in a conversation with the machine through text. The questioner is unaware of who or what is in the other space. If the questioner cannot distinguish between the machine and the human through the conversation, the machine is considered to have intelligence.

1.1.4 *Systems that act rationally*

This is a reasonable agent approach with an agent system that acts intelligently through a calculation model, defined by Schalkoff in 1990 and Luger and Stubblefield in 1993. The term "agent" literally means "representative" or "proxy." In the field of computer science, it refers to intelligent software or hardware that performs tasks on behalf of a person. Search engines and web crawlers that autonomously retrieve information are prime examples. A rationally agent is one that acts to get the best results and, in the case of uncertainty, the best-expected results. Rational behavior is the mechanism of inference to achieve set goals in a way that yields rational results. This is because it involves actions taken after careful thought processes, as opposed to impulsive or instinctive behavior. In conclusion, AI is a technology or science field that studies the methodology or feasibility of making such intelligence by implementing human cognition, reasoning, and learning from machines into computers or systems.

---

[2]The Turing Test: An experimental approach to determining whether a computer has human-like intelligence.

1.2 History of Artificial Intelligence

1.2.1 *The beginning of artificial intelligence*

The birth of AI began in 1943 when Warren McCulloch and Walter Pitts studied AI. The reasons for their study of AI were three motives: knowledge of physiology and neuronal function in the brain, formal analysis of propositional logic according to Russell and Whitehead, and the Theory of Computation of Turing. As a model of AI, they proposed a connection model in which neurons[3] process information by synapses.[4] A synapse refers to the junction between the axon terminal, which is the end of the axon that propagates stimuli outside the cell, and the gap between neurons where neurotransmitters are exchanged. It is marked on or off when another connected neuron stimulates one neuron. This model represents the working relationship between neurons, which theoretically proves that connecting the artificial neural network in the form of a net can mimic the simple function that operates in the human brain. The neuron is the first neuron model of the neural network conceptually defined in terms of AI as "proposition that provides sufficient stimulation." McCulloch and Pitts also argued that the concept of learning is needed for neuron-connected networks, and in 1949, Hebb defined it as "Hebbin learning rule" Later, Alan Turing announced a Turing test in 1950 on the feasibility of implementing the machine, and in 1951 on the Ferranti Mark 1 machine at Manchester University, Christopher Strachey and Dietrich Prince launched a chess program.

Subsequently, in 1959, Arthur Samuel, "a field of research that develops algorithms that allow machines to learn from data and execute actions that are not explicitly specified by code," is enough technological advancement to define and challenge machine learning. Achieved. Figure 1.1 illustrates the history of AI.

In the summer of 1955, John McCarthy, then associate professor of mathematics at Dartmouth College, decided to organize a group to clarify and develop thinking about thinking machines. In the early 1950s, there were various names for the field of "thinking machines:" cybernetics, automata theory, and complex information processing. However, instead of focusing on automata theory, John McCarthy chose a new name, AI. On September 2, 1955, the project was formally proposed by McCarthy, Marvin Minsky, Nathaniel Rochester and Claude Shannon.

The proposal is credited with introducing the term "AI." The 1956 Dartmouth AI conference gave beginning to the field of AI and gave succeeding generations of scientists their first sense of the potential for information technology to be of benefit to human beings in a profound way (Figure 1.2).

---

[3] Neuron: A processor with only simple computational functions designed based on neuroanatomical facts.

[4] Synapse: A connection point that transmits signals from one neuron to another.

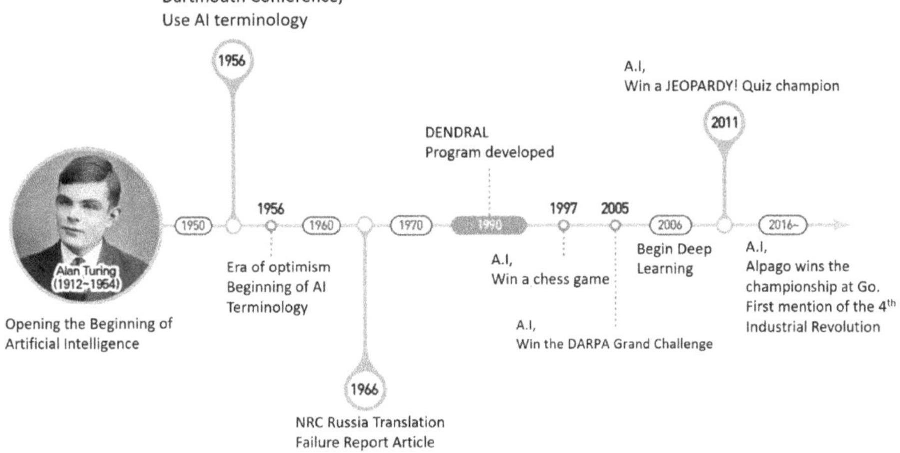

Figure 1.1 History of AI.

Figure 1.2 Dartmouth conference participants.

Proposal

A Proposal for the Dartmouth Summer Research Project on AI

In the summer of 1956, we propose conducting a 2-month, 10-person AI research project at Dartmouth College in Hanover, New Hampshire. The research is driven by the conjecture that all facets of learning and various characteristics of intelligence can be theoretically described with precision to enable machines to simulate them. Based on this conjecture, the study aims to progress in several areas. The focus will be on enabling machines to use language, create abstractions and concepts, solve problems that are currently reserved for humans, and develop methods for self-improvement. We believe that by making a deliberate effort to establish a group of carefully chosen scientists to collaborate over the summer, substantial advancements can be achieved in one or more of these areas.

The proposal discusses computers, natural language processing, neural networks, theory of computation, abstraction and creativity. However, after Dartmouth's workshop, AI did not achieve much, but later it became a direct opportunity to study various areas of AI, and in fact, it set the stage for continuous research at MIT, Carnegie Mellon University, Stanford University, and IBM in the United States.

1.2.2 *The early days of artificial intelligence*

The development of computers and programming tools for AI began to have primitive but simple computing power. At this point, Nowell and Simon have been very successful in developing a system called the General Problem Solver[5] that can change many of the factors in the calculations to reduce the difference between the desired and expected results. GPS was a program that modeled the process of humans solving problems, unlike Logic Theorist, which was based on logic. Logic Theorist was the first program to employ a non-numerical approach to thinking, combining search, goal-oriented behavior, and the application of rules. It modeled the reasoning process of humans who think logically.

GPS employs the Means-Ends Analysis method and the use of recursive procedures within the system. Means-Ends Analysis is a technique that instills confidence that the application of an operator has a purpose if it is applied during the process. GPS became the first program in AI capable of thinking in a manner similar to humans. The cognitive science approach, suggesting that AI and human cognitive abilities will eventually be similar to computers, is the foundation of interdisciplinary research in fields such as biology, linguistics, psychology, and computer science. McCarthy, who announced the beginning of AI, made three critical achievements when he moved from Dartmouth to MIT in 1958.

First, developing the representative LISP of AI program language. Second, Time Sharing[6] was introduced, which began to solve the problem of the expensive computer using problems. At the time, the fee for using a computer was high, so those other users could take turns using one computer at the same time for a short time. Third, in 1958, he published his first book, "Programs with Commonsense," which led to the development of the first complete AI program called "Advice Taker."[7]

---

[5] General Problem Solver (GPS): A program created in 1959 by John Cliff Shaw, Herbert Simon, and Alan Newell that can simulate the human problem-solving process with a computer program.

[6] Time-sharing system: A form of operating system that allows a computer to be used by multiple users simultaneously.

[7] Advice Taker: A system that interacts with its environment to acquire and reason about knowledge.

1.2.3 *The artificial intelligence slump*

From the outset, AI researchers predicted success, and even as early as 1957, Herbert Simon stated that "Machines will be capable, within twenty years of doing any work a man can do." However, at that time, AI was limited to solving only simple problems. In almost all cases, however, the early systems had to choose from a broader range of problems and hit limitations when they became more difficult. The reasons are, first, early AI programs only succeeded in simple grammatical manipulation because they had little knowledge of the subject. The National Research Council in the United States financed the translation of the Russian language, but it was difficult to resolve the ambiguity of the language and to understand the content of the sentence. Eventually, in 1966, the Advisory Committee issued a report saying that "there is no machine translation of the general scientific text and no immediate outlook." The government's financial support for academic translation projects was then canceled. Second, the problems that AI was trying to solve were, and are, complicated. The initial AI programs were very simple and could discover specific solutions to problems, but complex AI programs were not possible.

Therefore, at the time, it was hoped that a better computer machine or memory would be available. However, as the Computational Complexity Theory developed over time, it was found that solving a problem by one program did not linearly scale in time and space, depending on the difficulty of the problem. Third, there were limitations in the basic structure used to generate intelligent behavior. For example, according to a 1969 book by Minsky and Papert's Perceptron, learning impossible facts proved to be a problem in itself. Over the years, the Back-Propagation[8] learning algorithm for multi-layer neural networks was introduced to solve this problem.

1.2.4 *Activator of artificial intelligence*

There were several reasons why AI technology faced difficulties in advancing, but one of the main issues was that the problem-solving methods were just general-purpose search methods. Therefore, there was a lack of knowledge about the problem to be solved, which inevitably led to a decrease in performance when dealing with complex issues. This approach is called "weak methods," which, although common, would have been difficult to solve for larger or more difficult problems. The solution was to use more powerful domain-specific

[8] Back propagation: An algorithm that updates each variable by propagating the error from the output layer to the input layer by comparing the output value of the model with the expected value and updating the weights in the direction that reduces the error.

knowledge that makes it easier to deal with cases that typically occur and allows for more significant inference. Therefore, if someone tries to solve a difficult problem, they realize that they can only explain it efficiently by using someone who already knows the answer — in other words, expert knowledge. Buchanan's "DENDRAL" program announced at this time is a prime example. In 1965, DENDRAL emerged as the first practical application of AI. Developed by Edward Feigenbaum in the field of chemistry to assist geneticist Joshua Lederberg in the study of extraterrestrial life, DENDRAL was designed to infer the structure of organic molecules. It successfully applied the concepts of expert systems, converting the expertise, experience, and knowledge of human experts in the field of chemistry into logical rules to solve problems. In other words, it employed a technique of analyzing the "know-how" of experts, converting it into rules and expressing it within the framework of expert systems. DENDRAL proposed the most likely plausible candidate structures for new and unknown chemicals, and their performance was competing with human experts in some classes of organic materials. The importance of DENDRAL is probably the first successful knowledge-intensive system.

1.2.5 *The rise of artificial intelligence*

Following the success of DENDRAL and the achievements of expert systems, AI began to regain recognition. Companies that experienced successful applications of expert systems started to adopt and apply them in various domains. R1, which was utilized at Digital Equipment Corporation (DEC), was a commercially successful expert system designed to assist in coordinating orders when introducing new computer systems. Until 1988, DEC's AI group employed an additional 40 expert systems, and by 1986, it was reported that they were able to save approximately $40 million annually. Since then, neural networks have begun to flourish and it can be seen that neural networks have been applied in various fields. At the end of the 1970s, neural networks were abandoned, but work continued in other fields. Following that, Dr. John Hopfield developed the backpropagation algorithm, leading to the widespread adoption of neural network models. This was inspired by physical spin models, and the Hopfield network introduced a difference from existing algorithms by using fixed weights to compute complete information. In other words, applying neural networks to the network to predict communication amounts in advance, minimizing communication load, and operating efficiently. Recently, a great deal of research has been conducted on intelligent agents with intelligence in agent theory.

In particular, smart mobile systems in information retrieval systems are very active, and progress is being made and applied to numerous industrial operations around the world.

1.3 Classification of Artificial Intelligence

AI is divided into strong AI and weak AI from a philosophical point of view. Strong AI is the study of creating computer-based AI that can think and solve problems on its own, while weak AI cannot think or solve problems on its own. In other words, strong AI is capable of self-study, judgment, and execution, while weak AI is based on rules or predefined algorithms.

1.3.1 *Strong artificial intelligence*

1.3.1.1 *Artificial General Intelligence*

In the term "Artificial General Intelligence," "general" should be understood as meaning "universal" rather than "ordinary." In other words, unlike AI that can be applied only under specific conditions, it refers to AI that can be used and purposed towards all situations. Artificial General Intelligence, unlike weak AI, means to learn and do things others do, despite having never tried before. It is a form where self-learning by watching, listening, and imitating is possible. This is what artificial general intelligence, which is the goal that AI researchers are aiming for to make machines behave like human beings, is all about. Ultimately, a key goal in AI research is to develop machines with intelligence capable of successfully performing any intellectual task that humans can do. This vision is a significant theme in movies and futurology, reflecting the desire to achieve advanced AI that can emulate various aspects of human cognitive abilities.

1.3.1.2 *Artificial consciousness*

Artificial consciousness is one of strong AI, and is also called machine consciousness or synthetic consciousness. It can be called the next form of artificial general intelligence. If AI analyzes and understands ordinary objects, Artificial Consciousness goes beyond it to imitate emotions, the ego, creativity. It actively uses these elements, and judges whether they are suitable for the environment or manipulates itself, performing tasks it deems necessary, even those not commanded. In essence, artificial consciousness can be described as AI capable of performing all actions based on its own judgment. Moreover, the general criterion is self-awareness. The literal meaning of "conscious" is stated as "stimulating one's environment, one's existence, senses, and thoughts." According to the 1913 Cambridge English Dictionary, "conscious" is defined as "being aware that a particular thing or person exists" while Webster's Dictionary defines it as "possessing knowledge, whether by internal, conscious experience or by external observation". While self-awareness is crucial, it might be challenging to objectively verify. Therefore, artificial consciousness implies a sentient being similar to or surpassing human intelligence, as opposed to weak AI, which is merely a tool executing given tasks on a computer. It may be thought that AI can reach artificial consciousness when it develops, but at this point, it should be considered as a hypothesis.

In the end, whether or not it is possible to implement a consciousness that thinks like humans mechanically, and whether a human-like machine can be built theoretically can be a significant ethical debate.

1.3.2 *Weak artificial intelligence*

Weak AI is defined as research in creating computer-based AI that cannot think or solve problems on its own but addresses specific issues through defined learning. Although they are not really intelligent, nor possess intelligence, developing programs that mimic intelligence using a set of predefined rules seems to be intelligent behavior. Research in the field of strong AI is what today's AI researchers aim for, but it can still be considered close to impossible at this point. A significant amount of progress has been made in the field of weak AI, depending on the point of view of the goal. Google's AWS, DeepMind's AlphaGo and IBM's Watson are all rule-based AI, in other words, weak AI. Optimized AI can be described as AI technology that solves problems or handles tasks in a specific domain. Weak AI requires input of algorithms, basic data, and rules. It is characterized by the need for guidance, meaning it must be supervised. With the implementation of "learning" based on guided inputs and over time producing better outputs, programs that surpass or show similar performance to humans in limited domains have emerged. However, this still falls short of replicating a small fraction of human learning capabilities, making it challenging to view these entities in the same light as humans. Nevertheless, the development direction of weak AI can be considered highly positive. This is because there is no inherent reason for AI to be confined to imitation of human capabilities. With already achievable functionalities, AI has the potential to far surpass human abilities, especially in the aspect of solving given problems, which might even excel beyond strong AI in certain aspects. Today's weak AI is actively used in medicine, management, education, and services, and there is a limit that only a limited part of human cognitive ability can think and operate, unlike strong AI that can think for itself like a human.

Computers possess overwhelming advantages compared to humans in terms of calculation speed, memory capacity, accuracy, and diligence. Already, computers exhibit exceptional performance in tasks such as voice and image processing, augmented reality (AR), virtual reality (VR), and the metaverse. They can handle astronomical amounts of data, recall it rapidly, and provide uninterrupted services to thousands of users 365 days a year. The vision of ubiquitous computing proposed by Mark Weiser is becoming a reality. The current state of AI involves the implementation of relatively intelligent behaviors or decisions based on predefined algorithms and vast datasets. While AI can autonomously discover rules to solve problems, understanding why it made specific decisions remains challenging. Additionally, AI is confined to solving problems within limited scopes. In this context, the majority of AI created by humans can be classified as weak AI.

1.4 Practice Questions

Q1. What is AI?

Q2. Prof. Kathleen McKeown divided human thoughts and behaviors into four systems. Describe the four systems.

Q3. Describe the proposal for the Dartmouth workshop.

Chapter 2

Machine Learning

2.1 Definition of Machine Learning

IBM notes that machine learning is a field of artificial intelligence (AI) and computer science that focuses on the use of data and algorithms to mimic the way humans improve accuracy progressively through learning. Meanwhile, SAP states that machine learning is a subset of AI (AI) that focuses on training computers to learn from data and improve through experience, instead of explicitly programming them for learning and improvement. To summarize, machine learning is a field of AI that focuses on developing algorithms and techniques to enable computers to learn. It emphasizes representation and generalization. Representation refers to the evaluation of data, while generalization pertains to the handling of data that is not yet known. In other words, representation can be understood as the analysis and evaluation of the results using data whose origin is unknown, and generalization can be described as the process of making this generated data usable. The term "machine learning" was first used by Arthur Samuel, an IBM researcher in the field of AI, in his paper "Studies in Machine Learning Using the Game of Checkers." The machine is a computer that can be programmed and can be used as a server. Machine learning has been studied in three approaches. The first is the neural model paradigm. Neural models start from perceptron[1] and are now leading to deep learning. The second is the learning paradigm of symbol concepts.

This paradigm uses logic or graph structures instead of numerical or statistical theories and has been the core approach of AI from the mid-1970s to the late 1980s. The third is the intensive paradigm of modern knowledge. The paradigm, which began in the mid-1970s, started with the theory that knowledge that has

[1] Perceptron: A type of artificial neural network, an algorithm that models the learning function of the brain.

already been learned should be reused without the neural model of starting learning in a new state. These three methods are very representative models in AI research. However, entering the 1990s, the focus shifted from the traditional approach, which emphasized the learning methodologies of computers, to more practical machine learning research aimed at solving real-life problems, which has become the mainstream.

In the 1990s, the paradigm of machine learning was closer to computer based. Data mining,[2] which analyzes data from a statistical perspective, and much of it theoretically shared, and the ease of securing digital data due to the rapid spread of high-performance computers and the spread of the Internet also had a lot of influence on this movement. Arthur Samuel describes machine learning as a field of study that gives computers the ability to learn without being explicitly programmed. Following this, Tom Mitchell[3] provided a more precise definition of machine learning. Tom Mitchell said, "A computer program is said to learn from experience E with respect to some class of tasks T and performance measure P, if its performance at tasks in T, as measured by P, improves with experience E." For example, when a computer is trained to recognize letters, the machine classifies the newly typed letters (T) and experiences learning from pre-processed data sets (E). To do (P) is to say that the computer "learned." Ultimately, it is a way to improve the performance of a particular task through experience. Machine learning is often mixed with data mining, probably because it uses the same methods in data mining, such as classification and clustering. So, let's briefly look at the differences between data mining and machine learning. Machine learning involves using algorithms like classification and clustering to discover phenomena and characteristics in data, and to find predictive values for new data in order to solve problems. From a statistical perspective, the use of algorithms like classification and clustering to solve problems or to discover phenomena and characteristics in data can be referred to as "data mining." The mix of machine learning and data mining began in the 1990s when computer scientists drew on cases handled in statistics in their search for more efficient solutions to practical machine learning research. The main difference between machine learning and data mining is that data mining focuses on discovering attributes, which are characteristics of data, and machine learning focuses on prediction using attributes known through training data.

---

[2] Data mining: The process of processing and exploring large amounts of data to obtain valuable information.

[3] Tom Mitchell (1988): A computer program is said to learn from experience E with respect to some task T and some performance measure P, if its performance on t, as measured by P, improves with experience E.

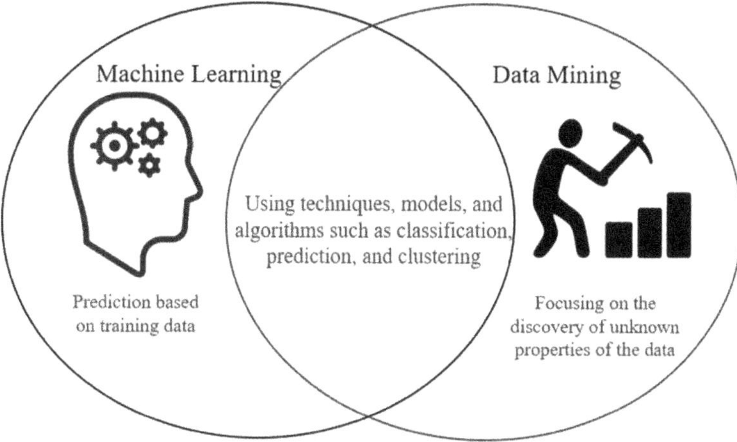

Figure 2.1 The difference between machine learning and data mining.

Figure 2.1 shows that machine learning and data mining use the same algorithm, but computer science focuses on prediction, and machine learning focuses on pattern discovery.

2.2 Classification of Machine Learning

Machine learning distinguishes between supervised learning and unsupervised learning, depending on whether or not the learning data has a label.[4] The biggest difference between supervised and unsupervised learning is whether labels are provided. Labels can be considered as objects (like people, cups, cats, dogs, etc.). This involves defining the attributes of the training data for learning. When labels are provided, it is classified as supervised learning, whereas the method without providing labels is classified as unsupervised learning. Figure 2.2 shows the classification of supervised and unsupervised learning.

Supervised learning, as the name suggests, involves training the machine using data that has a known answer. It means teaching the machine by giving input values and informing it about the corresponding labels for those values. For example, this method of learning involves showing photos of people and animals and telling the machine "This is a person and this is an animal."

Therefore, it is easy to determine whether the machine has correctly identified the answer or not. In supervised learning, there are two main types: classification and regression. Classification is a typical example of supervised learning, where given data is categorized according to predefined categories. Regression involves

[4] Label: Symbols and characters used to manage, process, or identify items, records, files, etc.

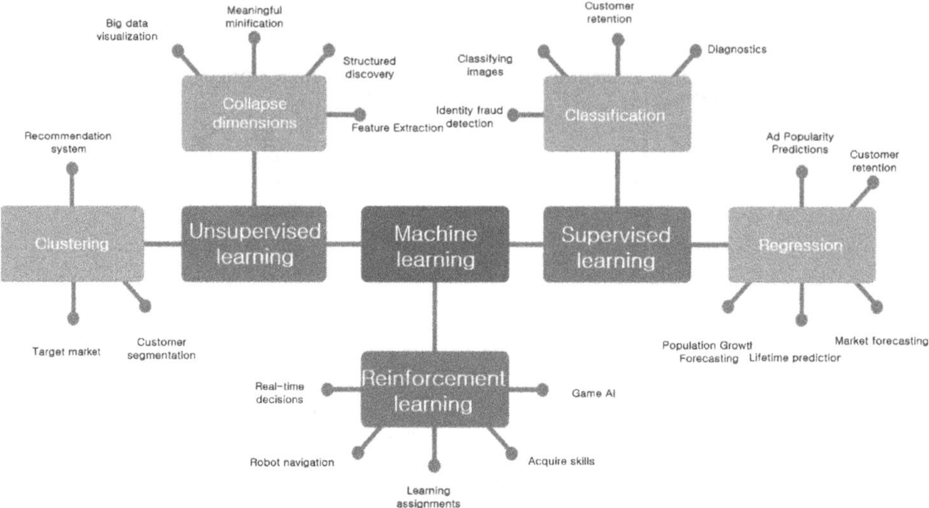

Figure 2.2 Supervised learning and unsupervised learning classification.

Table 2.1 Classification and regression in supervised learning.

Classification	Binary classification	Classifying some data into one of two categories.
	Multiclass classification	Classifying some data into one of several possible values.
Regression	Predicting values based on the characteristics of some data (the result can be a real number).	

predicting continuous values based on the features[5] known as predictor variables of some data, and it is often used for predicting patterns, trends, or tendencies.

Table 2.1 provides an easy-to-understand distinction between classification and regression.

To further explain supervised learning, we first recognize objects by using our senses, such as seeing or hearing them. Thus, to enable computers to recognize objects like humans, we need to provide them with images or sounds. If we provide photos, these photos are considered data for learning, and objects in real life or in the photos such as "cup," "desk," "bicycle," "cat" are referred to as labels. These objects, defined by humans upon viewing the objects or photos, become lessons learned by the computer when it reads labeled photos. Conversely, if the input data does not have labels, it is considered unsupervised learning because the

[5]Feature: In the field of machine learning, a characteristic of data that makes it better at classifying or predicting the value of something.

computer has not received any guidance from humans. Therefore, there are two types of learning: supervised learning, which involves teaching by presenting something, and unsupervised learning, which involves learning solely from phenomena without any presentation. Supervised learning includes classification and regression, while unsupervised learning encompasses various models like clustering, *K*-means, and dimensionality reduction.

2.3 Supervised Learning

Supervised learning involves learning from labeled data. Since the outputs are clear based on the inputs in the training data, this learning method involves using a given model to train on the data, progressively reducing the difference between the predicted values and the actual values when new data is input. In essence, supervised learning is a machine-learning technique that trains models using data with assigned labels. Supervised learning is primarily divided into regression and classification, with regression analysis being a notable example. Both classification and regression are types of supervised learning models, sharing the commonality of learning from input data with labels. The difference between classification and regression lies in the fact that classification has fixed outcome values, while regression can yield any value within the range of the dataset.

Thus, the outcome of classification is one of the labels included in the training dataset, while the outcome of regression is an arbitrary value calculated using a function determined by the training dataset.

Figure 2.3 represents the overall process of supervised learning. In supervised learning, the correct answers are provided during the learning process. The dataset contains variables, where the variable to be predicted is called the dependent variable, and the variables used to predict the dependent variable are known as independent variables. Here, the input is the Feature of the data, and the output is the respective label or Target variable. The goal of supervised learning is to train a mapping function[6] that can predict outputs for new input data. This mapping function is represented by a mathematical model or algorithm that uses input data as inputs and generates output data as outputs, minimizing the error between predicted and actual outputs. This error is typically measured using a loss function,[7] which calculates the difference between the predicted and actual outputs. Supervised learning can be used for various tasks like classification and

[6] Mapping function: A one-to-one correspondence of individual elements in a series or dataframe to a specific function.

[7] Loss function: A function for comparing the difference between the algorithm's predicted value and the actual correct answer when supervised learning.

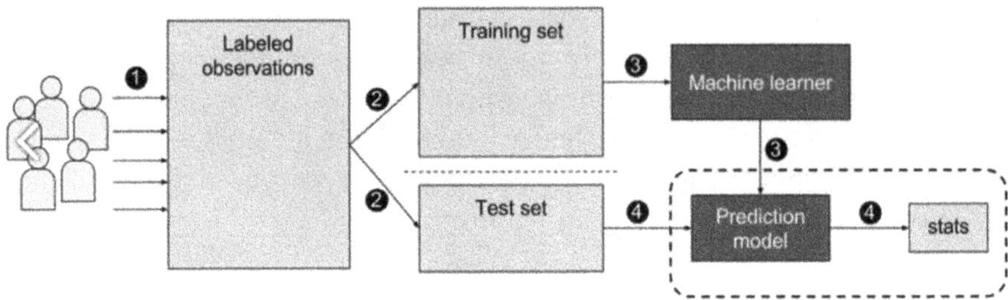

Figure 2.3 Supervised learning.

regression, where the goal of classification is to predict discrete labels or classes[8] for the input.

For example, classification may involve distinguishing images as a dog or a cat, whereas regression deals with continuous values of the dependent variable (predicted value). For instance, predicting a student's exam score based on study hours, forecasting the rise or fall in house prices due to factors like proximity to a subway station or school district, or predicting a company's transaction volume or sales can all be examples of regression.

Ultimately, a regression problem can be defined as predicting through real-number variables where the outcome values are continuous. The difference between classification and regression is that in classification, the outcome values are fixed within the training data, whereas in regression, the outcome values lie within the dataset's range. In other words, the result of classification is one of the labels included in the training dataset, while the result of regression is an arbitrary value calculated by a function determined by the training dataset. The differences between regression and classification are illustrated in Figure 2.4.

Table 2.2 presents a comparison between classification and regression.

More specifically, a classification model aims to identify the group to which new input data belongs, after being trained on labeled training data, hence the result of a classification model must be one of the labels from the training data. For example, if there is a dataset labeled with bicycle, cat, dog, book, or desk, then the result of the classification model must be one of these five labels.

Therefore, if there are two groups to be classified, it is called binary classification, and if there are three or more groups, it is called multinomial classification.

In statistics, Regression Analysis is a method of analysis that constructs a model between two variables for observed continuous variables and then measures its fit. Thus, a regression model uses labeled training data to express the

---

[8]Class: In object-oriented programming, a kind of framework that defines variables and methods to create a particular object.

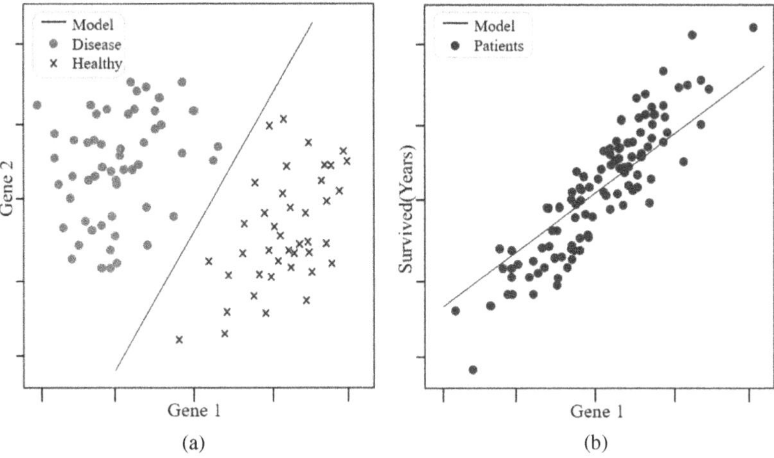

Figure 2.4 Comparison of (a) classification and (b) regression.

Table 2.2 Comparative analysis of classification and regression.

Basis of comparison	Classification	Regression analysis
Foundation	Discovery of models or functions where the mapping of objects occurs within predefined classes	Conception of models where the mapping of objects is done through values
Prediction in data	Discrete values	Continuous values
Algorithms	Decision trees, logistic regression, naive Bayes, *K*-nearest neighbors, support vector machines, etc.	Regression trees, random forest, linear regression, etc.
Method of calculation	Measurement of accuracy	Measurement of mean squared error

relationship between features and labels through a function, identifying specific patterns and predicting continuous values within those patterns. Therefore, regression analysis is also referred to as a predictive model for estimating continuous values. A regression model can also be applied in cases like classification, where it predicts several categorical outcome values. In statistics, such cases use a regression model known as logistic regression. Logistic regression, utilizing the "S"-shaped logistic or sigmoid function, shows excellent performance in binary classification within the [0, 1] input data range. In supervised learning, the main models are classification and regression, and related algorithms include *K*-nearest neighbors (KNN), support vector machine (SVM), decision tree, linear regression, logistic regression, random forest, neural networks, among others, which are some of the representative methods to be explored.

2.3.1 *Classification*

Classification refers to the process of recognizing, differentiating, and understanding a concept or subject, implying that a subject is in a category, and usually has a specific purpose. So classification is essential for all kinds of interactions concerning language, speculation, reasoning, decision-making, and the environment. In machine learning, classification is a representative technique of data analysis, which classifies objects having multiple attributes or variables into one of a predetermined group or class. The classification model is divided into various models according to algorithms. Among them, KNN, SVM, and decision tree models are representative.

2.3.1.1 *K-nearest neighbors*

The KNN model used in the classification is one of the most intuitive and simple of the machine learning models. This algorithm classifies new input data into specific values by finding the K-nearest data points to the current (new input data) and classifying it into the category that appears most among these K points. The KNN algorithm is a non-parametric method[9] used for both classification and regression, and in both cases, the input consists of the k nearest training data points within a certain space. KNN can be considered as part of the learning and classification process, where, given existing training data and a new data point x, the result for x is determined using the data points nearest to x. By observing data on a graph, it can be determined whether the problem should be addressed through classification or regression. This means it can be applied using either classification or regression methods.

As can be seen in Figure 2.5, (a) can clearly appear as a problem of classification. In Figure 2.5, (b) may seem like a problem of regression. In any situation, it's about finding the nearest K data points to the new data and determining its characteristics. The object is classified by the majority as the one assigned to the most common item among its KNN. Therefore, it can be advantageous to measure K as an odd number. In Figure 2.5, the characteristics of new data are determined by the majority rule, based on the number of nearest K neighbors. This means, as in Figure 2.6, if $k = 3$, the characteristics are of triangles, and if $k = 8$, it also takes on the characteristics of triangles, i.e., the characteristics of Class B.

The case of Figure 2.7 involves a regression problem. Calculating the distances to determine which point the new data is closest to, $X2$ ends up having the closest three data features. In KNN Regression, the output is the feature values of

[9]Non-parametric method: In statistics, a statistical test that makes no assumptions about parameters and calculates probabilities directly from the given data, regardless of the shape of the population.

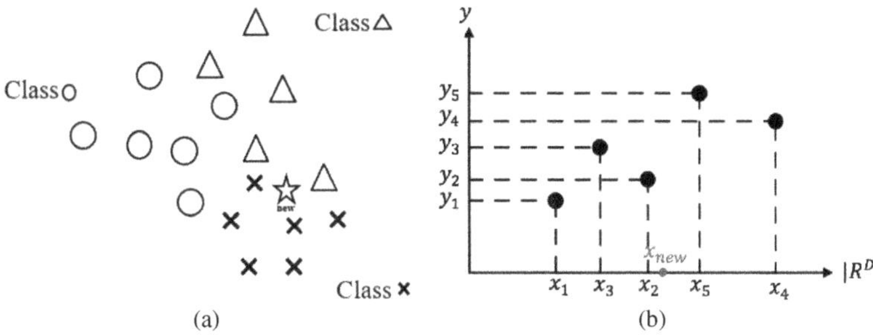

Figure 2.5 Various methods of KNN: (a) classification and (b) regression.

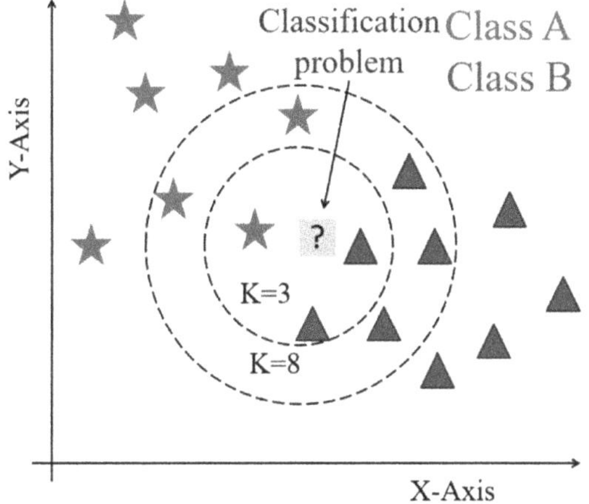

Figure 2.6 KNN algorithm.

the object. The value *y*(xnew) in this case is the average of the values held by the KNN, and the formula is as shown in Figure 2.7.

Let's provide an example for KNN regression, which is described as the average of KNN. People want to predict preferences for paintings when visiting an art gallery. Let's set the neighboring data for a painting with $K = 3$ to preferences of 3, 5, and 9, respectively. We will also assume distances from the new data point arbitrarily. If we represent them as [preference, distance from the new data], with values [9, 5], [5, 10], and [3, 2], the average of these preferences would be 5.66. However, the data point with the closest distance is [3, 2]. Even though its preference is lower, the proximity of the new data point might be considered more important. There are issues when determining distance. Before selecting *K*, a prerequisite is addressing the issue of distance metrics. The issue with distance metrics involves defining the standard for distance, typically done using metrics such as Euclidean distance or Manhattan distance. Euclidean distance is a method

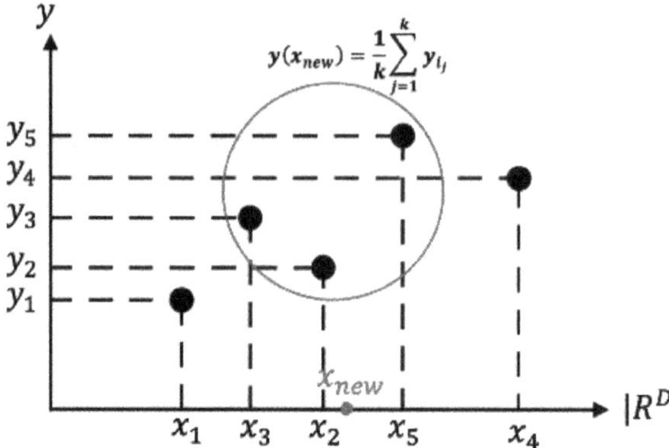

Figure 2.7 KNN regression.

used to calculate the distance between two points, and the formula is shown in Figure 2.6. The closer the distance, the more similarity is exhibited between the data points. The calculation method for Euclidean distance is as shown in (2.1).

$$d_{\{E\}}(x, y) = \sqrt{\{(x-y)\cdot(x-y)\}} = \sqrt{\{(x_{\{1\}} - y_{\{1\}})^2 + (x_{\{2\}} - y_{\{2\}})^2\}} \quad (2.1)$$

Simply put, in terms of coordinates, the units might represent noticeable distances, but those units could be related to things like money or weight. In other words, the distance on the coordinates can vary depending on the units used. Therefore, when defining the KNN algorithm, it is crucial to establish standards regarding the units of the data. Manhattan distance is a method for measuring the distance between two points. This is calculated as the sum of the distances moved along each coordinate axis. Specifically, the Manhattan distance between two points (x_1, y_1) and (x_2, y_2) is the sum of the absolute values of $|x_1 - x_2|$ and $|y_1 - y_2|$. The absolute values are used because distance cannot be negative. Therefore, Manhattan distance involves calculating the horizontal and vertical distances between two points separately and then summing them to find the total distance.

2.3.1.2 *Support vector machine*

SVM is a supervised learning model in the field of machine learning, used for pattern recognition and data analysis. It is applicable to both classification and regression analysis. Originally developed for binary classification, SVM is now widely applied across various domains in AI. The reason SVM is gaining attention is its ability to provide an optimal hyperplane boundary in classification, addressing issues that arise in classification problems. It is advantageous in minimizing

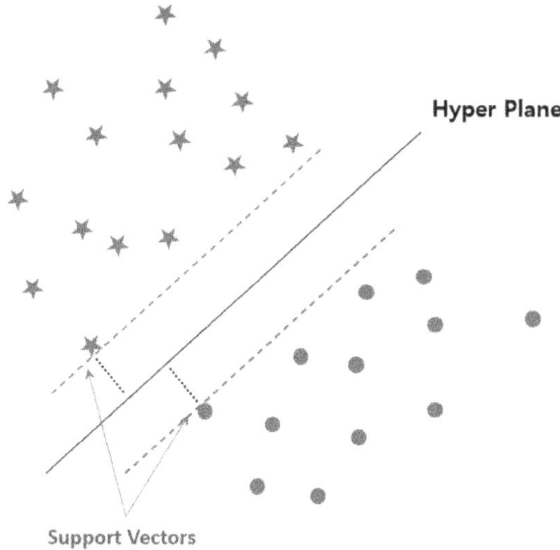

Figure 2.8 Support vector machine.

the empirical risk based on the principle of minimizing structural risk, even when dealing with small amounts of training data, making the interpretation of results easier.

A hyperplane is a space that is one dimension lower than its surrounding space. In three dimensions, it becomes a two-dimensional plane, and in two dimensions, it reduces to a one-dimensional line. In the case of two dimensions, if we separate the boundary with a line, it looks like Figure 2.8. Looking at Figure 2.8, we can classify the data into two categories: stars and circles. The line drawn to separate this data is called the hyperplane, and the data points closest to the hyperplane used to draw it are called support vectors. The distance between the hyperplane (boundary) and the support vectors is referred to as the margin.

In SVMs, the goal is to find the best boundary among numerous possible hyperplanes that can separate the two sets of data. In this context, finding the best boundary means identifying the straight line that maximizes the distance between the support vectors and the hyperplane. As the distance between the separating line and the data increases, it becomes more stable in effectively distinguishing between the two groups. In a two-dimensional plane, as shown in Figure 2.8, a straight line is determined by the equation $y = Wx + b$, where the slope W and y-intercept b influence the decision. The goal is to find W and b that maximize the margin. In Figure 2.9, there are two methods for finding the optimal hyperplane that separates two classes using support vectors and margins (could you clarify what a margin is?): hard margin and soft margin. The hard margin method is a strict way to determine the boundary that separates the two groups, making it

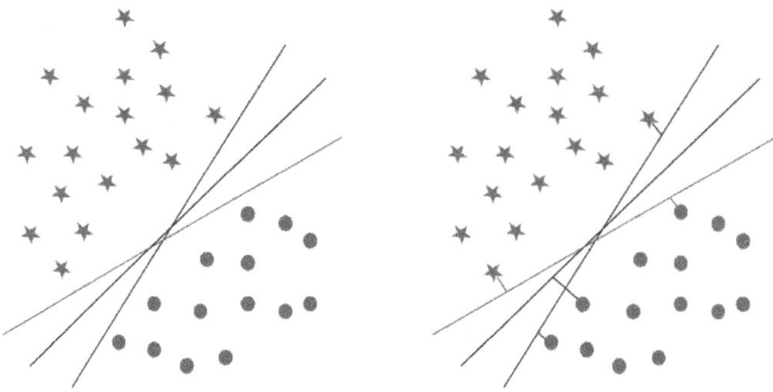

Figure 2.9 Hyperplane of SVM.

difficult to use when there are a few interfering elements. The term "hard" implies not allowing misclassifications and aiming for perfect classification, known as linear separation. Indeed, it means not allowing errors in classification. The concept of a hard margin is applicable only when each data class is perfectly separated for linear classification, but in reality, such situations are rarely encountered. In such cases, to maximize the margin, some degree of tolerance for misclassifications needs to be allowed. In such scenarios, the term "soft" is used, and it is referred to as a soft margin. If it is not possible to linearly separate the given samples perfectly into two groups, misclassification errors are allowed. In any case, for effective classification, it is essential to have a wide margin. This involves allowing some flexibility around the boundary where the support vectors are positioned.

Referring to Figure 2.9 as an example, the ultimate goal of SVM is to find a straight line that separates the two groups. The objective is to find a hyperplane that distinguishes between the two groups, ideally placing the hyperplane in a way that maximizes the separation between the two groups. This is because a greater separation between the two groups allows for more confident predictions when classifying additional input data in the future. In SVM, it was mentioned that it can be applied to both classification problems and regression analysis. So far, we have explained the issues related to classification, and next, we would like to explore analysis related to regression. Regression problems can be viewed in a similar context to classification. By maximizing the margin with respect to the hyperplane, efficient analysis can be achieved. Figure 2.10 illustrates a diagram of regression analysis in SVM.

Figure 2.10 contains data within the E range. Observing the margins, (a) is wider than (b). As E increases, the likelihood of data points falling outside the hyperplane will decrease. The data point labeled as 1 in (a) has a wider margin compared to (b), relative to the hyperplane. Therefore, one would prefer to shape the model in the form of (a). In other words, a narrow margin implies a higher probability of misclassification.

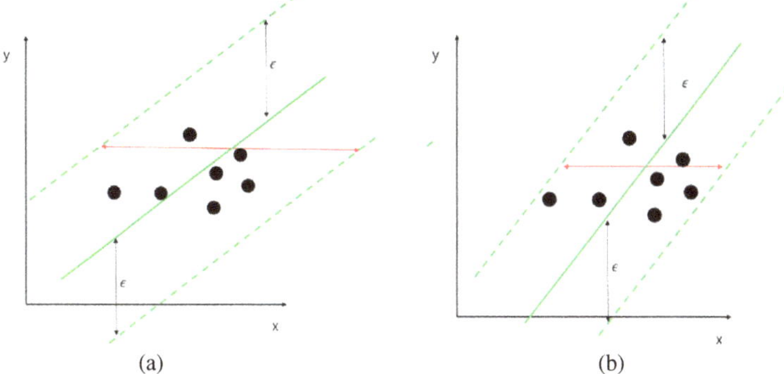

Figure 2.10 Regression problem's hyperplane. (a) Wide-margin hyperplane and (b) narrow-margin hyperplane.

2.3.1.3 *Decision tree*

A decision tree is a supervised learning model that can handle both classification and regression tasks. It automatically discovers rules in the data through learning, creating a tree-based set of classification rules. It is a predictive model that links input observations with target values, analyzing the data to represent the patterns between them as a combination of predictable rules. A decision tree is a common predictive modeling method used in statistics, data mining, and machine learning. It is a supervised learning methodology applied to problems where the goal is to predict or classify a dependent variable (Y) based on the relationships or scales of independent variables (X). This involves inputting observed values of independent variables into the model to classify or predict the dependent variable, making it a supervised learning approach. The primary reason for using a decision tree model is to determine which independent variables have the most significant impact when predicting or solving classification problems with the dependent variable.

Furthermore, for each independent variable, you can see the detailed criteria for predicting or categorizing based on which scale, and the intuitive interpretation allows you to identify key variables, present separation criteria, and explore linear and nonlinear relationships by considering interactions between variables. However, the downside is that it is unstable with small amounts of data, has a high incidence of overfitting, and can introduce errors near separation boundaries.

2.3.2 *Regression*

Regression analysis is a statistical method of analyzing continuous variables by fitting a model between two variables and measuring the goodness of fit. It is a statistical analysis method that assumes a mathematical model to find the functional relationship between variables and estimates this model from data on measured variables. To summarize, regression analysis is essentially a mathematical way of looking for tendencies or dependencies between variables. This means that regression is not a way to prove causality, but rather a way to implement a model where causality is assumed. Today, regression analysis is one of the most popular statistical models in use, having proven its usefulness in many fields, including economics, medicine, and engineering. In the following, we'll discuss a representative model for regression analysis.

2.3.2.1 *Linear regression*

Linear regression is a popular and representative regression algorithm that models the linear correlation of a dependent variable *y* with one or more independent variables (or explanatory variables) *X*. It is a regression analysis technique. A simple linear regression[10] with one independent variable is called a simple linear regression, and a multiple linear regression with more than one independent variable is called a multiple linear regression, i.e., it will contain more than one independent variable that adequately predicts the value of the dependent variable in a general linear equation. An example of a typical linear regression is shown in Figure 2.11.

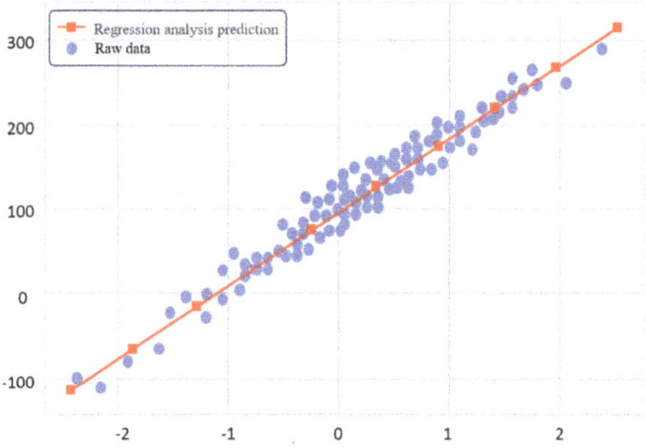

Figure 2.11 Linear regression.

[10] Simple linear regression: If you have a relationship between an independent variable *x* and a single dependent variable *y*, it is called simple linear regression, and if you extend it to multiple independent variables, it is called multiple linear regression. Almost all problems in the real world involve multiple independent variables, so when we talk about linear regression, it is usually multiple linear regression.

In the following, we'll discuss the different types of linear regression.

- **Simple linear regression:** Simple linear regression can be expressed as $y = wx + b$, and in machine learning, it is about finding the appropriate values of W and b, with W representing the weight multiplied by the independent variable x and b representing the bias against the constant term. The shape of the model will be a straight line or a plane, and in more than four dimensions, a hyperplane. Thus, a linear regression model is essentially a model that finds a straight line or plane (hyperplane) in the dimensional space representing the data that has the smallest error associated with it.
- **Multiple linear regression:** Multiple linear regression is represented by the expression $y = w_1x_1 + w_2x_2 + \cdots + w_nx_n + b$ and is influenced by multiple independent variables. In other words, it models a relationship in which two or more independent variables affect a dependent variable. This is usually done by using the method of least squares, which selects regression coefficients to build a model. For example, in a model that predicts the price of a house, it is not just a matter of zoning, but the number of rooms, number of bathrooms, size, etc. will change the prediction value, and in the part that predicts wages, you can analyze how many independent variables such as experience, age, external activities, etc. affect the price of a house and wages instead of just using education as a variable.

2.3.2.2 *Logistic regression*

Logistic regression is a model that allows for prediction and classification. It is a statistical technique used to predict the probability of occurrence of an event using a linear combination of independent variables. It is also a type of classification technique because the dependent variable is categorical data (no matter how much the value of the independent variable changes, the dependent variable is a value between 0 and 1, see Figure 2.12 and the result of the data is divided into a specific classification when the input data is given. The logistic regression model is shown in Figure 2.12. Logistic regression is used for categorical cases, as already mentioned. Linear regression is typically used when the dependent variable has a continuous normal distribution. If the dependent variable is categorical (yes/no, 1/0, pass/fail, purchase/non-purchase), logistic regression is used. When the dependent variable falls into two categories, it is called a binary logistic regression model. Unlike linear regression, logistic regression can also be seen as a type of classification technique because the dependent variable is categorical data and the outcome of that data given the input data falls into a specific classification.

Since not all data is linear in nature, logistic regression with multiple classes emerged. In Figure 2.12, based on the criteria for categorizing certain data, if it

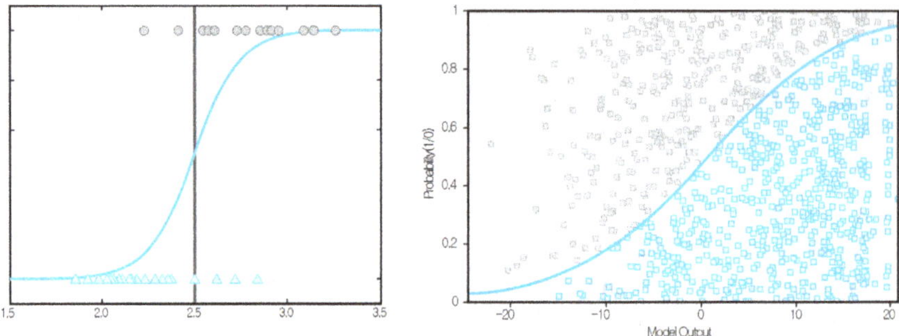

Figure 2.12 Logistic regression models.

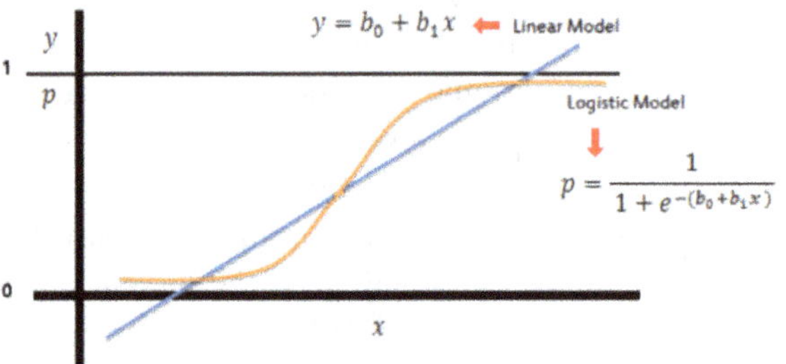

Figure 2.13 Odds and logit conversion.

exceeds the criteria, it falls within the classification, otherwise it falls outside the classification, so it is represented by 0 and 1. This is how we categorize whether an email is spam or not, or whether it is benign or malignant in terms of cancer.

Logistic models can solve problems in classification form by applying a sigmoid function to a linear model, as shown in Figure 2.12. However, unlike linear models, classification models cannot be interpreted intuitively through regression coefficients, which is why we use the concepts of Odds and Logit Transform, as shown in Figure 2.13.

Odds is the ratio of the probability of success to the probability of failure, expressed as "P = probability of success" and "$1 - P$ = probability of failure." Odds = 5 means that the probability of success is 5 times the probability of failure. Logit is a function that determines the probability of belonging to a class in the range $\pm\infty$, and log odds is a function that finds the probability through the value of the dependent variable in a transformation model (linear). The odds have two limitations: being within the range of $0 < \text{odds} < \infty$ imposes constraints, and the values of odds and probabilities exhibit asymmetry. Therefore, the logarithm function is applied. The process of applying a logarithmic transformation to the odds is called the logit transformation.[11] Through logit transformation, a linear equation is derived, allowing us to understand the changes in the logit based on the independent variable *X*. Logistic regression is an important technique in the fields of AI and machine learning, enabling the performance of complex data processing tasks without human intervention through learning. If a company gains valuable insights from business-related data, it can use these insights in predictive analysis to reduce operating costs, increase efficiency, and scale more quickly. For example, a company might discover patterns that lead to improving employee retention or designing more profitable products.

2.4 Unsupervised Learning

Unsupervised learning involves using machine learning algorithms to analyze and cluster unlabeled datasets. Unlike supervised learning, it doesn't provide predefined outputs in the learning algorithm. Instead, AI discovers specific patterns and correlations within the input dataset. Unsupervised learning is commonly used when information about the output is either unknown or impractical to obtain due to reasons such as insufficient data or high costs associated with collecting training data. For example, when a child is born, parents typically teach the child words like "dad" and "mom" as they begin to learn language. This method of teaching while providing guidance can be considered as supervised learning. In other words, supervised learning involves teaching a child things that humans already know, such as colors, numbers, or vocabulary.

On the other hand, unsupervised learning can be likened to letting the child solve problems and make inferences on their own. Typically, children learn on their own through creative activities such as imaginative play, writing, drawing, etc. Unsupervised learning exhibits the following characteristics:

1. Processes unlabeled data.
2. Enables users to perform more complex processing tasks.
3. Is more challenging for prediction.
4. Can be used to discover the fundamental structure of the data.
5. Occurs in real time.

---

[11] Logit transformation: Odds can be calculated as (probability that *k* wins the game/probability that *k* loses the game) = $p/(1 - p)$, so knowing the odds, we can calculate the inverse of the probability that *k* wins the game. However, if *k* wins once and loses six times, then the odds of *k* winning are 0.17. Conversely, if *k* wins six times and loses one time, the odds of *k* winning are 6. In other words, if the number of wins is less than the number of losses, it will have a value between 0 and 1, but if the opposite is true, it will have a value between 1 and infinity. This is not symmetric, so to solve this problem, we usually use logit, which is log(Odds), not Odds.

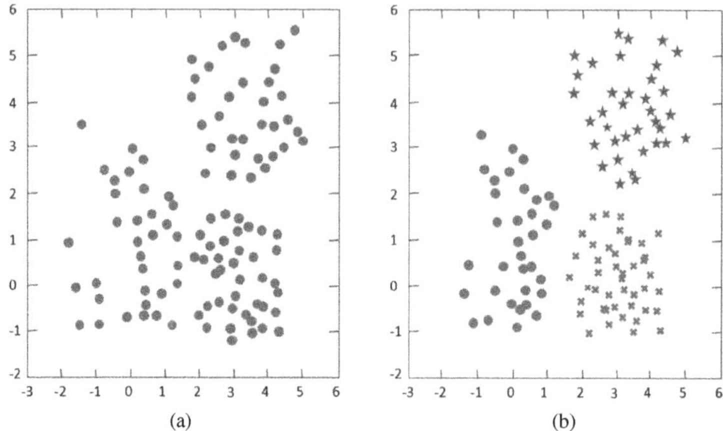

Figure 2.14 Cluster models in unsupervised learning. (a) Raw data and (b) clustering results.

Unsupervised learning is a method of training a computer without providing labels for the given data, aiming to discover hidden features or structures within the data. As previously explained, supervised learning classification involves creating a discriminant based on labeled training data, and using this discriminant to classify or predict newly input data. Unsupervised learning involves classifying data without labels, aiming to find groups of data distributed based on similar characteristics or conducting independent component analysis (ICA).[12] The goal is to discover groups of data with similar features and perform ICA to identify independent components. Therefore, unsupervised learning is used when there is no prior knowledge about specific outcomes, but the objective is to gain meaningful insights from the resulting data.

It is often utilized to explore hidden features or structures in unspecified intrinsic problems or datasets. One of the representative methods is clustering. Cluster models are essential when the training data lacks labels, making it crucial to understand how the input data forms groups. In other words, the fundamental principle of a clustering model is to analyze the characteristics of unlabeled data and group together data with similar features. Except for the absence of labels, clustering shares the same purpose as the classification model in supervised learning. Figure 2.14 illustrates an example of clustering.

[12] Independent component analysis (ICA): A computational method for separating a multivariate signal into statistically independent subcomponents. Each component is a non-Gaussian signal and consists of components that are statistically independent of each other. ICA is a specialized method for separating blind signals, for example, isolating the voice of a particular person from a conversation of several people recorded in a room.

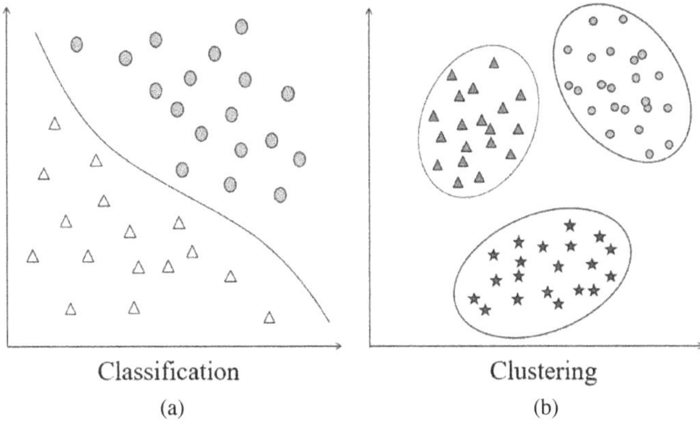

Figure 2.15 Comparison between (a) classification and (b) clustering.

In Figure 2.14, as shown in (a), without labels in the original data, it is not possible to know the characteristics of the data. However, in (b), when observing the clustering results, it is possible to see the phenomenon where similar data gathers into clusters. Analyzing data clustered in this way may reveal specific patterns, features, or structures. Looking at (b) in Figure 2.14, cluster models can sometimes resemble a classification format, and the distinction between classification and clustering may appear subtle. It involves classifying data without labels. Figure 2.15 compares classification and clustering.

The similarity between classification and clustering lies in their shared goal of data analysis, aiming to discover hidden patterns or structures. Both methods involve understanding the characteristics of the data and forming groups, employing similar approaches. However, the key difference between classification and clustering is that classification is an analysis technique used to predict the group of new data when labels are available, whereas clustering aims to understand the inherent characteristics of the data when the information about the categories of each object within the clusters is unknown, meaning there are no labels. Cluster models can be broadly categorized into two main types. The first is the smoothing[13] or partition-based clustering method, and the second is the hierarchical clustering method. Partition-based clustering in clustering involves specifying the number of clusters initially, dividing the data into K clusters, and then modifying the centroids using various criteria while reclassifying the clusters. Various techniques fall under partition-based clustering, such as K-means (mean, for continuous data), K-medoids (median, for continuous data), k-modes (a combination of mean and mode, for mixed data), DBSCAN (density-based analysis, used for

[13] Smoothing: Converting highly fluctuating values in time series data to values with smoother variations.

outlier detection), and more. This method involves initially designating each observation as a single initial cluster and then gradually combining clusters, two at a time, until all clusters become one. An advantage of hierarchical clustering is that it doesn't require pre-specifying the number of clusters, and it can be used for all clustering techniques, regardless of whether the data is text or numeric. However, it has the drawback of being computationally intensive since similarity needs to be calculated iteratively for every pair, making it challenging to use for large-scale datasets. The application fields of cluster models are very diverse. For example, distinguishing between human voices and noise to improve call audio quality on phones is a good example of discerning unlabeled data. In the medical field, cluster models are used when differentiating between disease and patient groups in clinical trials. In marketing, cluster models are employed when segmenting customers. Recently, they have found applications in marketing for market and customer segmentation. In social network services like Facebook, cluster models are used to create communities for users with similar interests. Google uses cluster models in its news service to group news articles on the same topic into the same category. In the medical field, spatial cluster analysis is used in epidemiological investigations to understand the distribution areas and transmission paths of specific diseases. In the promotion field, clustering is utilized when segmenting customers. Furthermore, in the field of statistics, there is a trend towards utilizing various clustering algorithms and methodologies for data analysis. Clustering is a widely used machine learning method in various fields, and we will now explore the commonly used *K*-means algorithm and DBSCAN clustering algorithm.

2.4.1 *K-means algorithm*

The *K*-means clustering algorithm is an algorithm that groups given data into *K* clusters. The *K*-means algorithm minimizes the variance of the distance difference for each cluster. It is one of the simplest unsupervised learning methods that clusters unlabeled data. In simple terms, *K*-means involves predefining the number of clusters and iteratively updating the cluster means while clustering the points closest to the centroids. In the case of *K*-means, the number of clusters (*K*) needs to be predefined because, in unlabeled data, we don't know how many clusters exist. The value of *K* in *K*-means represents the number of clusters. Choosing an appropriate *K* value is crucial. Two common methods for determining the *K* value are the Elbow method and Silhouette analysis. The Elbow method involves gradually increasing the cluster number *K* while performing clustering using the *K*-means algorithm. The sum of squared errors[14] (SSEs) is calculated for each iteration, and

[14] SSE: The sum of the squared distances between each data point and its cluster centroid; the smaller the SSE value, the better the clustering performance.

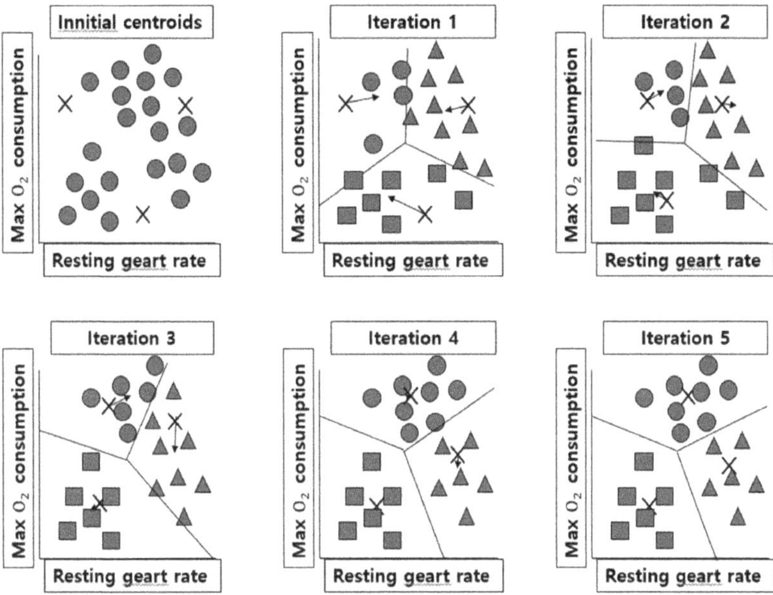

Figure 2.16 Process of *K*-means clustering.

a graph is plotted to choose the optimal *K* value where the SSE starts to level off. Silhouette analysis indicates how efficiently the distance between clusters is separated. Efficient separation means that the distance between different clusters is large, and data within the same cluster are close to each other. Once the number of clusters to be classified is determined, initial centroid values are assigned to each cluster. The meaning of the centroids used in *K*-means is the virtual group representing the cluster, and each learning data serves as a reference point to determine which cluster it belongs to. Initial centroids are arbitrarily chosen and get updated during the learning process, gradually moving towards the optimal center for each cluster. Figure 2.16 illustrates the steps of performing *K*-means clustering.

The core of the *K*-means algorithm is to find the optimal centroids. The concept of finding the optimal centroids is intuitive and straightforward. In other words, for each data point, calculate the distance to the *k* arbitrarily chosen initial centroids and consider the closest centroid as the center of the cluster. Once *k* clusters are determined, compute the average coordinates of the training data points belonging to each cluster, and set these averages as the new centroids. Then, recalculate the distances from all training data points to the newly defined *k* centroids and assign each data point to the cluster with the closest centroid. This process is repeated iteratively, updating the distances, until there is no change in cluster assignments or when a predefined number of iterations is reached. In algorithmic terms, this can be expressed as follows:

1 Determine the number of clusters ($k = n$) and set n random centroids.
2 Calculate the distance from each data point to the n centroids and assign the closest centroid as the center of its cluster.
3 Compute the average coordinates of the training data points for each cluster and set them as the new centroids.
4 Repeat steps 2 and 3 for each data point until there is no change in cluster assignments.
5 If no data point changes its assigned cluster, the learning process is complete.

The *K*-means model performs data computations for learning very quickly. When the scale of the learning data is n, most other algorithms often have computational complexities of $O(n^2)$ or $O(n\log n)$, while *K*-means has a complexity of $O(n)$. Ultimately, from the perspective of software engineering, it is represented as the algorithm with the lowest complexity when calculating algorithm complexity. Therefore, one of the advantages of the *K*-means algorithm is that it is relatively easy to understand and implement. Additionally, it operates relatively quickly and works well with large datasets, making it the most popular clustering algorithm. However, as a drawback, it requires the definition of means for application, and the algorithm's output can vary depending on the random initialization of centroids. The most significant drawback is that it assumes a circular shape for clusters, making its applicability relatively limited. Additionally, it requires specifying the number of clusters to be identified.

2.4.2 *Density-based algorithm*

DBSCAN is an algorithm for grouping together closely located data points in multi-dimensional data based on density. While *K*-means clustering creates clusters using the distance between clusters, density-based clustering, on the other hand, groups regions where observations are densely concentrated. Essentially, two parameters are required: the minimum cluster size and the maximum distance between clustered points.

The minimum cluster size represents the minimum number of data points needed to form a single cluster, influencing the determination of small clusters that can be considered negligible. Here, there are core points and border points. A data point is considered a core point if the number of neighboring points around it is greater than or equal to the "minimum cluster size." Core points play the role of the center of a cluster and, along with neighboring points, belong to the same cluster. A core point is the central element in a cluster, whereas a border point represents the maximum distance between data points within a cluster. This value restricts how far apart data points within a cluster can be. All data points within a given cluster must have distances from each other that are less than or equal to the

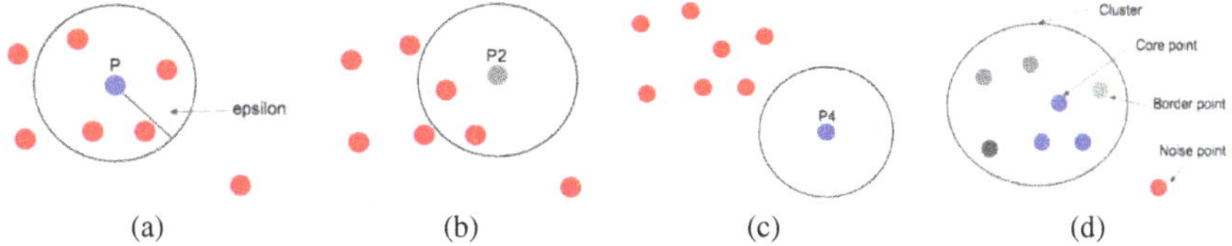

Figure 2.17 DBSCAN clustering. (a) The cluster centered around point P, (b) border points, (c) cluster including P, and (d) each points.

"maximum distance between clustered points" value. This ensures that data points within a cluster maintain a consistent density and proximity. Figure 2.17 illustrates clustering using DBSCAN. If we consider the distance epsilon from the reference point and set the minimum number of points within this radius, minPts, as minPts = 7:

1 In (a), considering the epsilon distance from the reference point and the minimum number of points minPts as parameters, P becomes a core point forming a cluster since there are 7 points within its radius.
2 For P2, since there are only three points within the epsilon radius, it doesn't meet the minPts = 7 criteria to be a core point. However, it becomes a border point as it belongs to the cluster centered around the core point P.
3 P4, regardless of the center point, does not fall within the range satisfying minPts = 7. Therefore, it becomes an outlier that doesn't belong to any cluster and can be ignored as a noise point.

As explained in Figure 2.17, DBSCAN assigns clusters to data that are densely packed around specific points and treats data outside these dense areas as noise or border points.

In the *K*-means model, each data point determines its cluster by measuring the distance from *K* centroids and assigning itself to the cluster with the shortest distance. However, as shown in Figure 2.18, DBSCAN considers a group of data with a certain density as the same cluster, regardless of the concept of distance. Table 2.3 compares *K*-means and DBSCAN.

The advantages of DBSCAN include not requiring the pre-specification of the number of clusters and effectively excluding outliers. As mentioned earlier, the two parameters that need to be defined are the radius size and the minimum cluster size. The radius size represents the distance at which points are considered connected, and the minimum cluster size indicates the number of points needed to form a cluster. For example, if MinPts is set to 10, even if six points are well connected, they will not be considered as a desired cluster but rather treated as noise. In conclusion, DBSCAN is useful when dealing with irregular data, expecting a significant amount of noise and outliers, and facing situations where predicting and

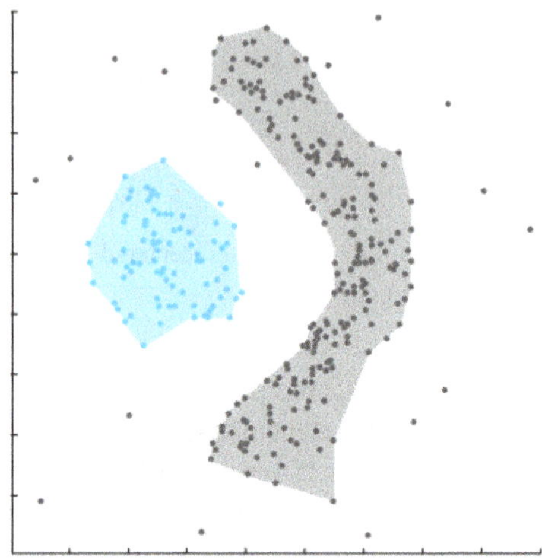

Figure 2.18 Density-based clustering algorithm.

Table 2.3 Compares DBSCAN, *K*-means.

Features	DBSCAN	*K*-means clustering
Shape of clusters	Clusters well when the shape of the clusters in your data is arbitrarily clustered	Clusters well when the shape of the clusters in the data is spherical
Number of clusters	No need to predetermine the number of clusters (density-based)	Need to predetermine the number of clusters to be clustered (centroid-based)
Outlier	Allows you to specify outliers that are not included in the clustering	All data is included in one cluster
Initial setting	Initial cluster state does not exist	Results vary greatly depending on initial centroid settings

interpreting data in advance is challenging. However, it comes with the drawbacks of high computational cost and difficulty in prediction and interpretation.

2.5 Reinforcement Learning

Reinforcement learning involves iteratively interacting with the environment, learning the optimal actions to take from the current state. Each action in a given state receives a reward from the external environment, and the learning process aims to maximize the cumulative reward over time. While the algorithms mentioned earlier performed learning on static datasets, reinforcement learning operates in a scenario where the agent[15] doesn't initially know which actions to take in a given state to maximize the reward. The learning process involves finding a strategy to maximize the actions leading to high rewards without prior knowledge of the optimal actions. In other words, reinforcement learning discovers the optimal actions through numerous trial-and-error iterations. The agent interacts with the environment, experiencing rewards and penalties. Through this iterative process, the agent learns how to maximize rewards. Additionally, the outcomes of actions may be expressed as delayed rewards over time. This encourages the agent to predict and consider the future consequences of its current actions. For example, in gaming, taking a certain action now may yield immediate rewards, but it could lead to losing the game in the long run. Instantaneous rewards may not necessarily lead to better decisions if not considering long-term goals. Therefore, the agent needs to weigh short-term rewards while striving for long-term objectives by consistently choosing actions known to be optimal. Indeed, due to its trial-and-error nature, reinforcement learning can be considered similar to human learning. The agent observes the environment and takes appropriate actions based on these observations. The agent then receives rewards from the environment based on its actions.

Figure 2.19 illustrates the iterative process of interaction with the environment to perform tasks that maximize the obtained rewards through the observation-action-reward cycle. The process of observing the state, acting, and receiving rewards is referred to as an experience.

Reinforcement learning has been influenced by various fields, with behavioral psychology and control theory playing a significant role. The concept of "trial and error" from behavioral psychology, describing how humans and animals learn, has been applied to machine learning. The agent learns by remembering the rewards and penalties for each action to make optimal decisions. Another influential field is optimal control, which emerged in the late 1950s. This theory focuses on optimizing the efficiency of dynamic systems for designing control devices. The optimization problem of dynamic systems involves making optimal decisions for each stage over time. Richard Bellman, an American mathematician, addressed the issue of discrete-time problems by introducing the Markov decision process model to solve this problem. Reinforcement learning is primarily used when making sequential decisions, as it involves learning through feedback received during unfolding events. Some argue that the learning process based on trial and error closely resembles human learning, making reinforcement learning the most representative model of AI. Reinforcement learning models are most effectively utilized in games and robotics. If a reinforcement learning model understands how

[15] Agent: A decision-making entity that interacts with a given environment (determining the state and rewards of the agent's actions, predicting the rewards it will receive after an action, and learning), deciding on actions and receiving rewards accordingly.

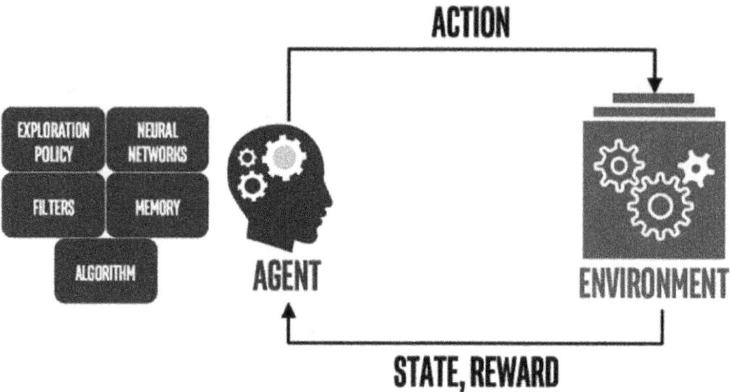

Figure 2.19 The forms of reinforcement learning.

the environment will change based on its actions, it can anticipate the changes before actually taking actions and execute optimal behavior. While planning provides the advantage of more efficient actions for the agent, if the model fails to accurately reflect the environment and errors occur, these errors can propagate into the agent's actions. Creating an accurate model becomes as challenging as building a good agent when such errors become more frequent. From a machine learning perspective, reinforcement learning is not purely based on mathematics but follows a control-based approach. While some view it as a form of supervised learning, there are also cases where it is categorized as unsupervised learning.

2.6 Procedures and Details in Applying Machine Learning

The goal of machine learning is to accurately output predicted values, whether it be for a regression analysis model or a classification of classes. The process to achieve this goal is outlined in Figure 2.20, and during the analysis, it is essential to consider the issues of underfitting and overfitting. Underfitting occurs when the model is too simple, failing to properly explain the input data and resulting in poor predictive performance.

Whether a model is underfitting or not can be judged based on the error, known as the cost, between the model's predictions and the actual values. In cases of underfitting, some improvement can be achieved through model selection and obtaining additional training data. Overfitting is when the model becomes overly tailored to the training data. The fundamental reason for overfitting is that many machine learning algorithms construct models in an inductive manner. Therefore, such issues can arise anytime machine learning methods are applied for analysis.

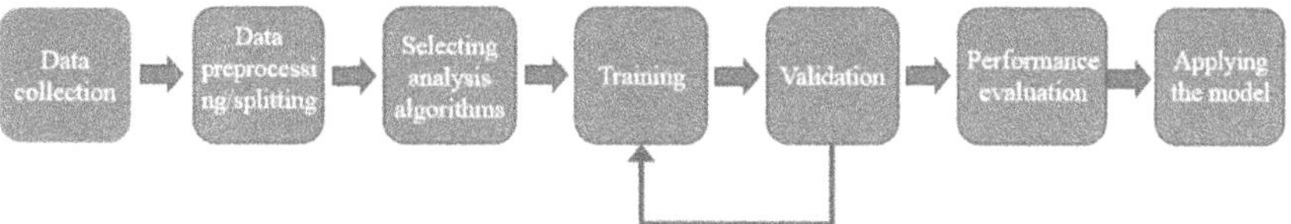

Figure 2.20 Procedure for machine learning implementation.

2.7 Practice Questions

Q1. What is machine learning?

Q2. Explain the difference between machine learning and data mining.

Q3. Describe supervised learning and unsupervised learning.

Q4. Describe the methods used in supervised earning and unsupervised learning.

Chapter 3

Model Validation and Evaluation

3.1 Splitting of Training, Validation, and Test Data

To model machine learning and validate the results, you need training data, validation data, and testing data. The three types of data are separated into the training dataset used for learning, the validation dataset providing an unbiased evaluation metric for the model applied to the training dataset when fine-tuning the model's hyperparameters, and the testing dataset providing an unbiased evaluation for the final model. There are two criteria for dividing the training, testing, and validation data. It can be determined by the total number of samples in the available data and the number of samples used in the actual model training. The ratio for dividing the training-testing-validation dataset is often determined on a case-by-case basis by the researcher. It is best to determine this through trial and error by training more models. A common practice is to set the data ratio as 5:2:3 and adjust the ratio based on each research environment.

Figure 3.1 shows an example where the collected data is divided into 70% training data and 30% test data. Then, 20% of the total training data is further divided for validation. The data is categorized into training data, validation data, and test data, and the researcher must decide the epoch, batch size, and iterations according to the intensity of the training. An epoch is a state where one complete training has been done on the entire training dataset, and batch size refers to the sample size that a model can process at once. It's inefficient to train the whole training data at once, so the data is divided into smaller parts for training. These smaller divided data are called mini batches. Iteration means repeating a calculation or computer processing procedure. For example, training 10 mini batches divided according to the batch size is called one iteration.

Figure 3.2 shows that when training data of 2,000 samples is divided into 200 mini batches, the batch size becomes 200, and 10 iterations complete one epoch. When determining the batch size, it should be considered that a larger batch size

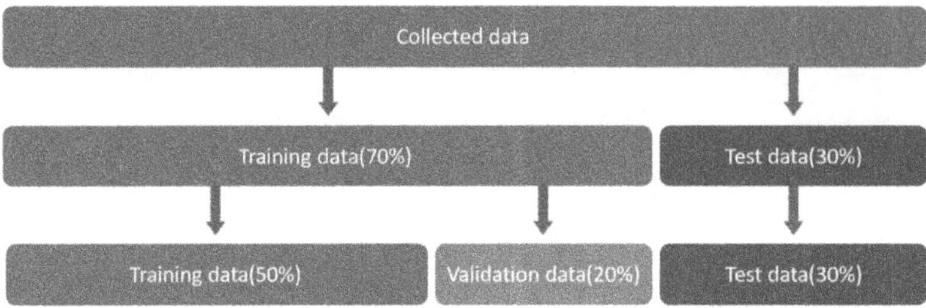

Figure 3.1 Splitting data for model validation.

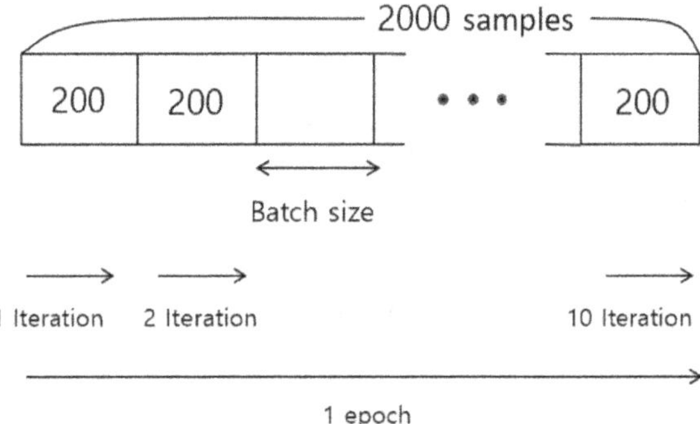

Figure 3.2 The relationship between batch size, iteration, and epoch.

reduces the time taken to build the model but also reduces the number of training iterations. Fewer training iterations may mean fewer adjustments to the weights to reach the minimum error value, which could result in lower predictive accuracy. This means that as the amount of data increases, the speed of training slows down, and issues like memory shortages can occur. In such cases, increasing the number of epochs to increase the number of trainings or raising the learning rate[1] can partially compensate for the reduced number of training iterations. Conversely, a smaller batch size may result in frequent weight updates, leading to unstable training, and increasing the number of epochs can increase the number of trainings but there's a risk of overfitting.

Therefore, an early stopping model, which can halt training before overfitting occurs, is needed. Early stopping involves periodically validating performance

[1] Learning rate: the amount or number of steps that are trained; a high training rate will result in faster results but may not produce the correct error values or cause overflows; conversely, a too-low training rate will result in slower results and more error values.

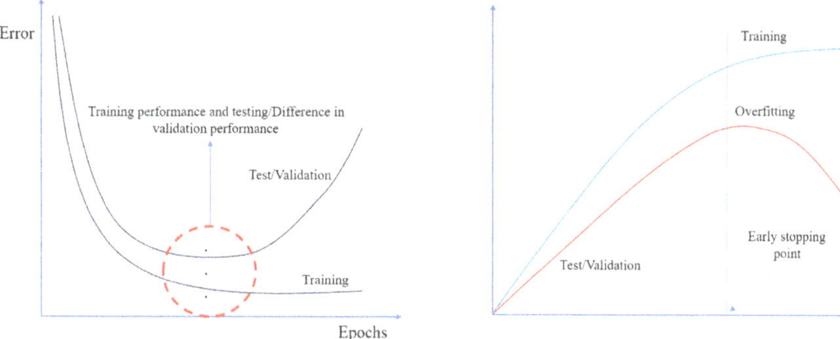

Figure 3.3 Early stopping before overfitting.

during training and stopping the training if there's no further improvement, indicating overfitting (Figure 3.3).

If training is terminated simply because there's no improvement in model performance, proper learning might not be achieved. Training should be terminated based on a sustained stagnation or decline, rather than temporary fluctuations. It's common to monitor whether performance continuously deteriorates over a certain number of iterations, and it's important to set a large enough number of monitoring iterations to prevent training from being too sensitively terminated. It's also important to decide on which performance criteria to base the termination. Typically, early stopping is based on model error, but accuracy or other performance metrics[2] can be used as well. To address these issues, adjustments in model selection, batch size, iterations, epochs, and learning rate should be made based on the data used for analysis, experience, environment, and the current predictive power level of the model, to ensure effective training and performance evaluation.

3.2 Evaluation Metrics for Classification Models

In artificial intelligence, machine learning involves training with multiple models. As these models have different algorithmic characteristics, there is a need for metrics to evaluate them. To evaluate models, the metrics differ between classification models and regression models. For classification models, metrics such as cross-validation, confusion matrix, accuracy, and precision are needed. For regression models, metrics include mean squared error (MSE), root mean squared error

[2] Scale: In artificial intelligence, evaluating performance, primarily by measuring the ability of a search algorithm. Decisions are mostly based on experience, but not completely, and computational methods such as concentration and branching coefficients are used.

(RMSE), and mean absolute error (MAE). We will now explore the methods of evaluating machine learning models.

3.2.1 *Cross-validation*

After creating an artificial intelligence model, it is necessary to evaluate whether the model is well-made. One method to evaluate a model's performance involves splitting the entire data into training and test data for assessment. This means training the model with training data and then evaluating its accuracy with test data. However, there is one weakness in this approach. If the performance of the model is evaluated and modified using a fixed initial test dataset, and this process is repeated, the model will eventually work well only on that one fixed test dataset. In such cases, the model becomes overfitted to a specific dataset and may perform poorly when predicting on different datasets. The solution to this is cross-validation. Cross-validation, as the name suggests, involves using all parts of the data for testing, not just a fixed portion, to validate the model's performance. This method is often used when there is little collected data or it is difficult to set up separate validation materials. To perform cross-validation, it is necessary to first decide how many folds to divide the data into.

If Four-fold cross-validation is performed as in Figure 3.4, the current data is divided into four sets, with one test data per fold. When dividing into four sets, the collected data can be divided using sequential, interleaving, or random sampling methods. To enhance the learning effect of cross-validation, it is recommended to make the data distribution of each dataset similar through random sampling and other methods. Whether the distributions are similar can be determined by drawing charts or performing descriptive statistical analysis, such as checking the mean and standard deviation. There is also a method to construct datasets with

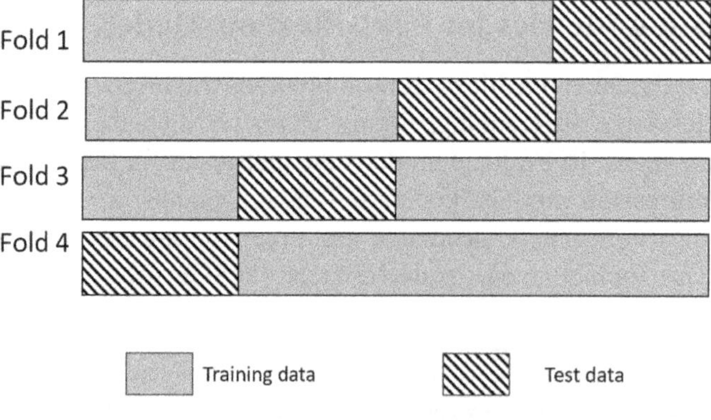

Figure 3.4 Cross-validation.

overlapping data among them. During training, one of the four datasets is set as the test data, and the remaining three are set as the training data. First, set the first dataset as the test data and the remaining three as the training data, perform training, evaluate performance, then set the second dataset as test data and the remaining three as training data, and perform training. This process is repeated four times. The final performance metric of the model is calculated as the average of the results from four performance evaluations. Cross-validation can be divided into five methods.

1. **Holdout method:** This method is known as the Holdout method or Holdout cross-validation. It involves randomly dividing the given training data into training and test sets at an arbitrary ratio. The most commonly used ratios are 8:2 or 7:3 for training to testing, and the advantage is that it involves only one iteration (training and validation), thus reducing the computational time burden. However, the disadvantage is that if the process of validating results on the test dataset and tuning model parameters is repeated, there is a high possibility that the model becomes overfitted to the test dataset.
2. **k-Fold cross-validation:** k-Fold cross-validation, the most commonly used cross-validation method, involves dividing the data into k data folds and assigning a different test data set in each iteration, thus forming a total of k "data fold sets (in the case of Figure 3.5, 4 training data + 4 tests)." Consequently, a total of k iterations are needed to train and validate the model. The validation results for each data fold set are averaged to obtain the final validation outcome.
3. **Leave-p-out cross-validation:** This method involves selecting p samples from the entire data (different data samples) for model validation. In other words, it extracts p data from the existing dataset and uses the remaining data as test data. For example, Leave-3-out cross-validation involves extracting three data points from the available data, as shown in Figure 3.6, and using the rest as training data. The remaining data is then used as test data to derive the validation results. This method also has the disadvantage of requiring a long computation time.
4. **Leave-one-out cross-validation:** Leave-one-out cross-validation, also known as LOOCV, refers to the case where $p = 1$ in the previously mentioned Leave-p-out cross-validation. It is widely used because it reduces computational time and can yield better results compared to Leave-p-out cross-validation. As the number of test datasets used for validation is smaller, the amount of data used

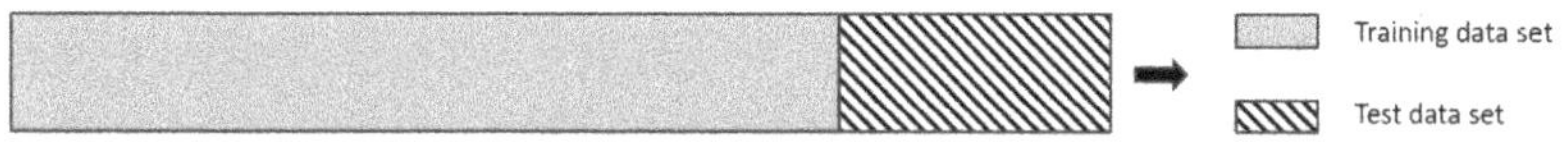

Figure 3.5 Holdout data configuration.

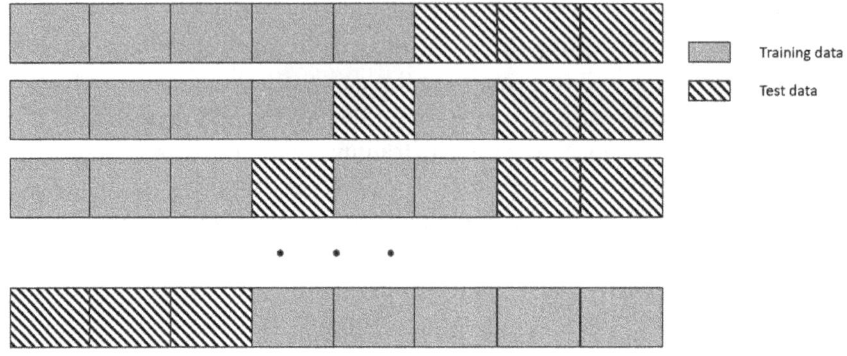

Figure 3.6 Leave-p-out cross-validation data configuration.

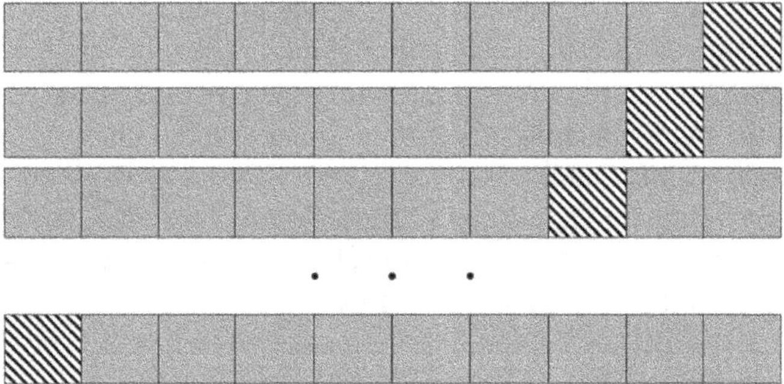

Figure 3.7 Leave-one-out cross-validation data configuration.

for model training increases. The advantage is that since only one data point is sacrificed for model validation, all other data can be used for model training. This is illustrated in Figure 3.7.

The advantage of cross-validation is that it allows the use of all datasets for evaluation, preventing data bias in evaluation and thereby avoiding overfitting to specific evaluation datasets. Additionally, based on the evaluation results, a more generalized model can be developed. However, the downside is the long duration of model training/evaluation due to the high number of iterations, but this is in comparison to the k-fold cross-validation method, and using other cross-validation methods appropriately can mitigate this disadvantage.

3.2.2 *Confusion matrix*

The Confusion Matrix is a table used to evaluate the prediction results of a classification model. Table 3.1 represents the confusion Matrix, which helps evaluate a model's performance in various aspects beyond just accuracy. This table is a

Table 3.1 Confusion matrix example.

	Output value		
	A	B	C
A_class	5	1	0
B_class	1	4	1
C_class	2	0	4

square matrix, displaying the values of all classes horizontally and vertically, with the algorithm's predicted results on top and the actual target values on the left. The element at (i, j) in Table 3.1 indicates whether a large number of input values in class i were predicted as class j by the algorithm. In this table, the values on the diagonal represent the correctly predicted cases, specifically the diagonal line starting from the top left to the bottom right. Table 3.1 contains three classes — A_class, B_class, and C_class — and the table is filled with the frequency of predicted and actual values for each case.

For example, for the A_class, it is classified into three categories based on whether the algorithm predicted it as A, B, or C.

As seen in Table 3.1, it can be observed that there are two misclassifications of C_class as A. To calculate the model's accuracy, divide the sum of these diagonal elements (number of correct predictions) by the total number of predictions. This calculated value represents the ratio of how accurately the model performed its predictions, i.e., the accuracy.

The confusion matrix is one of various methods to evaluate a model, offering an opportunity to deeply understand and improve the performance of machine learning algorithms. There are other methods used to evaluate machine learning algorithms, and we will continue to explore them.

3.2.3 *Accuracy metric*

Confusion matrices can be used not only to assess accuracy but also to perform many other types of analysis. A confusion matrix is an analysis tool for classification, predicting whether a sample belongs to one of two classes. The confusion matrix for classification consists of four elements:

1. **True Positive (TP):** The model correctly predicts positive; the actual value is positive (true).
2. **False Positive (FP):** The model incorrectly predicts positive; the actual value is negative → Type I error.[3]

[3] Type 1 error: Predicting that something should have happened when in fact nothing did.

3. **False Negative (FN):** The model incorrectly predicts negative; the actual value is positive → Type II error.[4]
4. **True Negative (TN):** The model correctly predicts negative; the actual value is negative (true).

Various metrics to evaluate the performance of a machine learning model can be derived from the confusion matrix. The metrics are shown in Table 3.2. In the table, the values on the diagonal represent correct answers, and the other values represent incorrect answers.

Accuracy is the ratio of correctly answered questions among the total number of problems, calculated as the sum of TP and TN divided by the total number of examples.

$$\text{Accuracy} = \frac{\text{TP} + \text{TN}}{\text{TP} + \text{FP} + \text{TN} + \text{FN}}$$

In binary classification, the model's performance can be distorted based on the composition of the data, so performance cannot be evaluated solely on accuracy, and accuracy as an evaluation metric should not be used for performance measurement in imbalanced label datasets.

Table 3.2 Classification performance metrics.

		Predicted		
		Negative (0)	Positive (1)	
Actual	Negative (0)	True Negative TN	False Positive FP (Type I error)	Specificity $= \frac{\text{TN}}{\text{TN} + \text{FP}}$
	Positive (1)	False Negative FN (Type II error)	True Positive TP	Recall, Sensitivity, True positive rate (TPR) $= \frac{\text{TP}}{\text{TP} + \text{FN}}$
		Accuracy $= \frac{\text{TP} + \text{TN}}{\text{TP} + \text{TN} + \text{FP} + \text{FN}}$	Precision, Positive predictive value (PPV) $= \frac{\text{TP}}{\text{TP} + \text{FP}}$	F1-score $= 2 \times \frac{\text{Recall} \times \text{Precision}}{\text{Recall} + \text{Precision}}$

[4]Type 2 errors: Predicting that something happened when it didn't, when in fact it did.

Table 3.3 Lease company model accuracy.

Model A	Installment delinquency	Installment payment
Predicted delinquency	TP(3)	FP(5)
Predicted payment	FN(7)	TN(85)

For example, consider a model created by a car lease company to classify customers who are likely to default on their installment payments from those who will pay normally the accuracy calculation for Model A would be as shown in Table 3.3.

$$\text{Accuracy} = \frac{\text{TP}(3)+\text{TN}(85)}{\text{TP}(3)+\text{FP}(5)+\text{FN}(7)+\text{TN}(85)} = 0.88 = 88\%$$

According to the formula, Model A shows an accuracy of 88%. The leasing company would have created the model to predict customers who are likely to default on their installments, aiming to prevent defaults in advance. However, out of the actual 10 defaulters [TP(3) + FN(7)], only 3 cases [TP(3)] were predicted as defaults. Is this really reflective of reality? This is likely not the result the leasing company would want. Therefore, there are two pairs of complementary measures that can interpret performance: Sensitivity and Specificity, Precision and Recall, and their definitions and examples are as follows.

$$\text{Sensitivity} = \frac{\text{TP}}{\text{TP} + \text{FN}}$$

$$\text{Specificity} = \frac{\text{TN}}{\text{TN} + \text{FP}}$$

$$\text{Precision} = \frac{\text{TP}}{\text{TP} + \text{FP}}$$

$$\text{Recall} = \frac{\text{TP}}{\text{TP} + \text{FN}}$$

Sensitivity is the proportion of actual positive examples that are correctly identified as positive, specificity is the proportion of actual negative examples that are correctly identified as negative, precision is the proportion of actual positive examples among those identified as positive, and recall is the proportion of actual positive examples among those identified as positive. We have explored the meanings of sensitivity to recall and explained how to analyze using these metrics with examples. Table 3.4 aims to predict cancer occurrence among 165 individuals.

Table 3.4 Cancer onset classification performance evaluation metrics example.

$N = 165$	Prediction NO	Prediction YES	
Actual result NO	TN = 50	FP = 10	60
Actual result YES	FN = 5	TP = 100	105
	55	110	

A total of 165 patients were tested for the presence of the disease. Of these, 105 were cancer patients, and 60 were not. Predictions were made for two possible classes (predicted) — "YES" for predicting the occurrence of cancer and "NO" for predicting no cancer. The results showed that, out of 165 cases, the confusion matrix predicted "YES" 110 times and "NO" 55 times. Based on the example, the meanings of each matrix are as follows:

- Accuracy: (TP+TN)/n=>(100+50)/165 = 0.91
- Sensitivity: 100/100+5=0.9523
- Specificity: 50/50+10=0.8333
- Precision: 100/100+10=0.909
- Recall: 100/100+5=0.9523

These are calculated as such. Accuracy indicates how correct the predictions are overall, and recall shows how many YES predictions are made for actual cases of cancer. Specificity indicates how many NO predictions are made for cases that are actually not cancer. When considering recall, it's the proportion of samples that are actually Positive which the model predicts as Positive. In other words, it represents how well the model captures the actual Positive data. A higher recall means the model is good at not missing Positive cases. However, a higher recall can increase FP, so precision should be considered alongside when evaluating a model's performance. Note that if FP decreases and FN increases, precision increases; if FP increases and FN decreases, recall increases. Precision and recall are metrics for evaluating the performance of classification models; high precision can distinguish falsehoods well but may miss truths, while high recall may not distinguish falsehoods well but won't miss truths. These two metrics are in a trade-off relationship, and it's important to decide which is more important depending on the situation. In cancer patient differentiation, it's better to have a high recall to not miss true cases, but for a judge in a trial, a high precision to distinguish falsehoods is preferable. Therefore, precision and recall should be appropriately judged according to the model's purpose and situation to evaluate and improve its performance. If precision and recall are important, can they be considered to evaluate the model with a single numerical expression? This is where the F1-Score comes from. F1-Score is calculated as follows.

$$\text{Fl} = 2 \times \frac{\text{Precision} \times \text{Recall}}{\text{Precision} + \text{Recall}}$$

The F1-score is the harmonic mean[5] of precision and recall, having values between 0.0 and 1.0 where higher is better, and it tends to perform much better when precision and recall are close to each other.

3.3 ROC Curve

The receiver operating characteristic (ROC) curve is a graph used to evaluate the performance of classification models, primarily used in binary classification. The ROC curve is a useful tool for evaluating and comparing the performance of models, representing changes in the false positive rate (FPR) on the *X*-axis and the true positive rate (TPR) on the *Y*-axis as the model's decision threshold is continuously altered, forming a curve connecting (0,0) and (1,1). This allows for visual confirmation of the changes in FPR and TPR that result from altering the model's classification threshold. For example, representing cases where cancer patients are predicted to have cancer and those that are not is as follows.

- **TPR:** The ratio of correctly predicting a case as 1 when it is 1 (Sensitivity), diagnosing cancer in cancer patients.
- **FPR:** The ratio of incorrectly predicting a case as 1 when it is 0 (1 – Specificity), diagnosing cancer in healthy individuals.

The ROC curve is used to intuitively determine which model (classifier) performs well by plotting the ROC curve for various models to choose those with high sensitivity and high specificity. At this time, the model with a graph that is most skewed towards the upper left corner can be considered to have the highest performance. This is illustrated in Figure 3.8.

The ROC curve is useful even in cases of class imbalance, where one class has significantly more or fewer samples than another. Class imbalance refers to a situation where the number of samples in one class is much more or less compared to another class. In such situations, rather than judging a model's performance solely by accuracy, a more careful evaluation can be done considering the ROC curve

[5] Mean: The reciprocal of the average of the reciprocals of individual data, often used to calculate the average of a ratio, averaging over the dimensions of the reciprocal, and taking the reciprocal again to return to the value of the original dimension.

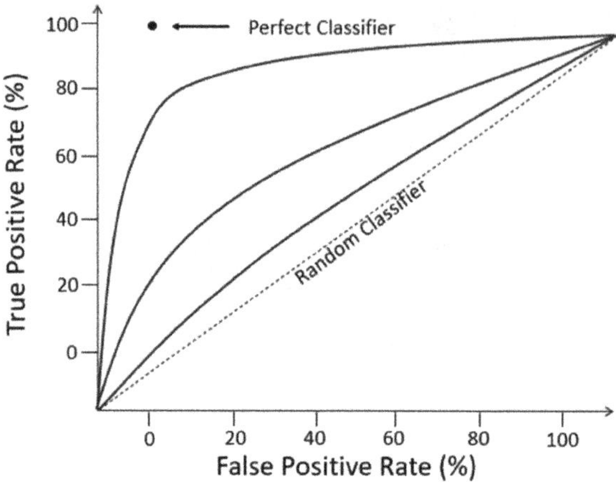

Figure 3.8 ROC curve.

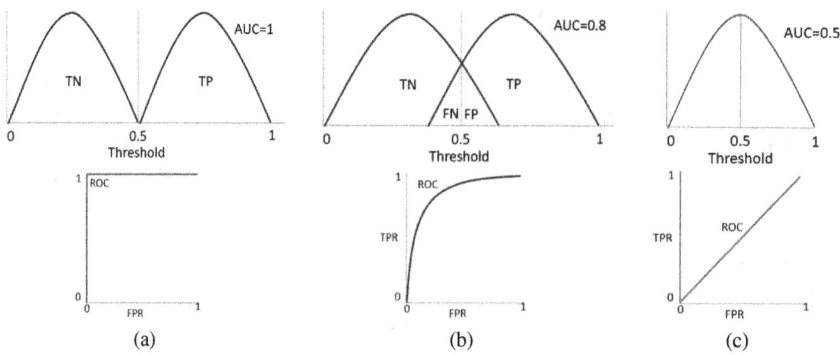

Figure 3.9 AUC model measurement curve. (a) Case of AUC = 1. (b) Case of AUC = 0.8. (c) Case of AUC = 0.5.

and AUC value. The area under the ROC curve is known as area under the curve (AUC), a numerical criterion for performance evaluation; the closer it is to 1, the closer the graph is to the upper left corner, indicating an overall good performance of the model. AUC values range from 0 to 1, with values closer to 1 indicating better classification performance. Figure 3.9 illustrates the measurement of a model's performance using AUC.

(a) The curve where two curves do not overlap at all indicates an ideal classification by the model. (b) The curve where two curves overlap represents a case with errors. The set value can minimize or maximize error values depending on the threshold. This implies that the model has an 80% probability of distinguishing between positive and negative. (c) Represents the worst-case scenario for a classification model. An AUC of 0.5 means that the model is incapable of distinguishing between positive and negative. In this case, the model's predictions should be inverted, or another model should be considered.

3.4 Metrics for Regression Models

Regression analysis is fundamentally a model that predicts continuous real values. Therefore, confusion matrices used in classification models cannot be used to evaluate the performance of regression models. To evaluate the performance of regression, one must examine the continuous real values of LOSS[6] and minimize errors. In machine learning, error is referred to as loss, cost, error, and the function defining the error is known as loss function, cost function, or error function.

Let's learn about loss and loss functions here. Loss functions compare the difference between actual values and model predictions, serving as an indicator to evaluate model performance. They serve as a standard to determine how similar the predicted values are to the actual values when arbitrary input values are processed through the model. The difference between the predicted and actual values constitutes the LOSS, and learning progresses by reducing this. Loss functions are used to reduce loss, and the model's loss is minimized by finding the point where the loss function is minimized through the gradient descent process. Therefore, loss functions are not fixed and can be varied as per the conditions described below.

- How to treat outliers?
- Which machine learning algorithm to use?
- Whether differentiation is possible to facilitate gradient descent?

These considerations can help in reducing the loss. Let's look at the metrics for evaluating loss:

- **MAE:** The average of the absolute differences between the model's predictions and actual values. It's intuitive and easier to interpret, more robust to outliers than MSE, but it doesn't indicate whether the model underperforms or overperforms due to the absolute values.
- **MSE:** The sum of the squared differences between the model's predictions and actual values. It's sensitive to outliers due to the squaring.
    - **RMSE:** The square root of MSE, makes interpreting the error metric easier by converting it back to units similar to the actual values.

---

[6]Loss: Also called error, it represents the difference between the predicted value minus the actual value.

- **Mean Absolute Percentage Error (MAPE):** MAPE converts MAE to a percentage. Like MAE, it's robust against outliers but shares the same disadvantages.
- **Root mean squared logarithmic error (RMSLE):** A log-transformed version of RMSE. Unlike RMSE, it measures relative error and penalizes underestimates more than overestimates.
- **R2 score:** A performance evaluation metric for regression models, showing how well independent variables explain the dependent variable. It indicates how well the regression equation predicts when a model is created.
- **Adjusted R2 score:** The coefficient of determination naturally increases with more dependent variables, but as the number of independent variables increases, the coefficient also increases, making it difficult to evaluate regression models based solely on the coefficient of determination.

Therefore, to compensate for this characteristic, the adjusted R2 Score is used, considering the size of the sample and the number of independent variables.

3.5 Imbalanced Datasets

Datasets require careful consideration from the point of creation. A quality dataset evenly contains an equal number of positive and negative data. However, most cases are not so, which can lead to problems in learning. This results in imbalanced datasets. Imbalanced Data occurs when the target/output variable is categorical and there is a significant difference in the number and proportion of observations per category, such as a normal to abnormal ratio of 90%:10%, or even more severely imbalanced like 99%:1%. In practice, it's not uncommon to encounter imbalanced data where the abnormal observations of interest are much less frequent than normal ones.

- For example, there are many challenging imbalanced datasets, such as telecom companies dealing with retained versus churned customers, credit card companies differentiating between normal and fraudulent transactions, and medical data comparing patients with and without certain conditions.
- In imbalanced data, if a classification model is trained with majority class dominating and minority class in fewer numbers, there's a high risk of creating a useless model that cannot properly classify the critical minority class.

For instance, if the ratio of majority class (normal) to minority class (abnormal) is 99%:1%, and the trained classification model classifies all data as "normal (majority class)," the accuracy will be 99%. At first glance, it may seem like a well-made model, but since it fails to properly classify the "abnormal (minority class)" category of interest, it becomes useless. There are mainly two approaches to address data imbalance: resampling and cost-sensitive learning. Resampling is an effective approach to solve data imbalance, including two main methods: under-sampling and oversampling. Undersampling reduces data from the majority class to balance with the minority class, while oversampling increases the data of the minority class to achieve the same goal. However, there are certain caveats to consider.

- Collecting more data for the minority class is needed but can be challenging in reality.
- Under sampling, which involves deleting samples from the majority class to balance numbers, leads to information loss and is therefore used when there is abundant data.
- Over sampling is used when data is scarce, increasing the risk of overfitting but is generally more commonly applied.

Cost-sensitive training (or penalized model) allows for giving more weight to the minority class, making it possible to achieve better results than traditional training methods on datasets with data imbalance. This approach, which rebalances by giving different loss values to each class at the algorithm level, can enhance the model's generalization performance by penalizing the head-class or giving more weight to the tail-class.

3.6 Overfitting and Underfitting

The ultimate goal of machine learning is to predict the test dataset using a model trained with the training dataset. To create a model that performs well, it is necessary to ensure that the developed model is neither overtrained nor undertrained. Therefore, when conducting analysis using machine learning, it is essential to consider the issues of overfitting and underfitting. Overfitting[7] is a problem that occurs in machine learning models when they consider too many aspects of the dataset during training. This results in excessively complex learning. While the model may show high performance on the training dataset, it may not accurately predict or classify new data. This happens because if the model learns too much data, it ends up learning dataset noise and inaccurate features as well. As a result, high variance is observed when evaluating with test data, leading to inaccurate predictions. In other words, the model becomes overly dependent on the training data and does not perform well on new data. Therefore, it is necessary to prevent overfitting. To reduce overfitting, methods such as diversifying, adjusting, and

---

[7] Overfitting: Optimizing for training data that deviates from the overall characteristics.

normalizing the training dataset can be used, along with prioritizing features based on their importance and eliminating elements that do not affect the prediction outcome. Additionally, various methods such as reducing the complexity of the model, applying dropout, and using early stopping should be employed to solve the problem. Underfitting[8] refers to a situation where the model fails to overfit or learn generalization properties, resulting in large errors in both training and test data, indicating the model's simplicity and poor predictive power. It's a case where the model trains well on the training dataset but performs poorly on the test dataset, indicating a lack of generalization to new data. Underfitting can be caused by the model being too simple compared to the data, insufficient data compared to the model, or unrefined and noisy training data. To resolve underfitting, increasing the complexity of the model by reducing dropout, increasing the number of features, and removing noise from the dataset are effective methods. Additionally, increasing the number of epochs to extend the training sessions is another method. Figure 3.10 shows overfitting, overtraining, and an appropriate state.

Figure 3.10(a) appears to approximate the data on the left relatively well, but the rest of the data is scattered away from the regression graph, with data in the area of large variables distributed towards the bottom right, indicating that the regression graph is not fitting properly as it continues to ascend. Figure 3.10(b) shows a graph that passes through all the data across the entire range of variables. It doesn't show a perfect fit for all the data, but it can be predicted to reasonably fit new data that comes in. For Figure 3.10(c), if the error between the training data and the generated model is calculated, it would be almost zero. The regression graph passes through all points, and it can be judged that the model's performance is better as it creates a graph that satisfies the training data more than Figure 3.10(b). In Figure 3.11, to determine which model is better, one test data point is added to compare the error with the model, and the sample data is represented with

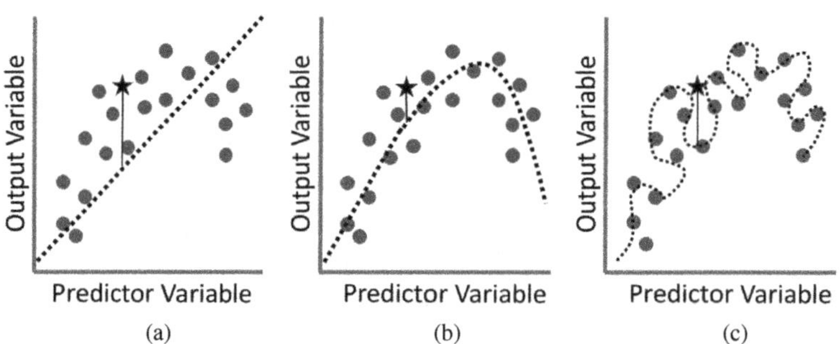

Figure 3.10 Examples of (a) underfitting, (b) fitting, and (c) overfitting.

[8] Underfitting: Occurs when a model is too simple to learn the intrinsic structure of the data.

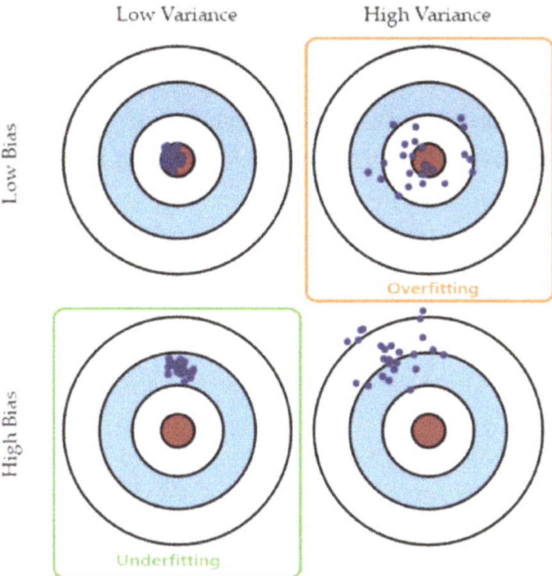

Figure 3.11 Bias and variance.

an asterisk. The added sample data did not significantly deviate from the trend of the entire training dataset in Figures 3.10(a)–3.10(c). However, looking at the error with each generated model, Figure 3.10(a) shows a significant error, Figure 3.10(b) shows little difference, and Figure 3.10(c), despite having no error with the training data, shows a large error with the sample data. Underfit models provide inaccurate results for both training and test datasets, thus exhibiting high bias. On the other hand, overfitting increases variance, making it sensitive to noise in the training data, which may lead to poor generalization on test data.

Bias can be seen as a measure of how far from the actual values it is, based on the perceived accuracy of the training data, and variance is a measure of how much the data is dispersed. Underfitting indicates an inability to get close to actual values and a concentration of predictions in specific areas, showing that learning has not occurred properly, and the prediction results, though close to the actual values, are widely dispersed. In other words, the predictive accuracy for untrained elements is low.

3.7 Practice Questions

Q1. What are the main purposes of splitting data into training, validation, and test sets in the context of machine learning model evaluation?

Q2. How does cross-validation benefit the process of evaluating machine learning models, and what problem does it help mitigate?

Q3. Describe the importance of addressing overfitting and underfitting in machine learning models and provide two strategies for dealing with these issues.

Chapter 4

Data and Source Collection for Machine Learning

4.1 Kaggle

Established in 2010, Kaggel hosts predictive modeling and analytics competitions, where data scientists from around the world and companies participate in ongoing contests to determine who can classify or predict more efficiently and effectively, with prize money at stake. Figure 4.1 shows the homepage of Kaggle, a website for data science and machine learning.

The competition topics vary with each event. As the winning algorithms are accessible, researchers in the field must refer to this site. For researchers interested in a specific topic but facing challenges in obtaining data, Kaggle provides uploaded datasets for preliminary testing, allowing them to formulate a research direction before proceeding. Kaggle offers a diverse range of datasets, including those tailored for competitions and others for researchers to test and analyze, such as text, images, and sound. Figure 4.2 shows a partial list of Kaggle datasets.

A partial list of Kaggle datasets includes examples.

- **Titanic: Machine learning from disaster:** A beginner-friendly dataset for predicting survivors of the Titanic. It's used to predict survival based on information like gender, age, ticket class, and more.
- **House prices: Advanced regression techniques:** A dataset for predicting house sale prices. It includes a number of house characteristics (location, size, number of rooms, etc.), making it perfect for practicing regression analysis.
- **MNIST handwritten digit database:** A large database of handwritten digit images. It is widely used to practice image processing and classification algorithms.

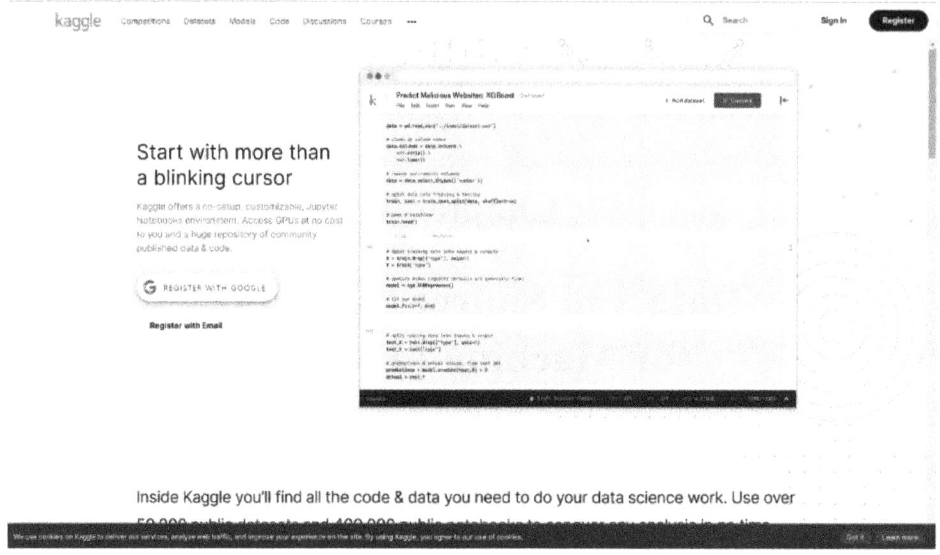

Figure 4.1 Kaggle site homepage.

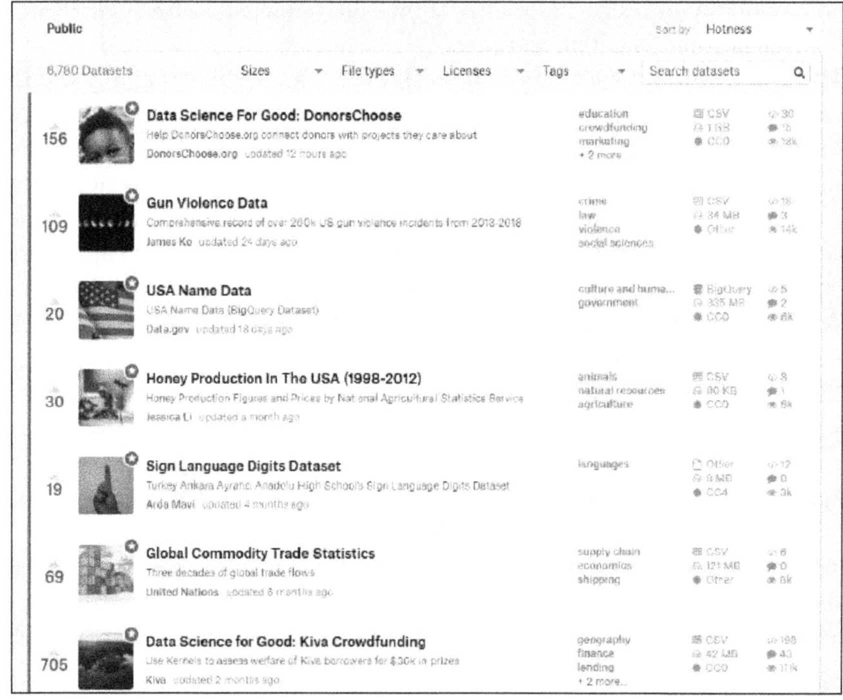

Figure 4.2 Partial list of Kaggle datasets.

- **Google landmark recognition:** A dataset for recognizing images of famous landmarks around the world. Used for research to advance image recognition and processing techniques.
- **COVID-19 open research dataset challenge (CORD-19):** A dataset that collects academic papers, articles, and more related to COVID-19. It was organized to help research and analyze information about the pandemic.
- **Common voice:** A large speech dataset in multiple languages from Mozilla. It contributes to the development and improvement of speech recognition systems.

In addition to these, Kaggle offers tons of other datasets across a variety of fields, including finance, healthcare, education, natural language processing, and more. Each dataset can be used to solve a specific problem, practice a skill, or gain new insights, and is considered a valuable resource by the data science and machine learning community.

4.2 GitHub

GitHub is a web service by Microsoft based on Git, a distributed version control software. It provides features that support source code hosting and collaboration. Users can remotely manage Git repositories and utilize web services for issue tracking and collaboration. Figure 4.3 is GitHub homepage.

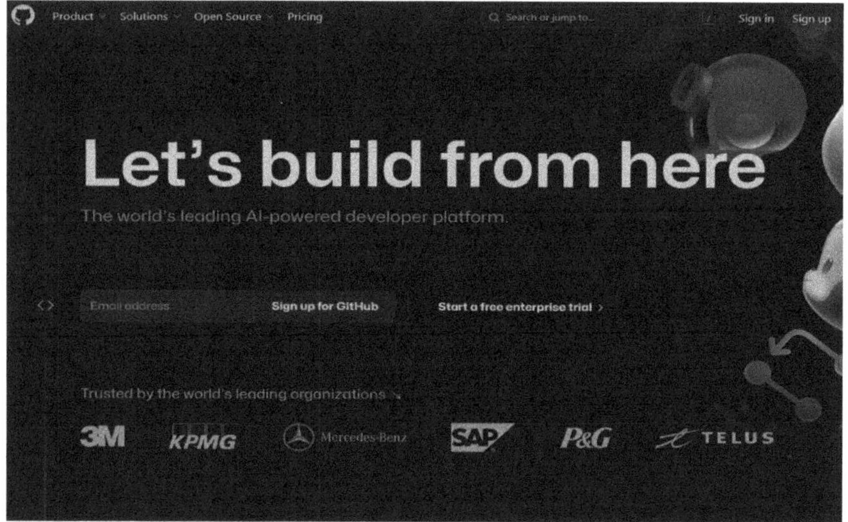

Figure 4.3 GitHub homepage.

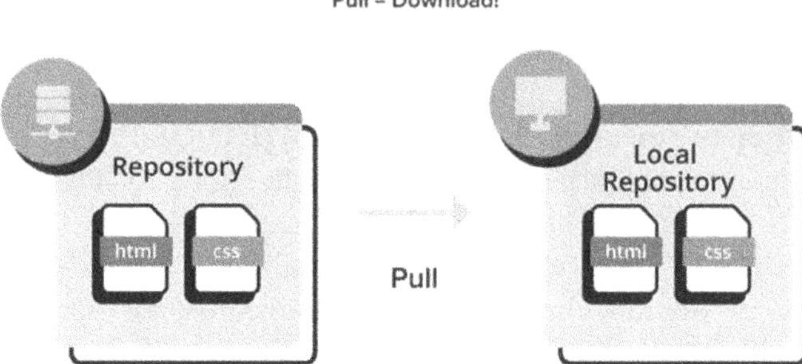

Figure 4.4 GitHub page.

With features like forking to copy repositories and pull requests to request code changes, GitHub is well-suited for collecting source code related to data manipulation rather than data collection. As a result of these functionalities, GitHub is renowned as a central hub for various open-source projects. It hosts JavaScript framework Vue.js, container tool Docker, web framework Ruby on Rails, machine learning library TensorFlow, Python data analysis library Pandas, Facebook's JavaScript framework React, information visualization library D3, and serves as a platform for development and collaboration.

As shown in Figure 4.4, the core functionality of GitHub, much like downloading from the Cloud, lies in hosting Git remote repositories. Through this, users can securely access Git repositories both in their local development environment and online. A repository is divided into two types: the Local repository, which is the version of the project repository stored on one's own computer, and the Remote repository, which is the version of the project repository on a server other than one's own computer. The Remote repository is useful for collaboration within a team, allowing for the sharing and review of project code. Additionally, it provides the advantage of merging with the Local version of the project and applying changes. In addition to repository hosting, GitHub offers functionalities such as issue tracking, pull requests, source code browsing, wikis, insights, and more. It provides not only individual repositories but also features for managing repositories at the team level.

4.3 UCI Machine Learning Repository

The UCI Machine Learning Repository is hosted and managed by the Center for Machine Learning and Intelligent Systems at the University of California, Irvine. Figure 4.5 is UCI Machine Learning Repository site.

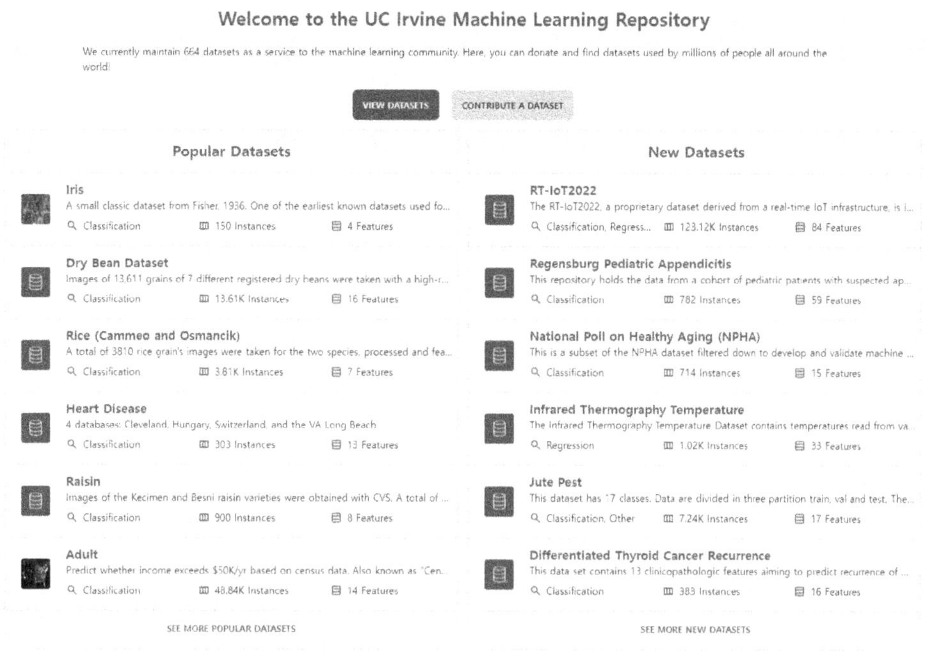

Welcome to the UC Irvine Machine Learning Repository
We currently maintain 664 datasets as a service to the machine learning community. Here, you can donate and find datasets used by millions of people all around the world!
VIEW DATASETS
CONTRIBUTE A DATASET
Popular Datasets
Iris
A small classic dataset from Fisher, 1936. One of the earliest known datasets used fo...
Classification
150 Instances
4 Features
Dry Bean Dataset
Images of 13,611 grains of 7 different registered dry beans were taken with a high-r...
Classification
13.61K Instances
16 Features
Rice (Cammeo and Osmancik)
A total of 3810 rice grain's images were taken for the two species, processed and fea...
Classification
3.81K Instances
7 Features
Heart Disease
4 databases: Cleveland, Hungary, Switzerland, and the VA Long Beach
Classification
303 Instances
13 Features
Raisin
Images of the Kecimen and Besni raisin varieties were obtained with CVS. A total of ...
Classification
900 Instances
8 Features
Adult
Predict whether income exceeds $50K/yr based on census data. Also known as "Cen...
Classification
48.84K Instances
14 Features
SEE MORE POPULAR DATASETS
New Datasets
RT-IoT2022
The RT-IoT2022, a proprietary dataset derived from a real-time IoT infrastructure, is i...
Classification, Regress...
123.12K Instances
84 Features
Regensburg Pediatric Appendicitis
This repository holds the data from a cohort of pediatric patients with suspected ap...
Classification
782 Instances
59 Features
National Poll on Healthy Aging (NPHA)
This is a subset of the NPHA dataset filtered down to develop and validate machine ...
Classification
714 Instances
15 Features
Infrared Thermography Temperature
The Infrared Thermography Temperature Dataset contains temperatures read from va...
Regression
1.02K Instances
33 Features
Jute Pest
This dataset has 17 classes. Data are divided in three partition train, val and test. The...
Classification, Other
7.24K Instances
17 Features
Differentiated Thyroid Cancer Recurrence
This data set contains 13 clinicopathologic features aiming to predict recurrence of ...
Classification
383 Instances
16 Features
SEE MORE NEW DATASETS

Figure 4.5 UCI machine learning repository.

Each dataset retrieves its own web page listing all the detailed information, including relevant publications, related to the investigated details. Therefore, as the datasets are freely accessible in ASCII and CSV formats, it serves as a valuable resource for collecting data in the fields of machine learning and data science. As one of the oldest and most accessible collections of datasets, it plays a pivotal role in the development and experimentation of machine learning techniques and methodologies. The repository offers a diverse range of datasets used for various tasks such as classification, regression, clustering, among others. These datasets range from simple toy datasets, ideal for teaching and initial experiments, to complex real-world data. One of the key characteristics of the UCI Repository is its simplicity and ease of use. Each dataset comes with a detailed description, including information about the dataset's background, attributes, and any relevant papers that have used or referenced it. Over the years, the repository has significantly grown both in the number of datasets it offers and in its reputation within the machine learning community. The repository has become a benchmark for evaluating and comparing algorithms using its datasets. It is widely cited in numerous research papers and is considered an essential resource for empirical analysis in machine learning. In addition to its utility for education and research, the UCI Machine Learning Repository also serves as a historical archive for the machine learning community. It reflects the evolution of the field, showcasing how datasets and the challenges they present have developed over time. From simple datasets like the famous Iris dataset to more complex and high-dimensional data, the repository offers a window into the history and development of machine learning.

4.4 Google Dataset Search

Google Dataset Search is the ultimate search engine for researchers, data journalists, and data enthusiasts to find the datasets they need for their work and is the ultimate tool for finding publicly accessible datasets on the internet. Figure 4.6 shows the Google Dataset Search website.

With its extensive coverage and user-friendly interface, this search engine indexes datasets from a variety of sources, such as universities, governments, and organizations, making it possible to access a vast amount of data from a single search. Its unique features distinguish it from other search engines, making it the go-to choice for anyone looking for datasets online. The platform covers a diverse range of disciplines, including environmental science, social sciences, and government data, as well as education datasets. This makes it an indispensable resource for interdisciplinary research. Users can easily access datasets on a variety of topics, such as climate change, economic trends, healthcare data, public transportation patterns, and more. Additionally, each indexed dataset is accompanied by key information that provides valuable context and relevance. This section contains

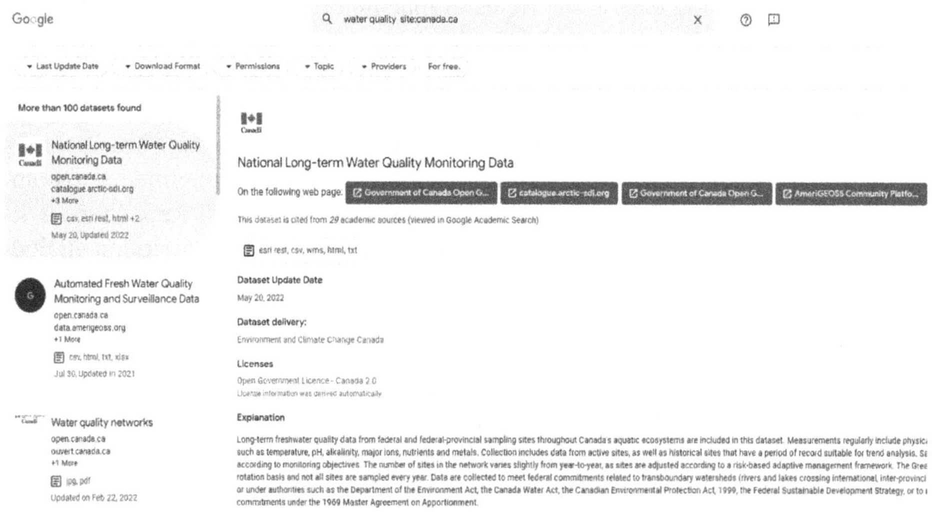

Figure 4.6 Google dataset search.

the title of the dataset, the name of the provider, a brief description, and occasionally the geographical area that the data covers.

4.5 AWS Public Datasets

AWS Public Datasets is a collection of rich open datasets provided by Amazon Web Services. The purpose of this service is to facilitate easy access and analysis of large datasets for researchers, data scientists, developers, and educators. AWS hosts data collected from various fields, offering an environment to leverage data without the need for complex infrastructure. Figure 4.7 shows the AWS Public Dataset.

AWS Public Datasets include data from diverse domains such as geographic information systems (GIS), life sciences, meteorology, space science, and more. For instance, Earth observation data, genomics research data, and meteorological data are made available to users. These datasets are widely utilized for public research, innovative projects, and educational purposes. By simplifying the storage and management of datasets, AWS enables users to focus more on data analysis and value creation instead of spending time on complex data management. Additionally, leveraging AWS's cloud computing resources allows users to process large datasets quickly and efficiently. Another significant feature of AWS Public Datasets is its user-friendly nature. Through the AWS Management Console, users can easily discover the datasets they need and utilize them with a variety of AWS analysis and processing tools. This streamlines the data analysis and processing workflow, allowing users to concentrate more on their domain expertise.

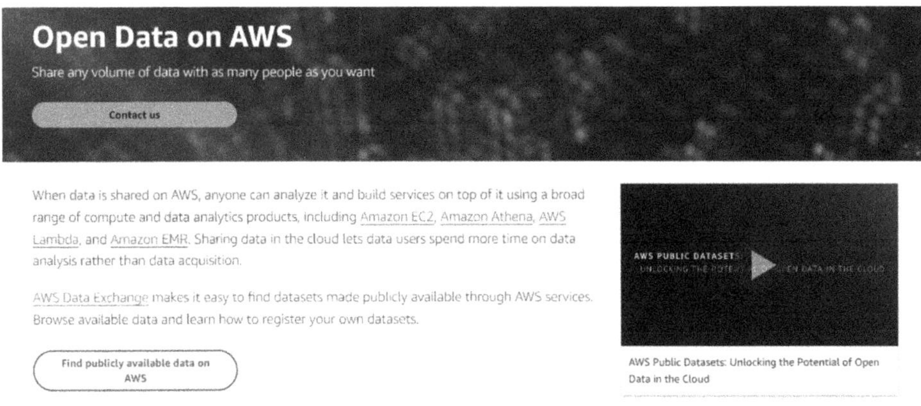

Figure 4.7 AWS public datasets.

4.6 Data.gov

Data.gov is an online data portal operated by the US government, providing access to vast amounts of public data held by various government agencies. Designed to enhance public transparency and promote citizen engagement, this platform encompasses datasets collected by multiple institutions. Figure 4.8 shows the Data.gov website.

The purpose of Data.gov is to foster innovation through data, improve the efficiency of government operations, and deliver valuable information to citizens. Data.gov offers data on various topics, including environment, health, education, science, energy, security, and international relations. These datasets provide valuable information for researchers, policymakers, developers, journalists, and the general public. One of the main advantages of this portal is its user-friendly interface and search functionality. Users can easily find desired datasets through keyword search, category-based exploration, or institution-specific filtering. Additionally, most datasets are available for download, and support for API access enables developers and researchers to seamlessly integrate them into their projects or studies. Furthermore, Data.gov plays a significant role as an educational resource, allowing students and teachers to conduct learning and research based on real-world data. This helps students develop skills in data science and analysis, gaining valuable experience in solving real-world problems.

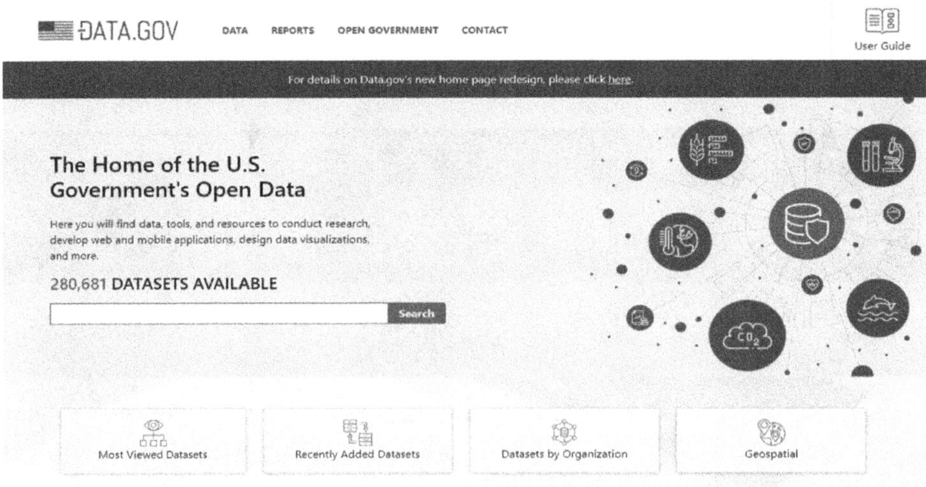

Figure 4.8 Data.gov.

4.7 Stanford Large Network Dataset Collection

The Stanford Large Network Dataset Collection (SLNDC) is an extensive compilation of large-scale network datasets provided by Stanford University. Figure 4.9 shows the Stanford Large Network Dataset Collection website.

Designed for research in complex networks and systems, it serves as a vital resource for researchers in various fields such as social science, computer science, mathematics, and statistics. SLNDC plays a crucial role, particularly in the areas of graph theory, complex network analysis, system modeling, and big data analysis. The datasets include various types of networks such as social networks, internet networks, biological networks, and infrastructure networks. Examples include online social media interactions, internet connectivity structures, protein interaction networks, and air traffic networks. These datasets can be utilized to study phenomena such as structural characteristics of networks, dynamic behaviors, information propagation, and community formation. One of the key strengths of SLNDC lies in the diversity and scale of its data. This collection includes network datasets with hundreds to thousands of nodes, making it highly suitable for large-scale network analysis and experiments. Additionally, the datasets are systematically organized, allowing researchers easy access and analysis. SLNDC also holds significant value as an educational resource, with these datasets being applicable for teaching network theory and analysis techniques in university courses.

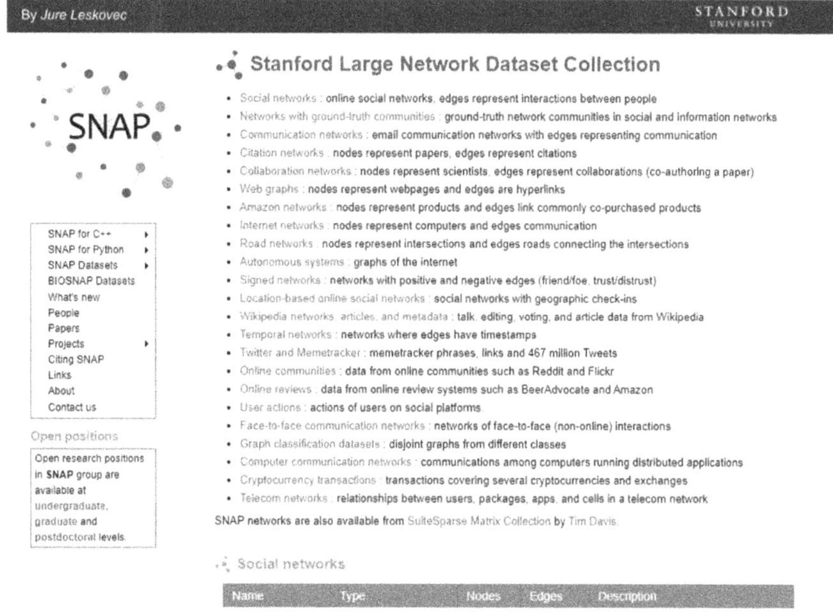

Figure 4.9 Stanford large network dataset collection.

Table 4.1 Link to the data collection site.

Name	URL
Kaggle	https://www.kaggle.com/
GitHub	https://github.com/github
UCI Machine Learning Repository	https://archive.ics.uci.edu/
Google Dataset search	https://datasetsearch.research.google.com/
AWS Public Datasets	https://aws.amazon.com/datasets/
Data.gov	https://catalog.data.gov/
Stanford Large Network Dataset Collection	http://snap.stanford.edu/

Students can apply theoretical knowledge in practical exercises using real-world complex network data. Table 4.1 provides example links to the introduced data collection sites.

4.8 Practice Questions

Q1. Describe the role of Kaggle in the machine learning community and how it supports both competition and research.

Q2. Explain how GitHub, despite being primarily a code hosting platform, contributes to machine learning data and source collection.

Q3. Discuss the significance of the UCI Machine Learning Repository to the development and experimentation of machine learning techniques.

Chapter 5

Deep Learning

5.1 Definition and Concept of Deep Learning

Artificial intelligence, as previously mentioned, is defined as understanding human judgment, behavior, cognition, and enabling machines to possess human intelligence. Recently, many people seem to confuse machine learning and deep learning in the field of artificial intelligence. Therefore, I would like to explain it in an easy-to-understand way. Artificial intelligence should be designed to enable machines to have the same level of intelligence as humans, and for this, they need to undergo learning, with machine learning being one area of this learning. Deep learning, which is part of machine learning, is a field that learns similar to neural networks, and Figure 5.1 represents this relationship.

In simple terms, deep learning can be defined as a field of machine learning that teaches computers to think like humans, based on artificial neural networks that enable computers to learn on their own through a high level of abstraction resembling the human brain. Neurons in the brain, which are specialized cells that transmit electrical signals, can be described as a collection of neurons, with approximately 100 billion of them and trillions of connections. These neurons are connected in a network, which is referred to as a neural network, and an artificially created network is called an artificial neural network. Figure 5.2 depicts a neural network with connected neurons.

Neurons receive signals in their connected dendrites and transmit signals in their axons. In between, there are synapses through which signals are transmitted by sending or storing information in electrical signals above the threshold value. Neurons are connected together, receiving signals in their dendrites and transmitting signals in their axons, with synapses between each neuron.

- **Branches:** The part that receives signals transmitted by cells.
- **Axon:** Transmits signals from one cell to another cell.

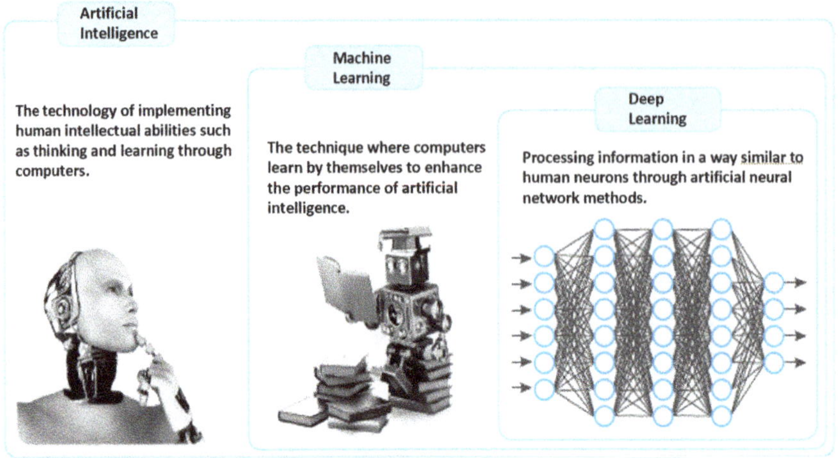

Figure 5.1 Relationship diagram of artificial intelligence, machine learning and deep learning.

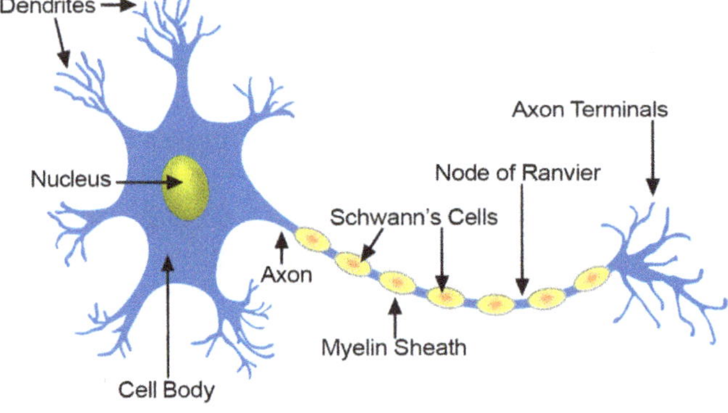

Figure 5.2 Neuron.

- **Synapse:** The part between the dendrites and axon terminals that is responsible for transmitting the strength of signals. Depending on the development of the synapse, the same signal can be transmitted strongly or weakly.
- **Cell body:** The organelle generates its output signal from the received signals through dendrites and transmits it to the next cell. In this process, the signal is transmitted in a nonlinear manner, not simply as the sum of input signals.

A neural network is a modeling of the basic unit of the nervous system, which is the neuron. In an artificial neuron (node), it receives multiple input signals and outputs a single signal. This is similar to the actual neurons sending out electric signals to transmit information. Only signals that exceed a certain threshold are

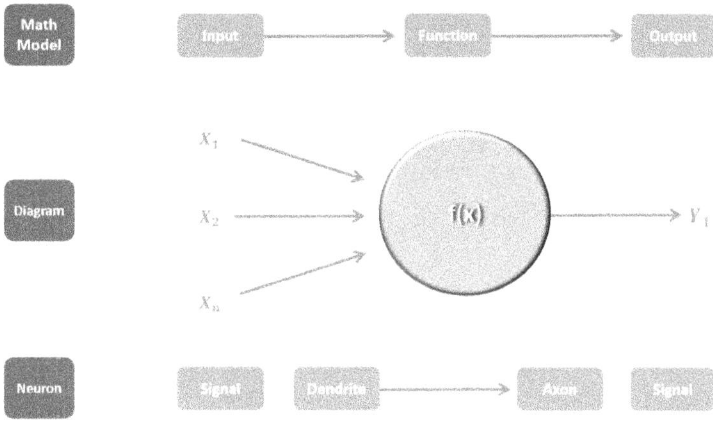

Figure 5.3 Neural network modeling example.

transmitted, and they are used to transmit or store information as electric signals. If we express this flow of artificial intelligence in mathematics and neurons, it would be represented as shown in Figure 5.3.

The operating principle of a neural network can be represented in the form of a function that takes input at one node and produces output. As shown below, one node can be seen as a function that receives input signals and delivers results. Therefore, if we receive an input signal x and output y, it can be represented as follows.

$$y = wx + b$$

Here, w represents the weight, and b represents the bias. For example, when an input signal x is received, it is calculated with a pre-assigned weight w, and if the resulting value exceeds the threshold value θ, it outputs 1. In other words,

$$wx > \theta \Rightarrow 1$$

When θ is moved to the left-hand side of the equation, we get $b = -\theta$ (Figure 5.4).

In the same way, the cases with two input signals and three input signals are as follows:

$$y = ax_1 + bx_2 + c$$
$$y = ax_1 + bx_2 + cx_3 + d$$

The formula above can be expressed as shown in Figure 5.5.

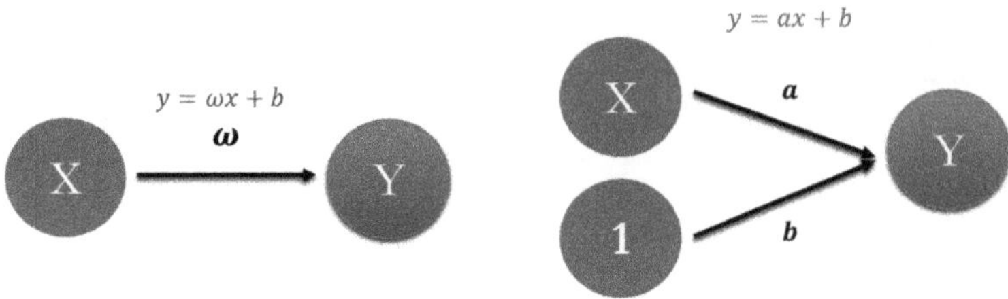

Figure 5.4 Bias.

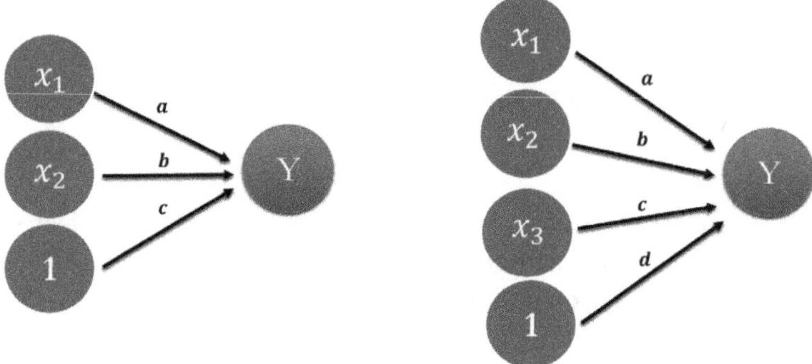

Figure 5.5 Multibias.

If there are two outputs, it can be represented using matrix multiplication as follows. For convenience, let's assume the bias is 0, and w_{ij} represents the weight connected from the ith node of the input layer to the jth node of the output layer (Figure 5.6).

Just as a neuron requires a stimulus above a threshold to respond, in an artificial neuron, when multiple input signals are given, they are calculated with pre-assigned weights, and if their total sum exceeds a predetermined threshold, it outputs 1, and if it does not, it outputs 0 (or −1). The function that determines this output is called the activation function. A typical example of an activation function is the sigmoid function.

$$f(x) = \frac{1}{1+e^{-x}} = \frac{e^x}{e^x+1}$$

This is an approximation of the step function or heaviside function, which is activated or deactivated based on a threshold value. In Figure 5.7, the sigmoid function is illustrated.

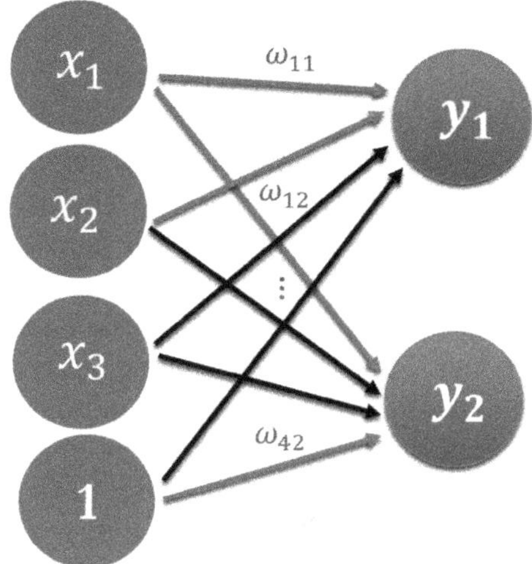

Figure 5.6 Bias weight.

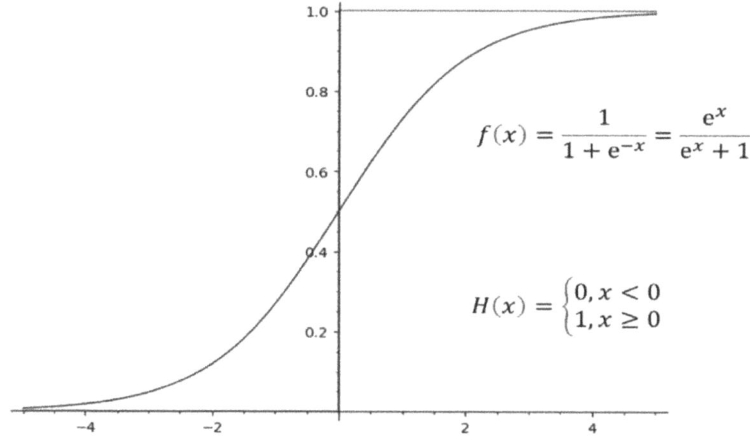

Figure 5.7 Sigmoid weight.

$$H(x) = \begin{cases} 0, x < 0 \\ 1, x \geq 0 \end{cases}$$

Deep learning technology is a widely used technology already by global IT companies such as Google, Facebook, and Amazon. It is known for its excellent performance in pattern recognition, image and voice recognition, natural language processing including machine translation. Deep learning is divided into perceptron, multilayer perceptron (MLP), and the current deep learning.

5.1.1 *Perceptron*

The origin of the neural network can be traced back to the perceptron, proposed by Frank Rosenblatt in 1958. The perceptron, as an early neural network model capable of learning, can be seen as a primitive neural network. It is a classifier that categorizes input data into one of two categories. The perceptron simulated the way neurons transmit information through electrical signals in a statistical manner, with the concept of applying weights to each of its n inputs. A single-layer perceptron consists of an input layer and an output layer. The input layer is where data is entered, and the entered data is transmitted to the output layer neurons, where it is output according to the activation function. The output layer can be composed of an arbitrary number of neurons in the design of the perceptron, and Figure 5.8 illustrates a conceptual diagram of an output layer composed of a single neuron.

In Figure 5.8, each input value is weighted and then goes through a net input function before reaching the activation function stage. The use of a function in the activation stage can be said to enhance the expressiveness of the model. Improving the model means increasing its complexity, which plays a crucial role in solving nonlinear problems. By using an activation function, the output value for an input does not remain linear, thus transforming a linear classifier into a nonlinear system. To elaborate, an activation function is a function that decides whether to activate the output value, and if activation is needed, it assigns an activation value. In other words, it is designed to activate under specific conditions. For example, let's assume there is the following equation:

$$x0(0.1)w0(0.3) = 0.3$$
$$x1(0.5)w1(0.5) = 0.25$$
$$xn(0.7)wn(0.7) = 0.49$$

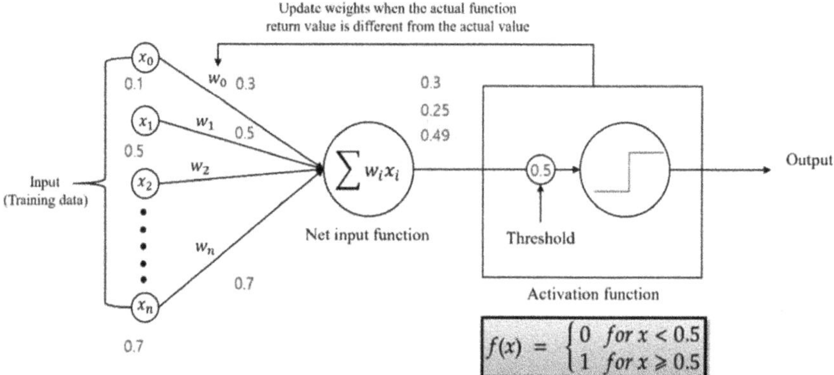

Figure 5.8 Perceptron concept diagram.

The value of the net input function is fed into the activation function based on a threshold of 0.5. If the predicted value is greater than 0.5, it outputs "1," or if it is less than or equal to 0.5, it outputs "0." Therefore, if the threshold for activation is exceeded, activation occurs; if not, it remains inactive. The condition for activation varies depending on the activation function used. Commonly used activation functions include the sigmoid function, the tanh function, and the ReLU function, each with different threshold values. Figure 5.9 illustrates these activation functions.

The term "sigmoid" means "S-shaped" and in the equation, "e" in e minus x squared refers to the natural constant, which has a value of approximately 2.7182... It is a real number that takes a real number value as input and expresses it as a value between 0 and 1. The "tanh" function is similar to the sigmoid function, but it takes a real number value as input and expresses it as a value between –1 and 1. The "ReLU" function is the most commonly used activation function. It outputs the input as it is if the input is greater than 0, and it outputs the input if it is less than or equal to 0. After the operation of the activation function, as shown in Figure 3.4, the weighted sum of signals is calculated by assigning weights to each of the n input signals. If the total sum of the signals exceeds a threshold value, it outputs a value of 1; otherwise, it outputs either 0 or 1. In other words,

$$w_1x_1 + w_2x_2 + w_3x_3 + \cdots + w_nx_n > \text{Threshold value}$$

If it is "1," otherwise it becomes "0." The weights serve as parameters that determine how much influence the input value will have on the output value by distinguishing the boundaries of the threshold value. In other words, the input value is entered into each neuron along with its respective weight, and is transformed into a single value through the linear input function. Although the perceptron is the beginning of artificial neural networks, like any other beginning, it started to show limitations in solving problems with very simple XOR models.

Figure 5.10 XOR operation, as shown in the picture, is a logical operation where the output is 0 when the input values are the same, and 1 when they are different. However, when trying to model this operation, simple linear

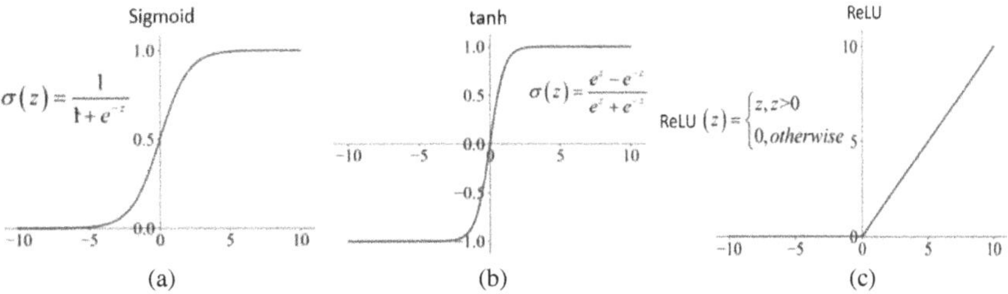

Figure 5.9 Activation function. (a) Sigmoid function, (b) tanh function and (c) ReLU function.

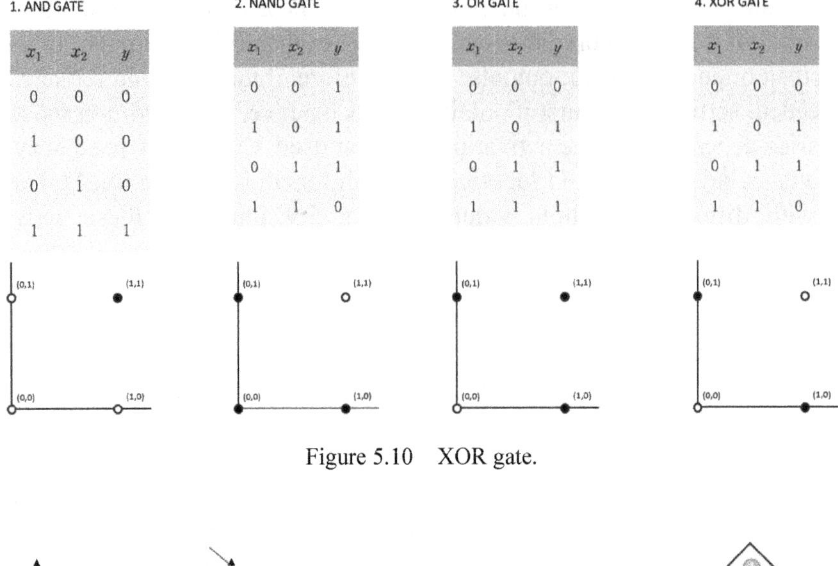

x_1	x_2	y (AND)
0	0	0
1	0	0
0	1	0
1	1	1

x_1	x_2	y (NAND)
0	0	1
1	0	1
0	1	1
1	1	0

x_1	x_2	y (OR)
0	0	0
1	0	1
0	1	1
1	1	1

x_1	x_2	y (XOR)
0	0	0
1	0	1
0	1	1
1	1	0

Figure 5.10 XOR gate.

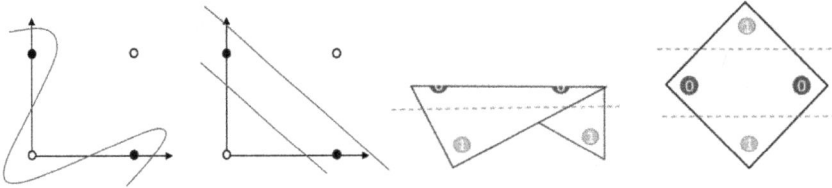

Figure 5.11 XOR classification.

classification methods alone cannot appropriately separate these patterns. Looking at Figure 5.10, the part that cannot be classified is the XOR section. The method to linearly separate this section is shown in Figure 5.11.

The core challenge of the XOR problem is that it cannot be separated linearly, and a complex network structure is required. Additionally, the application of a nonlinear activation function is necessary, and it was difficult to solve the problem with limited data. Due to this complexity, the solution to the XOR problem was not easy initially. Therefore, concepts such as MLP and nonlinear activation functions were introduced to overcome this problem. The introduction of this approach laid the foundation for neural networks and deep learning technology in the field of artificial intelligence.

5.1.2 *Multilayer perceptron*

The multilayer perceptron was introduced to solve the disadvantage of perceptron that cannot learn simple tasks like XOR, and the solution is surprisingly simple. As shown in Figure 5.11, if we draw curves or divide it into two using straight

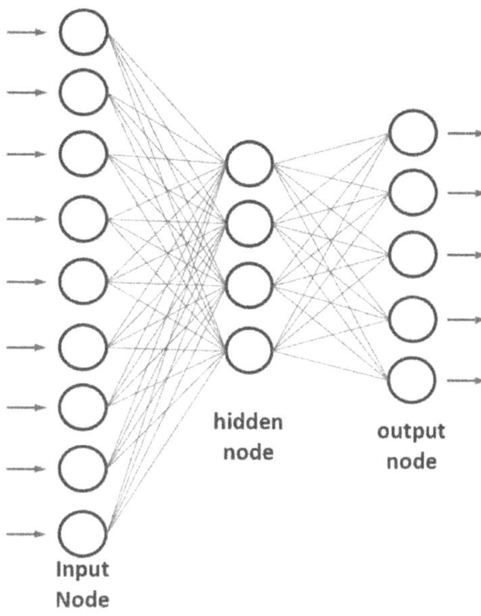

Figure 5.12 Multilayer perceptron concept diagram.

lines like the AND or OR operations, classification becomes possible. So, by applying both AND and OR operations, it can be solved. In other words, by adding a hidden layer[1] to the single-layer perceptron consisting of the input layer and output layer and calculating the weighted sum in the hidden layer as with the existing output layer, the problem can be solved. Not only one but several intermediate and hidden layers were included. Therefore, by adding one or more intermediate layers, it became a multilayer perceptron, and it was observed that the classification ability improved as the number of hidden layers, also known as hidden layer 1 or hidden layer, increased. To solve more complex problems than the XOR problem, multilayer perceptron can have numerous additional hidden layers. The number of hidden layers can be two or even dozens, depending on the user's preference. Figure 5.12 illustrates the concept of a multilayer perceptron.

The introduction of the multilayer perceptron made it possible to solve not only the XOR problem but also more complex problems that cannot be classified linearly. However, as the number of hidden layers increases, the number of weights also increases, making learning difficult. The traditional gradient descent method uses gradients to update weights, but updating the weights in multiple layers requires a significant amount of computation and memory. To solve this, the backpropagation algorithm was devised. Until now, the operation of the multilayer perceptron was a feed-forward neural network (FFNN), moving from the input

---

[1] Hidden layer: Any layer that lies between the input and output layers.

layer to the output layer. This feed-forward network propagates the input values to derive predicted values, calculates the error with the target values, and then uses the backpropagation algorithm to update the weights to reduce the error with the actual values.

The backpropagation algorithm propagates errors from the output layer to the input layer and updates the weights of each layer to overcome difficulties. Subsequent research in the field of deep learning has continued using various methodologies, but with the re-examination of learning methods based on control theory, new artificial neural networks have emerged. The Boltzmann machine, proposed by Geoffrey Hinton and Terry Sejnowski, improved upon the existing Hopfield network by combining it with neural network algorithms, serving as a powerful computational device for large-scale parallel processing and a stochastic recurrent network that allows learning based on its internal structure, capable of solving various combined problems. Through such methods, many disadvantages of the multilayer perceptron have been resolved, enabling the use of unlabeled data and unsupervised learning, which is expected to solve many of the longstanding issues.

5.2 Types of Artificial Neural Networks

Artificial neural networks come in various types, categorized by algorithms and theories, with complex multiinputs, directional feedback loops,[2] unidirectional or bidirectional flows, and diverse layers. However, the commonly referenced artificial neural networks are generally divided into deep neural network (DNN), convolutional neural network (CNN), and recurrent neural network (RNN) for explanation.

5.2.1 *Deep neural network*

The DNN include several hidden layers between the input and output layers, allowing them to model complex nonlinear relationships similar to general artificial neural networks. Utilizing multiple hidden layers allows for more sophisticated processing of input signals than using just one hidden layer.

The model of DNN is illustrated in Figure 5.13. To learn complex data, artificial intelligence technology capable of handling intricate learning through multiple hidden layers is needed. Not only do perceptron fail to solve XOR operations, but they are also limited in handling complex learning, necessitating DNN models that can consist of dozens to hundreds of hidden layers.

[2] Feedback loop: The repeated impact of an effect (output) on a cause (input), reinforcing or sustaining an outcome.

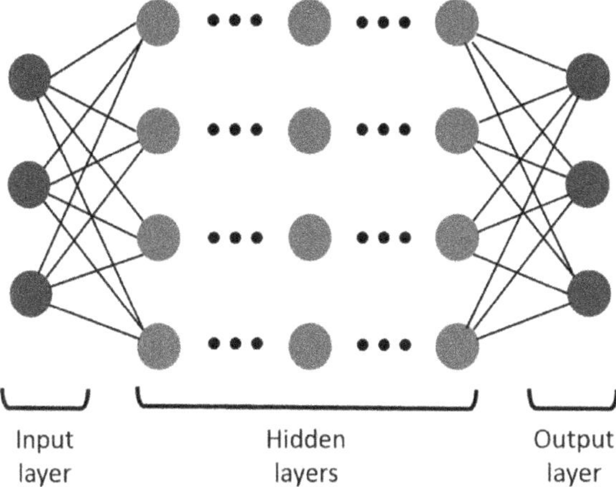

Figure 5.13 General DNN model.

DNNs are used by adjusting each weight so that the desired output is produced for the same input layer, which, while slow, has the advantage of yielding stable results. However, many problems can arise when trained in a naive[3] manner, among which overfitting and high time complexity are common issues. There are methods to solve these problems, but recently, the use of dropout regularization has led to the random omission of some units in the hidden layers during training. This method helps to address rare dependencies that can occur in the training data.

The high time complexity is due to backpropagation and gradient descent being suitable for finding local optima,[4] but the issue is that they take a long time for training. DNN is frequently used in fields like classification and numerical prediction, image processing, speech recognition, and natural language processing, and are especially useful in areas like image training and character recognition.

5.2.2 *Convolutional neural network*

The CNN technique is a method that addresses issues arising in conventional neural networks when processing multidimensional data such as images or videos. DNNs primarily use one-dimensional data. While it's feasible to train with text or numerical data without much loss, when an image is inputted, it is flattened into a

[3] Naïve: Naïve Bayes assumes that each measure is independent, which is why it is called naïve.

[4] Local optima: A local optimal solution is recognized as the best solution only within a certain interval, while a global optimal solution is recognized as the best solution in the entire search space, and the goal is to find the global optimal solution in many areas of optimization problems.

one-dimensional line of data. This process leads to the loss of spatial information in the image, making feature extraction or learning inefficient and limiting accuracy. Therefore, CNNs take images as raw input, maintaining the spatial/local information while forming layers of features. Traditional filtering techniques used fixed filters to process images. The basic concept of CNNs is "Let the elements of filters represented by matrices be automatically trained to be suitable for data processing." For instance, when developing an image classification algorithm, we could use filtering techniques to enhance classification accuracy. However, one issue is that filters to be used in the algorithm must be determined through human intuition or repetitive experimentation. In such situations, using CNNs, the algorithm can automatically learn filters that maximize image classification accuracy, thereby improving classification accuracy. For example, suppose we're given an image for recognizing a pig, and we want to create a model that can determine whether it's a pig. In that case, a key feature like the pig's snout could be an important point. Therefore, it becomes important for the model to discern whether the pig's snout is present in any given image. However, in the whole image of the pig, the snout is relatively a small part. Hence, it would be more efficient for the model to look at the snout part rather than the whole image.

This is where CNN comes into play. CNN neurons don't need to look at the whole image to recognize patterns (like the pig's snout); they can identify it as a pig by just observing the features. Figure 5.14 illustrates the feature extraction of the pig's snout.

Additionally, in Figures 5.14(a) and 5.14(b), it can be observed that the pig's snout is located in different parts of the image. Figure 5.14(a) shows the pig's snout located towards the middle right, and Figure 5.14(b) shows it at the bottom middle of the image, which highlights the importance of identifying and processing distinctive parts of the image for efficient learning, rather than the whole image. To illustrate, let's take the example of an image of the handwritten number "8," which is one of the samples extracted from the MNIST dataset. Upon closer inspection, it's composed of pixels in an 18×18 grid. The input values of the image data can be represented as an 18×18 matrix. In other words,

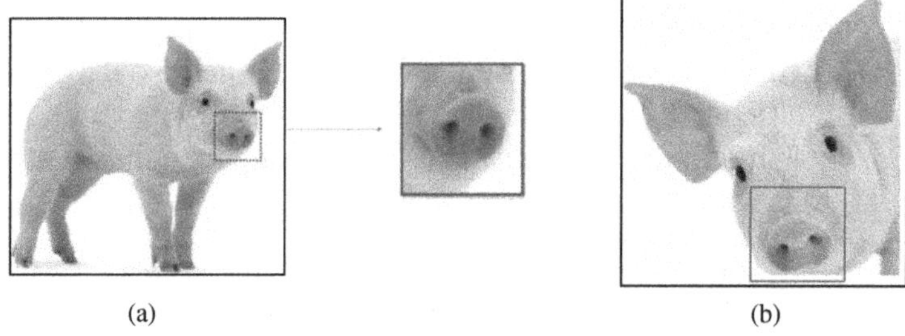

Figure 5.14 Image features from a CNN. (a) Positioned towards the middle-right. (b) Positioned in the middle bottom.

a two-dimensional image can be represented as a matrix, and the input value for the CNN is the image represented in matrix form. This is depicted in Figure 5.15.

Let's assume an image input value represented by a 5 × 5 matrix, similar to Figure 5.15, and in the CNN, there are filters (kernels). In the above example, the filter is of size 3 × 3. Simply put, this one filter is applied over the entire image input value. In essence, the goal is to repeatedly apply the same filter to every area of our input image to find and process patterns.

CNN are a type of multilayer perceptron designed to use minimal preprocessing, consisting of one or more convolutional layers and the usual artificial neural network layers on top. The layer structure of a CNN includes an input layer, an output layer, and several hidden layers in between. These layers perform operations that alter the data with the intention of learning the unique features of that data. The most common types of layers include the convolution layer, the activation or ReLU layer, and the pooling layer. The convolution layer passes the input image through a series of convolution filters, with each filter activating specific features in the image. The ReLU (Rectified Linear Unit) layer maps negative values to 0 and leaves positive values unchanged, enabling faster and more effective training. Only activated features are passed to the next layer, which is also referred to as activation. The pooling layer performs nonlinear downsampling, simplifying the output by reducing the number of parameters the neural network needs to learn (Figure 5.16).

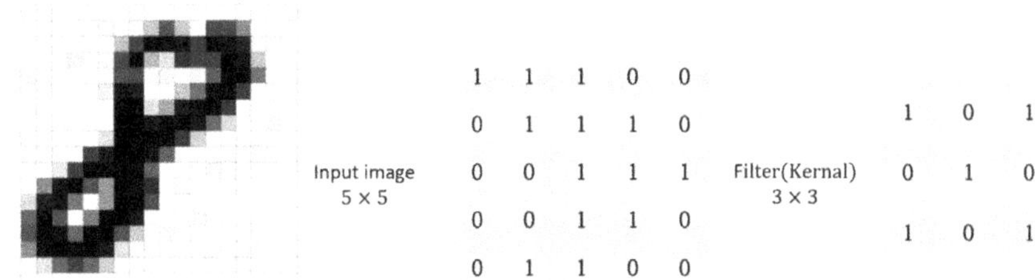

Figure 5.15 Matrix representation of sample data.

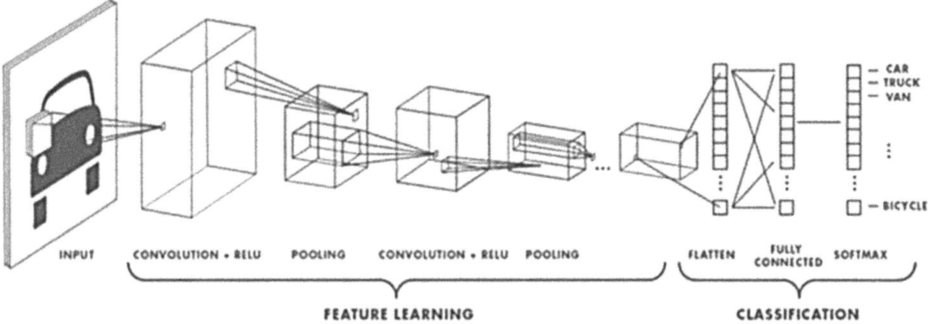

Figure 5.16 CNN model structure.

5.2.3 *Recurrent neural network*

RNN is a type of artificial neural network that uses sequential or time-series data for tasks like language translation, natural language processing, speech recognition, and image processing, dealing with data on a sequence basis. A sequence refers to an arrangement of elements, like words in a sentence. Models designed to handle sequences are called sequence models, and among these, RNN is one of the most fundamental sequence models in deep learning. Unlike previously learned neural networks, RNNs possess a "Hidden State," which allows them to reference previous data. This capability enables RNNs to understand the relationship between past and present data, facilitating the processing of continuous data. Previously learned neural networks process values from the hidden layers through activation functions, and these values proceed only in the direction of the output layer.

RNN is unique in that they not only send the results from the activation functions in the hidden layer nodes towards the output layer but also feed them back as input for the next calculation in the hidden layer nodes. For instance, in translation, the meaning of a sentence changes based on the order of words, and remembering which words were used previously can be helpful for predicting subsequent words. Figure 5.17 illustrates the basic structure of an RNN.

- x_t is the input at time step t.
- S_t is the hidden state at time step t, which is part of the network's "memory" and is calculated based on the previous time step's hidden state and the current input value at time step t.

$$S_t = f\left(Ux_t + Ws_{t-1}\right)$$

- The nonlinear function f is typically a tanh or ReLU function, and the initial hidden state S_{-1} for calculating the first hidden state is usually initialized to zero.

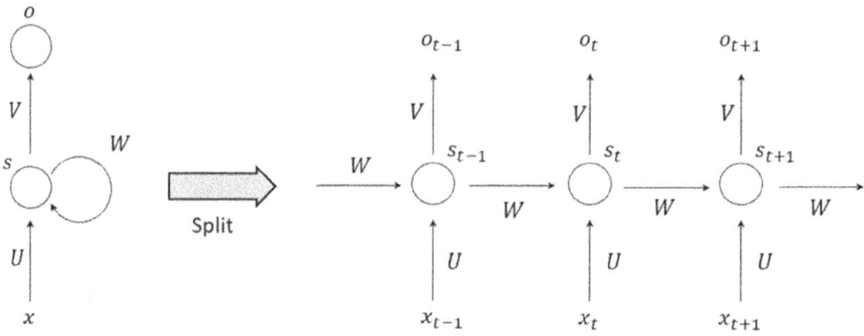

Figure 5.17 Basic structure of RNN.

- O_t is the output at time step t, which, for example, becomes a probability vector with as many dimensions as there are words when one wishes to predict the next word in a sentence.

Figure 5.17 represents an RNN unfolded to show its recurrent nature. It receives input values to produce output, and this output is then fed back into the network as input. The unfolded structure (on the right) illustrates this process at each time step, showing the inputs, outputs, and weights for each time step. The computation performed by the RNN layer is described by the following formula. In the formula, there are weights that transform the input value to the output value, and weights and biases that convert the RNN output to the output at the next time (t). As seen in the figure, the algorithm allows for the representation of information by accumulating past information onto current information through an internal recurrent structure, enabling continuous updating of information due to data circulation. Therefore, RNN is structured to handle input and output values regardless of the length of the sequence and is widely used in various applications such as NLP, speech recognition, and image captioning (annotating images with text).

5.3 Practice Questions

Q1. What is deep learning?

Q2. What is a perceptron?

Q3. Describe the types of artificial neural networks.

Chapter 6

Case Study

6.1 AlphaGo

AlphaGo, developed by Google's DeepMind Technologies Limited, is an artificial intelligence (AI) program for playing the board game Go. AlphaGo has a total of four versions: the version that defeated the 2-dan Go professional Fan Hui in October 2015, AlphaGo Lee that beat Lee Sedol 4:1 in March 2016, AlphaGo Master that won 3:0 against the 9-dan Go professional Ke Jie, and AlphaGo Zero, introduced in 2017 through Nature.

AlphaGo made history by becoming the first computer Go program to defeat a professional Go player in even games (without handicaps) using the traditional rules. In October 2015, it won five games out of five against the 2-dan French Go professional Fan Hui, who had won the European Go Championship three times. In March 2016, AlphaGo shocked the world by defeating the world's top-ranked Go professional, Lee Sedol 9-dan, in a five-game match with a score of 4-1, establishing itself as the "best existing AI." AlphaGo received an honorary 9-dan rank from the Korea Baduk Association and was registered as a guest player in the Korean Go scene, allowing participation in tournaments hosted by the Korea Baduk Association. While the possibility is low, it can participate in tournaments held by the Korea Baduk Association at any time. Figure 6.1 illustrates the AlphaGo system that was actually used.

Unlike other games such as chess, Go was considered much more challenging for computers to defeat humans. This is because the game has a significantly larger number of possible moves than chess, making the application of traditional AI techniques like brute force quite challenging. However, what allowed AlphaGo to succeed was not only the importance of data and computational power in AI learning but, more crucially, the algorithm.

The key was reducing the near-infinite possibilities, and AlphaGo was designed to make the most favorable choices among options through a trained

Figure 6.1 The actual AlphaGo system that was used.

deep neural network and Monte Carlo Tree Search. The neural network of the system was initially bootstrapped with expert knowledge from human gameplay. AlphaGo underwent training to mimic human play by attempting to match the moves of expert human players recorded in a database of about 30 million moves. Through the application of these techniques, AlphaGo was able to surpass human players in the game of Go.

6.1.1 *System configuration*

At the initial development, it was announced that the hardware utilized parallel computing with CPUs and NVIDIA GPUs. AlphaGo has two versions: a "Single version" running on a single computer with 48 CPUs and 4–8 GPUs, recording only one loss in 500 games (including "Crazy Stone" and "Zen"). The "Distributed version" uses multiple computers connected to the network, consisting of 1,202–1,920 CPUs and 176–280 GPUs. It was tested in both asynchronous[1] and distributed

---

[1] Asynchronous mode: Performing tasks in parallel, which means that even if a task is not finished, it will not wait for the next task to be performed.

Table 6.1 Configuration and performance.

Configuration	Search threads	No. of CPU	No. of GPU	Elo rating
Single [10, pp. 10–11]	40	48	1	2,181
Single	40	48	2	2,738
Single	40	48	4	2,850
Single	40	48	8	2,890
Distributed	12	428	64	2,937
Distributed	24	764	112	3,079
Distributed	40	1,202	176	3,140
Distributed	64	1,920	280	3,168

modes across varying numbers of CPUs and GPUs, with 2 seconds allocated for each move. The ELO ratings are shown in Table 6.1.

6.1.1.1 *AlphaGo match*

In October 2015, AlphaGo defeated European Go champion Fan Hui with a score of 5-0. AlphaGo used 1,202 CPUs and 176 GPUs in the match against Fan Hui. This corresponds to AlphaGo version 12, also known as AlphaGo Fan. These results demonstrate the capabilities of AlphaGo, which combines deep learning and reinforcement learning techniques, and are considered a compelling example showcasing the potential advancements in AI technology. AlphaGo showcased its ability to solve complex problems in the game of Go and develop strategies for future games in the field of Go.

6.1.1.2 *AlphaGo Lee*

In the match against Lee Sedol 9-dan in March 2016, AlphaGo used 48 TPUs instead of GPUs. At that time, the version of AlphaGo was only disclosed as being at the level of "Version 18," showcasing machine learning improvements in AlphaGo play. However, during a conference in 2016, Google introduced its custom-designed integrated circuit,[2] the Tensor Processing Unit[3] (TPU), and mentioned

[2] Integrated Circuit: A composite electronic device with a very small structure in which many electronic circuit elements are inseparably combined on a single substrate or on the substrate itself.

[3] Tensor Processing Unit (TPU): A processing unit used to quickly process machine learning workloads customized by Google.

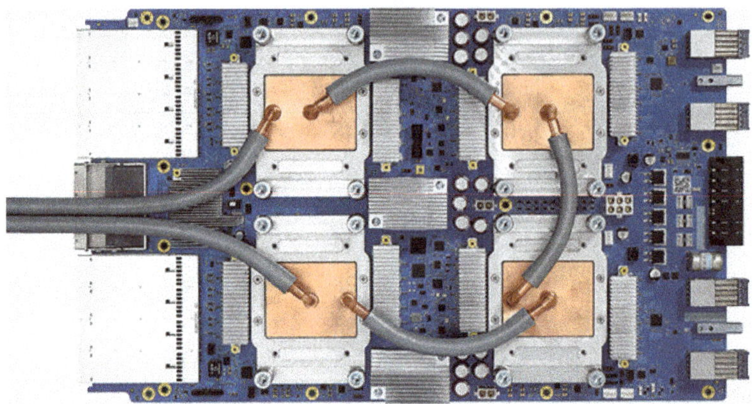

Figure 6.2 Tensor processing unit.

that AlphaGo, which played against Lee Sedol 9-dan, utilized TPUs. Figure 6.2 depicts an image of the TPU.

6.1.1.3 *AlphaGo Master*

It is a single version that uses four TPUs. In early 2017, it achieved a winning streak of 60 games against professional Go players in online games and secured a victory in a match against Ke Jie in May of the same year. AlphaGo Master, powered by the second-generation TPU module, was unveiled on May 17, 2017, at the Google I/O 2017 conference. The TPU modules used here consist of four TPUs, delivering a computational performance of 45 TFLOPS (45 trillion operations per second) per module, totaling 180 TFLOPS, and each module supports a memory bandwidth of 64 GB. Google claimed that the computational performance of TPU was 30 to 80 times higher than the latest CPUs at that time. While the previous version of AlphaGo made inferences based on what it had learned, AlphaGo Master can learn and infer simultaneously, reducing the training time to one-third of the previous duration. Additionally, with a reduced physical volume, the energy efficiency improved by approximately 10 times.

6.1.1.4 *AlphaGo Zero*

It is the final version of AlphaGo that uses four TPUs. Introduced through a paper titled "Mastering Chess and Shogi by Self-Play with a General Reinforcement Learning Algorithm" published in the scientific journal Nature on October 19, 2017, AlphaGo Zero learns and improves its strength solely based on the rules of Go without relying on human moves or supervised learning. Within 36 hours of training, AlphaGo Zero surpassed the level of AlphaGo Lee, remained undefeated against AlphaGo Lee in 100 consecutive games after 72 hours, and achieved a record of 89 wins and 11 losses against AlphaGo Master in a match 40 days later.

During this period, AlphaGo Zero engaged in 29 million self-play games to facilitate its learning process. The emergence of AI that does not require big data training is significant in addressing challenges in fields where utilizing AI was difficult due to the lack of big data, unlike Go. Since the conclusion of the century-defining match in 2017, AlphaGo has been involved in a project aimed at reducing Google's data center power consumption. Google's power consumption was exceptionally high, reaching 4.4 terawatts as of 2014, comparable to the energy consumption of 360,000 households in the United States. Through various efforts, Google successfully reduced cooling costs by 40%, leading to an overall power consumption reduction of around 15%. Ultimately, AlphaGo faced challenges of consuming significant power and requiring extensive computational systems, leading to the realization that learning from all data is not feasible. Consequently, the human-in-the-loop (HITL) approach is continuously being researched. While AlphaGo's success concluded with Go, its ability to transcend human capabilities and continuously improve through a broad learning capacity mirrors the human learning process, suggesting ongoing advancements in the future.

6.1.1.5 *AlphaZero*

It is a general AI algorithm applicable to board games such as chess, shogi, and Go. On December 7, 2018, they published a paper titled "A general reinforcement learning algorithm that masters chess, shogi, and Go through self-play" in the scientific journal Science. Similar to AlphaGo Zero, it doesn't require big data learning and builds its own dataset by repeatedly playing against itself. It took 2 hours to surpass Elmo, the winner of the 2016 shogi competition, 4 hours to outperform Stockfish, the 2017 chess champion, and 30 hours for AlphaGo Zero. On December 14, 2017 (US time), Dr. Huang Shih-chieh announced the conclusion of AlphaGo's journey and the allocation of all its resources to other AI developers.

6.1.2 *Algorithm implementation*

Until the emergence of AlphaGo, Go was considered the only analog game that computers couldn't surpass humans in. The Go AIs developed so far had a significant skill gap compared to professional players. So, how did AlphaGo manage to achieve such remarkable skill, surpassing existing algorithms and even defeating world-class professional players? Let's delve into the approach AlphaGo took to the game of Go and the process it underwent.

6.1.2.1 *AI approach strategy*

Let's explore how AI approaches a 1:1 match format board game. Suppose you've been tasked with creating an AI that wins board games, with no time constraints, allowing unlimited time for the program to consider all possible actions from the current state, looking as far ahead as possible and choosing the most optimal action. In the absence of time constraints, the most ideal method would be to use the Minimax Algorithm, considering all possible actions and anticipating the opponent's moves based on the actions taken. To illustrate the minimax algorithm, a tree structure diagram is provided in Figure 6.3.

The circled part represents my turn, and the squared part represents the opponent's state. Considering all possible actions for both me and the opponent, the score I can obtain after 4 moves is indicated on the "4" layer. The highest score I can achieve among these scores is +infinity, and it would be excellent if I could attain this score. However, a rational opponent would aim to minimize the score I can achieve. Therefore, they would take an action leading to the smaller score between +10 and +infinity, which is +10. On my turn, I would choose an action that maximizes my score. Following this thought process from layer 4 to layer 0, it becomes evident that the action I should take is not the left path to achieve the highest score of +10 or +infinity but the right path to obtain the highest score of "–7" when considering the opponent's actions as well. In 1:1 match games like Go, employing this approach is considered the most ideal, applying the minimax algorithm, meaning choosing the best move among the worst moves the opponent could offer. Ultimately, to create a powerful AI, a new algorithm is needed that efficiently explores the search space, minimizing the time used while achieving performance close to the minimax algorithm. Various methods, such as Alpha-beta Pruning and Principal Variation Search, have been proposed to find such

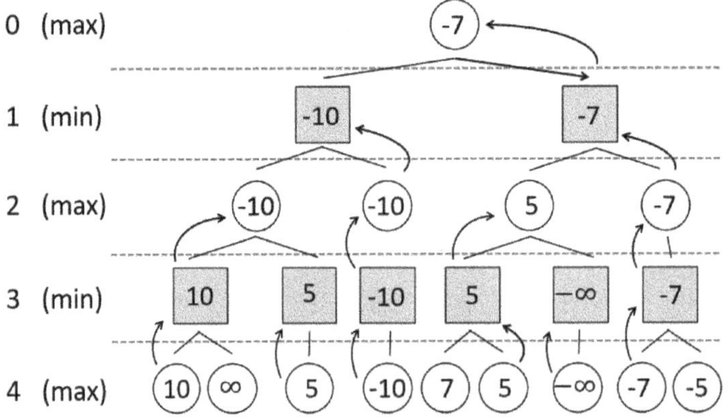

Figure 6.3 Minimax algorithm.

an approach. Now, the method I will introduce is Monte Carlo Tree Search, which has been widely used in board game-related AI and serves as the foundation for AlphaGo.

6.1.2.2 *Monte Carlo Tree Search*

In computer science, Monte Carlo Tree Search (MCTS) is an empirical search algorithm used primarily for decision-making, especially in games. This approach encompasses methods of solving problems using a probabilistic approach. Instead of directly addressing complex equations or exploring massive search spaces that are practically impossible to fully examine, this method employs randomness to approximate the solution. For instance, consider drawing a circle with a radius of 1 inscribed in a 2×2 square and randomly placing points inside the square. Calculate (the number of points inside the circle)/(the total number of points), and you have an approximation. By generating an immense number of points, the value will eventually converge to $\pi/4$. This is a typical example of the Monte Carlo method, using Monte Carlo principles to calculate the value of π. MCTS is an algorithm designed to explore the action tree in minimax by applying Monte Carlo techniques, allowing it to select good actions without exhaustively examining all states. Let's take a look at how MCTS constructs a tree using randomness and examine the basic form of MCTS (Figure 6.4).

The basic steps of MCTS consist of four stages.

1. **Selection:** Starting from the root, it moves towards the leaf nodes by choosing the most successful child. Typically, randomization is introduced at this stage to prevent the search from getting biased towards one side.

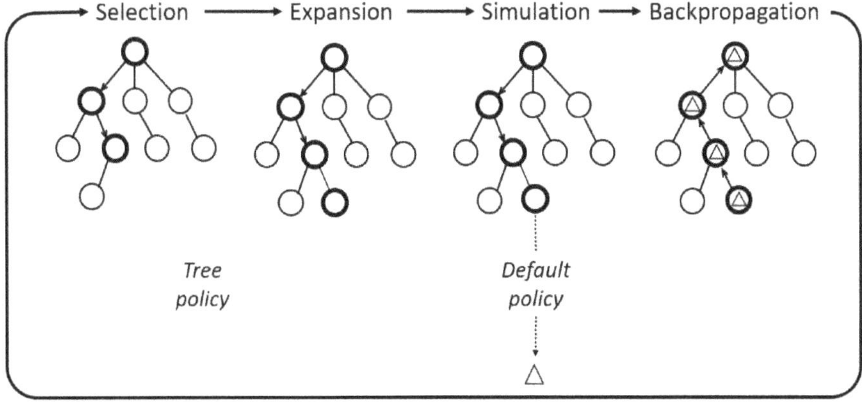

Figure 6.4 MCTS diagram.

2. **Expansion:** When reaching a leaf node, if the leaf node is not a final node that concludes the game, it generates the child nodes for the next stage and moves to one of the child nodes.
3. **Simulation:** Randomly playing the game from the current state, known as playout, to obtain the outcome regarding whether I win or lose by playing the game until the end.
4. **Backpropagation:** Based on the outcome of the current simulation, update the result values for the nodes along the path I took during the simulation.

In MCTS, repeating these steps as much as possible predicts the win rates for each action I can take. When deciding on my final action after exploring to a certain extent, I choose the action that has been simulated the most among the possible actions from the root. Surprisingly, it has been proven that repeatedly applying MCTS indefinitely converges to the results of minimax. However, the convergence speed of the basic MCTS is generally slow, so methods are needed to reduce the search space and efficiently find results. The ultimate result sought through convergence in MCTS is "how likely I am to win" at each node (game state) in the tree. Let's assume there is a theoretical function $v^*(s)$ that calculates the extent to which I can win the game given the current states. Although we cannot know the actual shape and operation of such a function, we can assume the existence of such a function. If MCTS is carried out indefinitely, the values of each node in MCTS will converge to the values of the theoretical win rate calculation function $v^*(s)$ for each state. Conversely, approaching this from the reverse, if there exists a function $v(s)$ that can approximate $v^*(s)$, it is possible to approximately obtain the desired results through MCTS without deep exploration. Moreover, by slightly modifying such a function, it could also provide values for "which action is better in the current state," allowing for more effective pruning in the MCTS tree. In other words, creating a function $v(s)$ that calculates the extent to which I can win from the current game state is a crucial problem. Previous researchers have approached MCTS and $v(s)$, leading to the creation of Go AIs like Pachi and Fuego. However, because the function $v(s)$ itself is extremely challenging to create, previous research failed to produce a powerful $v(s)$ function, ultimately resulting in limitations that could weaken AI performance. At this point, the DeepMind team at Google incorporated deep learning to greatly enhance the performance of creating the $v(s)$ function.

6.1.2.3 *AlphaGo pipeline structure*

Through the explanations so far, we have learned that AlphaGo effectively explores the state-action tree using the MCTS technique and utilizes a learned state evaluation function $v(s)$ through deep learning. However, the actual internal workings of AlphaGo involve a much more complex computation process than described so far. Moreover, it consists of multiple neural networks, not just one, forming a structure where each complements the others. Let's take a rough look at how the actual neural networks inside AlphaGo are composed.

AlphaGo has a network called the "policy network" designed to predict the next move on the board in the current state, foreseeing both where the opponent might place a stone and suggesting the optimal move for itself. In the paper, this is referred to as the "policy network," and it is structured in the form of a convolutional neural network (CNN). To train the policy network, AlphaGo first required a vast amount of data. To train the CNN, AlphaGo utilized a large dataset comprising around 30 million moves from the KGS Go server, providing data on "given a board state, where to place the next stone." This data was employed to train the policy network. Additionally, a "Rollout policy" network, which is less accurate but much faster at computing the same problem, was prepared and trained alongside the policy network. Policy network is also an outstanding network. Even without using MCTS, simply placing stones using only the policy network could defeat existing Go AIs with high winning rates. However, the network alone is not enough to operate the true AlphaGo. For MCTS exploration, a "value network," denoted as $v(s)$, calculating whether the player can win from the current state, was needed. The value network has a CNN structure similar to the policy network. However, while the policy network calculates a probability distribution for placing a stone in the final position, the value network provides a single value for the given states. To train this network, AlphaGo stored basic data from "self-play," collecting winning and losing states, and used them as training data for the value network, successfully completing the learning process. AlphaGo is considered to have conducted MCTS using policy network, rollout policy, and value network, with each network serving a specific purpose.

- **Policy network:** Originally, in the selection stage, MCTS used a method of choosing the most successful child node. However, AlphaGo's MCTS incorporates a method that moves more towards actions with higher probabilities from the policy network. This can be interpreted as AlphaGo implementing a form of human-specific "intuition" that traditional computer algorithms could not achieve. To prevent excessive focus on one direction, if a certain action is visited too frequently, AlphaGo encourages exploration in other directions by applying some penalty to that action.
- **Rollout policy:** While the rollout policy has lower performance compared to the policy network, it has the advantage of much faster calculations. Therefore, during the evaluation stage of MCTS, the simulation is conducted using this rollout policy network.
- **Value network:** Similarly, the value network is used during the evaluation stage. In the evaluation stage, scores are obtained using both the rollout policy and the value network. The final score for this state is calculated by combining the two scores in a certain ratio.

According to the paper, using a combination of the rollout policy and the value network together produced significantly better results than using them individually. After completing the processes of selection, expansion, and evaluation for a chosen state, a score is obtained. This score is utilized in the backup process to update the scores and visit counts of the traversed paths. After updating for a certain duration, the final decision is made by selecting the most frequently visited move from the available moves in the MCTS tree. The process of updating the MCTS tree, as described earlier, can fortunately be distributed for computation. Therefore, AlphaGo, in its final development, distributed computations across multiple CPUs and GPUs instead of relying on a single computer to perform MCTS. As a result, it achieved an exploration capability with moves that even professionals found incomprehensible.

6.2 IBM Watson

Named after IBM's first president, Thomas J. Watson, Watson is an AI computer system capable of answering questions posed in natural language. Developed through IBM's Deep QA project led by project manager David Ferrucci, Watson appeared on the famous quiz show Jeopardy[4] on February 14, 2011, serving as a kind of technology demonstration project, similar to Deep Blue. Interestingly, the first machine to achieve victory in a human-versus-machine competition was not AlphaGo but IBM's Deep Blue, which defeated chess world champion Garry Kasparov in 1996. Watson, an upgraded version of Deep Blue, went on to secure a triumph by surpassing quiz champions on the Jeopardy show in 2011. In 2004, IBM research team manager Charles Lickel observed people in a restaurant captivated by watching Ken Jennings' 74-game winning streak on "Jeopardy!" In 2005, IBM research team director Paul Horn assigned developers to Charles Lickel, instructing them to develop Watson, capable of defeating humans in Jeopardy! The actual development of Watson began in 2005, and it was a more challenging process than chess because it required real-time analysis of language. Nevertheless, Watson became the first AI to defeat humans. Figure 6.5 illustrates the Jeopardy! IBM "man versus machine" challenge.

IBM is expanding "Watson" into cognitive computing, combining AI and machine learning for inference and learning capabilities. This fusion of big data and AI involves the computer autonomously analyzing vast amounts of data and making judgments, learning and accumulating knowledge in human language to support human decision-making. IBM Watson applies advanced natural language

[4] Jeopardy: It is a television quiz show created by Merv Griffin on NBC (broadcast in the United States) that first aired on March 30, 1964 and ran until 1975, then was revived in 1978 and ran until 1979. It then returned to the air six years later in 1984, with a new host, AlexTrebek, who continues to host the show to this day.

Figure 6.5 Jeopardy! IBM "man versus machine" challenge.

processing, information retrieval, knowledge representation, automated reasoning, and machine learning technologies for unstructured question-answering in undefined domains. It continuously advances its expertise through machine learning, incorporating voice recognition, image recognition, and visualization technologies to interact in human language, learn, and evolve. AlphaGo, applying deep learning technology, excels in the game domain, while Watson shines in the health field. AlphaGo's core technology involves deep learning for clustering or classifying objects and data. AlphaGo implements a value network, calculating the probability of winning on the Go board, and a policy network, assigning scores to positions on the board, using deep learning technology. Google DeepMind utilized TensorFlow, a deep learning system, to create AlphaGo. TensorFlow is an open-source library for machine learning and deep learning, providing a versatile algorithm adaptable to various fields. In essence, TensorFlow serves as a well-known big data platform (Table 6.2).

Watson's core role lies in accumulating knowledge in specialized fields, such as medical terminology, to generate optimal data and support human decision-making. When healthcare professionals input various clinical information, Watson provides advice on a patient's condition and treatment options. Leveraging millions of diagnostic reports, patient records, medical literature, and other data sources, Watson autonomously evaluates and suggests the most promising treatment methods. It significantly shortens the time required for patient information interpretation and medical literature collection, from several days to just a few minutes, aiding in patient care. Over time, Watson has advanced its technological

Table 6.2 Illustrates IBM Watson versus Google AlphaGo.

IBM Watson	VS	AlphaGo
IBM	Development company	Google DeepMind
David Ferrucci	Developer	Demis Hassabis
Self-learning through machine learning	Learning method	Intensive learning using deep neural network technology
Won the Jeopardy quiz game in 2013	Record	Won the Go game with Lee Sedol in 2016
NLP	Characteristic	Decision-making system
Cloud	Using system	TensorFlow
Medical, finance	Main application field	Game, health

capabilities to analyze previously challenging unstructured data, including images, videos, MRI data, and patient movements.

6.2.1 *Operation of Watson*

Watson collects the knowledge needed to understand a specific field, also known as a corpus, following the guidance of experts. The process of building the corpus involves loading relevant documents into Watson. Content curation, involving human intervention, is essential to discard outdated, poorly rated, or irrelevant information for effective corpus construction. This is called content curation. Good content curation is the key to good learning. To learn how to interpret vast amounts of information, Watson undergoes training with experts in machine learning, partnering with professionals to train itself, acquiring the ability to learn optimal responses and identify patterns. Machine learning is a broad term applied to AI, allowing it to perform human-like reasoning. The comprehensive term encompasses AI projects such as CYC, initiated by Douglas Lenat in 1984, aiming to build a database of common sense. CYC, originally a proprietary project by Cycorp, Lenat's company, has a smaller version of its database distributed as OpenCyc under an open source license. For instance, statements like "All trees are plants" and "Plants eventually die" are included. When questioned about whether a tree is dead or alive, the inference engine[5] can derive a clear conclusion and provide accurate answers. Experts upload training data in the form of questions and answers that serve as validation material for Watson. Experts regularly review interactions between users and Watson, providing feedback to enhance

[5]It is the brain of the expert system, providing a methodology for reasoning about the information in the knowledge base and formalizing conclusions, and is responsible for finding answers in the knowledge base.

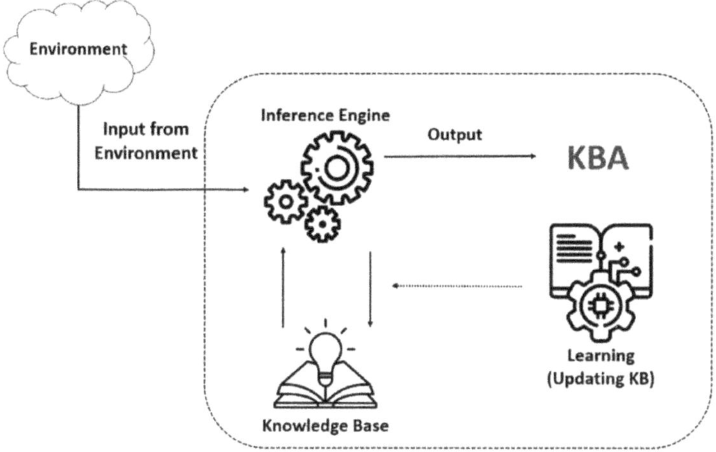

Figure 6.6 Knowledge-based agent in AI.

Watson's ability to interpret information. Figure 6.6 illustrates the knowledge-based agent in AI.

Such rules were created by humans. It was nearly impossible to manually create all the knowledge in the world due to the immense time required. However, in the case of Watson, such tasks were automated through natural language processing. Ultimately, Watson can continuously update with new information and adapt to changes in the knowledge and language interpretation of the field.

6.3 GPT

GPT stands for "Generative Pre-trained Transformer," an AI technology that learns vast data through machine learning and generates sentences based on this pre-training. When a user inputs a question as if talking to AI, GPT creates sentences "like a human" based on the learned data to provide answers. This serves as a conditional probability prediction tool fundamental to natural language processing, analyzing various text data to create appropriate sentences.

The most famous model is the GPT released by Open AI in 2018, which was trained with 117 million parameters. In 2019, Open AI released GPT-2 in four stages, with model sizes ranging from approximately 124 million to 1.5 billion parameters, ten times that of the previous version, capable of producing a page of text with human-like proficiency in just 10 seconds. Released in 2021, GPT-3, with its 175 billion parameters, marked an improvement of over 100 times the performance of GPT-2. As a result, GPT-3 is trained on a vast amount of text data, enabling it to learn a variety of language patterns and styles. Tasks that GPT-3 can perform include solving various language-related problems, random writing, basic

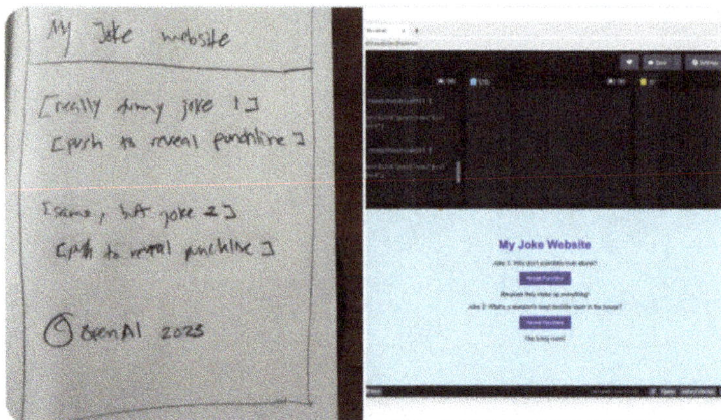

Figure 6.7 A website sketch converted into HTML source and implemented as an actual website.

arithmetic, translation, and simple web coding based on given sentences. Recently, the GPT-4 version was released, and it is currently available through a paid service called "Chat GPT Plus." GPT-4 features a multimodal capability to recognize and understand images and generate text information about those images. As an example, Greg Brockman, co-founder of Open AI, shared on his social media an image of a website created from a hand-drawn sketch converted to HTML source using GPT-4 (Figure 6.7).

The previous model, GPT-3.5, could process up to 3,000 words per request in English, but GPT-4 can handle up to 25,000 words. While GPT-3.5 could remember about 8,000 words during a conversation, GPT-4 can recall up to 64,000 words, equivalent to the length of a novel, to respond more accurately to user queries.

6.3.1 *Principles of GPT*

The principles of GPT largely divide into large language models (LLMs) and reinforcement learning from human feedback (RLHF). LLMs are subsets of AI trained on vast amounts of text data to generate human-like responses to dialogue or other natural language inputs. To generate these natural language responses, LLMs use multilayered neural networks as deep learning models to process, analyze, and predict complex data. However, they have limitations in accurately understanding human desires. Therefore, these limitations can be overcome by improving the technology through reinforcement learning with human feedback. RLHF is an advanced approach to AI system training that combines reinforcement learning with human feedback. It involves integrating human trainers' wisdom and experience into the model training process to create a more robust learning experience.

This technique uses human feedback to generate reward signals, which are then used to improve model behavior through reinforcement learning. The RLHF process is divided into four stages: initial model training, collecting human feedback, reinforcement learning, and iterative processes. Initially, the AI model is trained using supervised learning, where human trainers provide examples with labels for correct behavior. The model learns to predict the correct actions or outputs based on the given inputs. After the initial model is trained, human trainers participate to provide feedback on the model's performance. The feedback, based on quality or accuracy, ranks various model-generated outputs or actions and is used to create reward signals for reinforcement learning. The model is then fine-tuned using proximal policy optimization (PPO) or a similar algorithm that integrates rewards signals created by humans, continuously improving performance through learning from the feedback provided by human trainers.

Finally, the process of collecting human feedback and refining the model through reinforcement learning is iteratively repeated, continuously enhancing the model's performance.

6.3.2 *Examples of GPT application*

Open AI has released a new API based on GPT-3.5-turbo, an optimized version of GPT-3.5 with improved speed. The usage fee is set at $0.002 per 1,000 tokens or 750 words, which is ten times cheaper than GPT-3.5. Additionally, Open AI assured that all data entered by companies and users, including sensitive internal inputs and information, will not be used for model training, making it easier to integrate AI features into mobile apps, websites, products, and services. Below are some representative examples of its application.

Figure 6.8 shows "Q-chat" from Quizlet, a global learning platform. As a global learning platform, it has provided learning services to over 60 million students, and since 2023, it has used GPT-3 for various student learning cases like vocabulary and mock exams, in collaboration with OpenAPI. Thus, it has evolved into a fully adaptive AI tutor, engaging students with fun chatting experiences and adaptive questions based on learning materials.

Figure 6.9 depicts AI shopping assistant by Shopify. Shopify's customer app, Shop, used by 100 million users to find and purchase their favorite products and brands, has enhanced its new shopping assistant feature using Chat GPT. The shopping assistant provides personalized recommendation services based on customer requests when they search for products. The new AI-based shopping assistant in Shop scans millions of products to help buyers quickly find what they want or discover new products.

Figure 6.10 is "My AI" from Snapchat. This AI chatbot, equipped with Chat GPT, is trained to converse in a fun and lively tone, offering users a friendly and

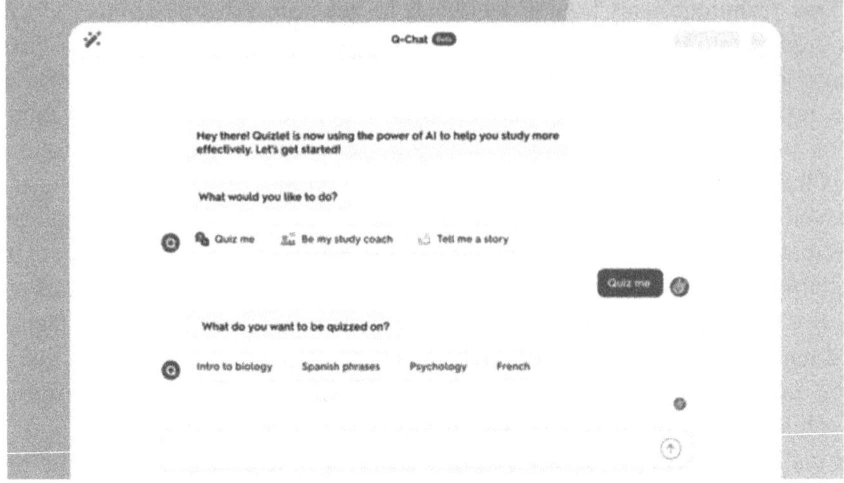

Figure 6.8 Q-chat.

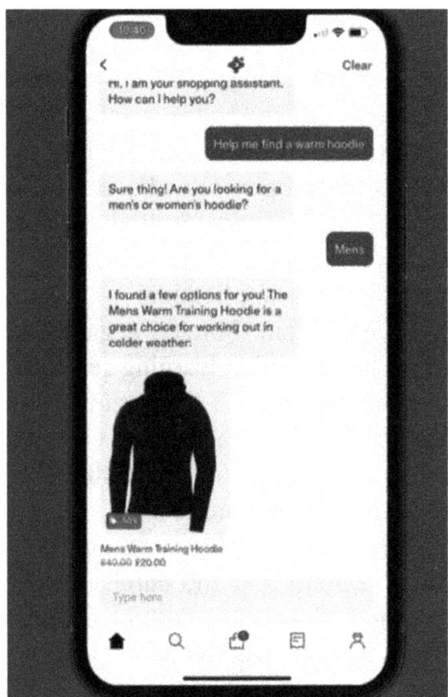

Figure 6.9 AI shopping assistant.

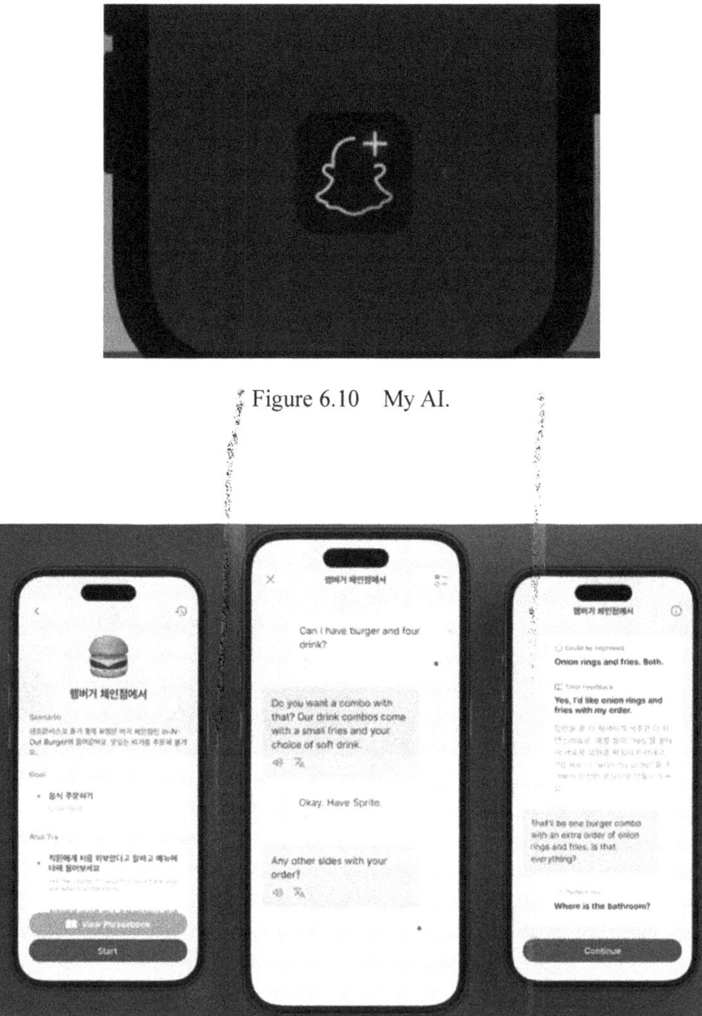

Figure 6.10 My AI.

Figure 6.11 AI speaking.

customizable chatbot environment. It includes features like providing birthday gift ideas or writing poems on specific topics.

Figure 6.11 is "AI speaking" from Speak. It is an AI-based language learning app that offers and uses learning paths for naturally speaking English. Using Whisper, it provides human-level accuracy to language learners of all levels, offering true open-ended conversational practice and precise feedback. Additionally, companies like Coca-Cola, SK Telecom, Samsung SDS, and DL E&C are applying Chat GPT in various fields.

6.3.3 *Try using Chat GPT*

Open AI's Chat GPT is divided into a free version and a paid version that utilizes GPT 4. First, let's go to the website at https://chat.openai.com/.

As shown in Figure 6.12, you can sign up for Chat GPT through the Sign-Up option. You can also access it with existing Google, MS, or Apple accounts.

The free version allows conversation through the GPT 3.5 engine and can be utilized in various ways, from simple everyday conversations to coding and document summarization. OpenAI's GPT has an open API, enabling its use as a module (Figure 6.13). Let's go to https://platform.openai.com/apps.

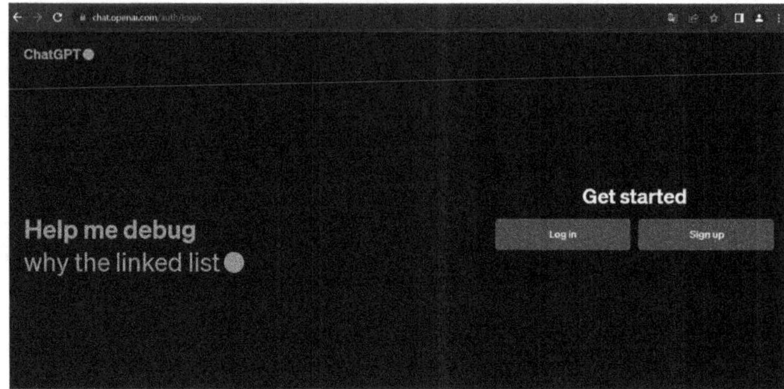

Figure 6.12 Chat GPT login page.

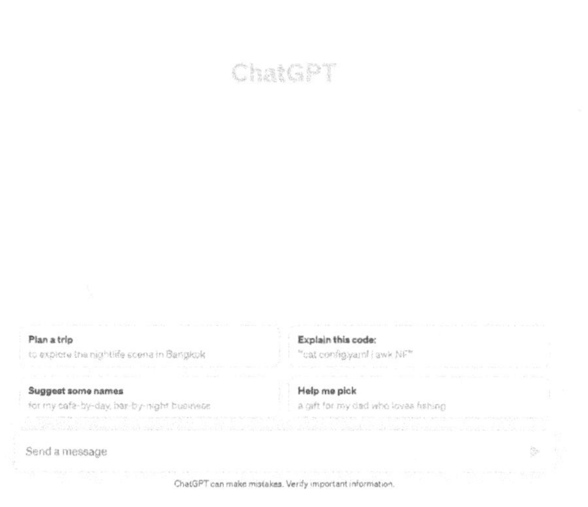

Figure 6.13 Chat GPT 3.5.

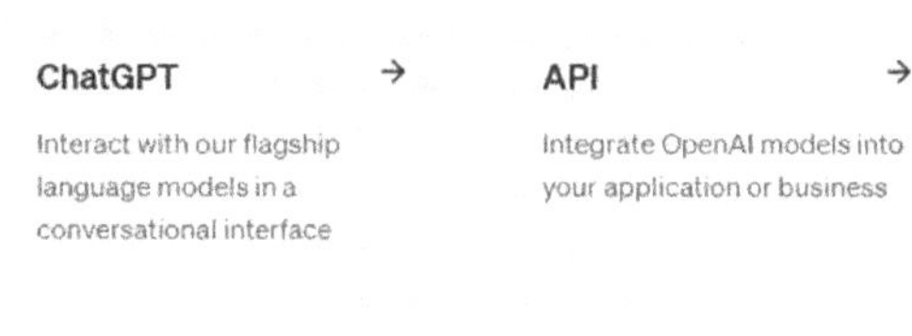

Figure 6.14 GPT API.

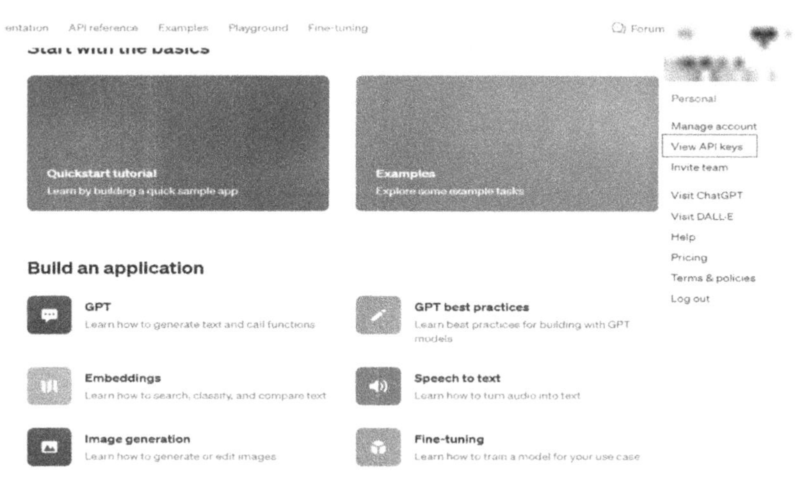

Figure 6.15 View API key.

If you are logged in the screen shown in, Figure 6.14 should appear. Here, click on the API.

As shown in Figure 6.15, clicking on the account at the top right will display a menu. Here, click on "View API."

As shown in Figure 6.16, you will see "Create New Key," where you can obtain and use an API key. The price for the API is based on usage, and the pricing policy and usage methods for the GPT API can be checked at the link below.

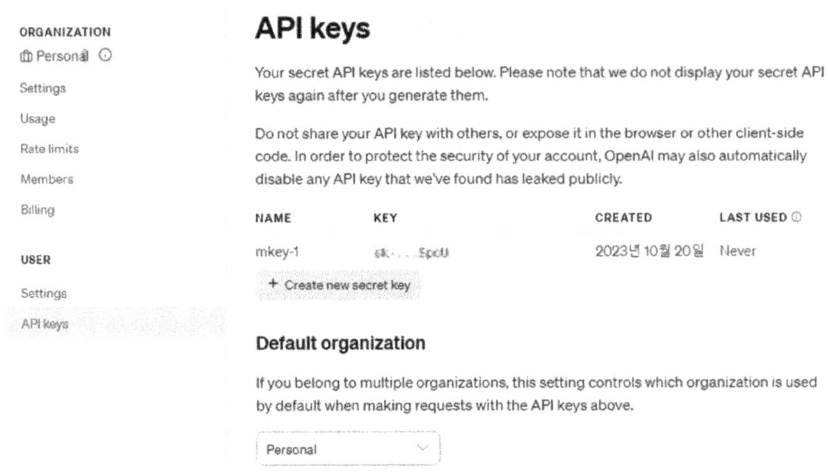
ORGANIZATION
Personal
Settings
Usage
Rate limits
Members
Billing
USER
Settings
API keys
API keys
Your secret API keys are listed below. Please note that we do not display your secret API keys again after you generate them.
Do not share your API key with others, or expose it in the browser or other client-side code. In order to protect the security of your account, OpenAI may also automatically disable any API key that we've found has leaked publicly.
NAME
KEY
CREATED
LAST USED
mkey-1
2023년 10월 20일
Never
+ Create new secret key
Default organization
If you belong to multiple organizations, this setting controls which organization is used by default when making requests with the API keys above.
Personal

Figure 6.16 Create new key.

6.4 Practice Questions

Q1. What is AlphaGo?

Q2. Describe the MCTS which is implemented as the AlphaGo algorithm.

Q3. Describe Watson, an AI system implemented by IBM.

Q4. What is GPT?

Part 2

Platform-Based Artificial Intelligence

Chapter 7

Amazon Web Service (AWS)

Amazon Web Services is a leading provider of cloud services with over 200 data centers worldwide. AWS offers cost savings, performance improvements, and innovative services to numerous customers ranging from businesses to government agencies, maintaining a leading position in the cloud industry. This platform provides a variety of services ranging from basic infrastructure services such as computing, storage, and databases to advanced technologies such as machine learning, artificial intelligence, and the Internet of Things. Customers can easily migrate applications to the cloud through AWS and leverage various platform features to build efficient business processes. AWS provides a wide range of services that can be utilized by large and small businesses, various industries, and public sector organizations, all accessible through a pay-as-you-go pricing model. In summary, AWS offers over 200 diverse services ranging from data warehousing to content delivery, allowing for quick utilization without initial costs.

Figure 7.1 shows a summary of the services provided by AWS in the form of an AI platform. Some key features are listed as examples below.

- **Amazon personalize:** A tool for recommendation services that learns from user data to provide recommendations that meet the user's or system's needs.
- Users can input their own data into personalize, and the service will learn from the data and provide responses according to the user's requests.
- **Amazon forecast:** A tool that provides a time series prediction service, offering predicted values for a specific future point based on various time series data.
- **Amazon lookout for metrics:** A service that detects abnormalities that may occur in existing business metrics. It identifies abnormal patterns in data using various algorithms.

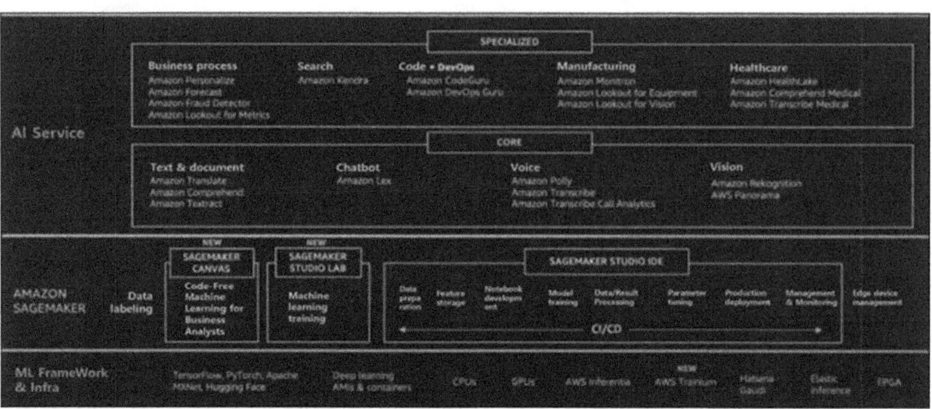

Figure 7.1 AWS platform.

- **SageMaker training:** It helps users effectively train their models by developing development code for user model training, or as a tool for model creation and training, allowing users to train models using their own data and algorithms.
- **Processing service:** An AWS service that supports data and result processing tasks, providing tools and services to quickly process and analyze large-scale data.
- **SageMaker Canvas:** A tool for business analysts that visualizes user data for analysis, allowing users to obtain analysis results without the need for complex code writing.

AWS provides the benefits of cloud computing, allowing users to avoid large hardware investments or time-consuming maintenance. Cloud computing is a computing service provided over the internet, offering computing power, database storage, applications, and other IT resources on-demand. Cloud service platforms like AWS provide users with all the necessary infrastructure and services, which users can provision and use in real-time through web applications. The main advantages of cloud computing are as follows:

1. **Removal of the need for capacity estimation:** Users can secure or reduce resources as needed in real-time.
2. **Cost reduction in the operation and maintenance of data centers:** Users can reduce the burden of maintaining infrastructure and focus on their core business.
3. **Fixed capital costs become variable costs:** No upfront payment is required as users only pay for the costs incurred while using the service.
4. **Economies of scale:** Cloud service providers reduce costs by providing services in large quantities and pass on these cost savings to users.

Therefore, by utilizing cloud computing, IT departments and developers can focus on core tasks instead of everyday infrastructure management. With the growth of cloud computing, various service models and deployment strategies have emerged, such as infrastructure-as-a-service[1] (IaaS), platform-as-a-service[2] (PaaS), and software-as-a-service[3] (SaaS). Understanding and effectively utilizing them is the key to effective cloud utilization.

7.1 Amazon Simple Storage Service (Amazon S3)

Amazon S3 boasts excellent scalability, data availability, security, and performance in the field of object storage services. By utilizing this service, users can securely store data for various purposes such as building a data lake, hosting websites, supporting mobile applications, backup and restore, archive storage, enterprise application management, IoT device integration, and big data analysis. Additionally, Amazon S3 provides rich management tools that allow users to control, structure, and configure data access, taking into consideration specific business requirements, organizational goals, and regulatory compliance.

7.1.1 *Creating an S3 bucket*

Next, let's see how to create an Amazon S3 bucket After logging in on the AWS homepage, navigate to the address https://s3.console.aws.amazon.com/. You will see the screen shown in Figure 7.2. Click the "Create bucket" button to start creating the bucket.

When you click on "Create bucket" in Figure 7.2, the screen shown in Figure 7.3 will appear. In Amazon S3, users can create buckets by directly setting the bucket name. However, this bucket name must be unique worldwide and cannot contain spaces or capital letters. Users can also select the region for the bucket. It is recommended to disable access control list (ACL) in the object ownership settings. If ACL is enabled and S3 bucket is used, another AWS account can become the owner of the bucket, posing a risk of inadvertently granting S3 bucket usage permissions to other users. This can result in unexpected charges, so caution is advised.

---

[1] Infrastructure-as-a-service (IaaS): Infrastructure-as-a-service includes the basic components of cloud IT, typically providing access to networking capabilities, computers, and data storage space.

[2] Platform-as-a-service (PaaS): Allows you to be more efficient by taking the burden of resource purchasing, capacity planning, software maintenance, patching, or any other monolithic tasks associated with running an application.

[3] Software-as-a-service (SaaS): Software as a service delivers a finished product that is operated and managed by a service provider. In most cases, when we talk about software as a service, we're referring to end-user applications.

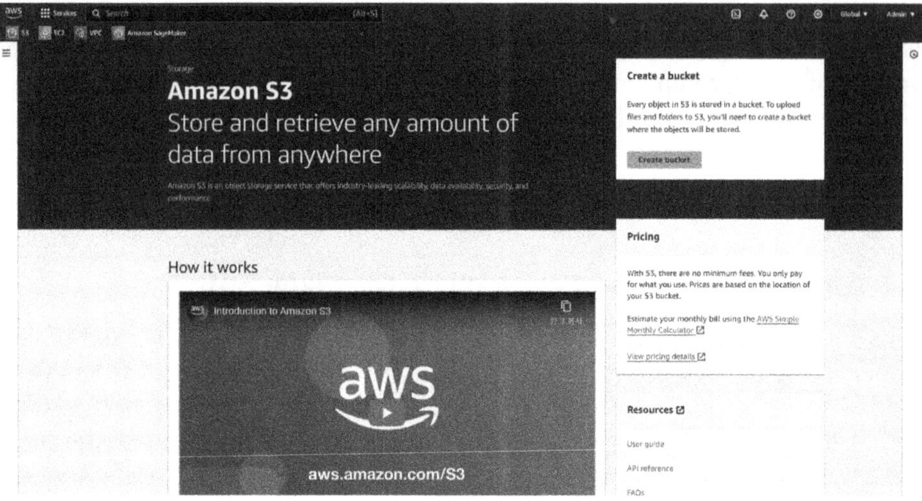

Figure 7.2 Amazon S3 screen.

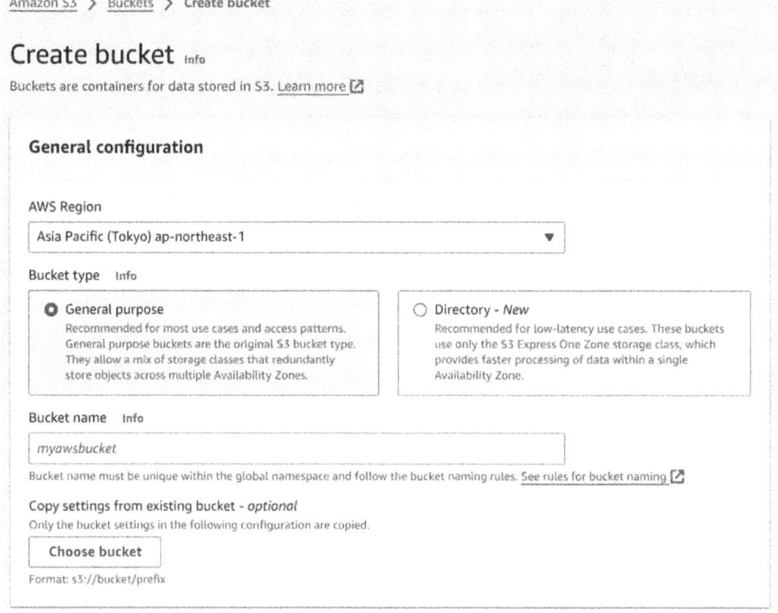

Figure 7.3 Creating an Amazon S3 bucket.

After setting the bucket name, you need to disable the ACL and then set up bucket versioning and default encryption. Customize the bucket versioning and default encryption according to the user's environment and click on "Create bucket" below to create the bucket.

Default encryption Info

Server-side encryption is automatically applied to new objects stored in this bucket.

Encryption type Info

- ◉ Server-side encryption with Amazon S3 managed keys (SSE-S3)
- ○ Server-side encryption with AWS Key Management Service keys (SSE-KMS)
- ○ Dual-layer server-side encryption with AWS Key Management Service keys (DSSE-KMS)

  Secure your objects with two separate layers of encryption. For details on pricing, see **DSSE-KMS pricing** on the **Storage** tab of the Amazon S3 pricing page.

Bucket Key

Using an S3 Bucket Key for SSE-KMS reduces encryption costs by lowering calls to AWS KMS. S3 Bucket Keys aren't supported for DSSE-KMS. Learn more

- ○ Disable
- ◉ Enable

▸ **Advanced settings**

ⓘ After creating the bucket, you can upload files and folders to the bucket, and configure additional bucket settings.

Cancel　Create bucket

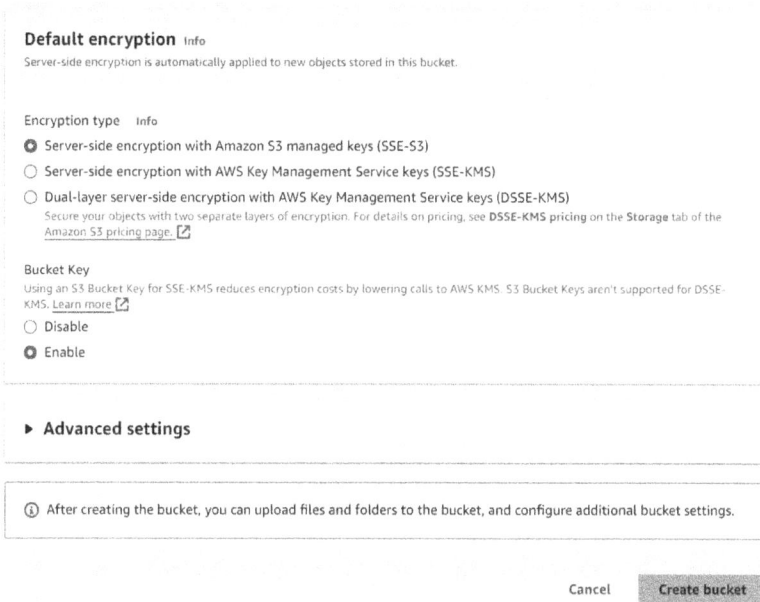

Figure 7.4　Amazon S3 bucket encryption settings.

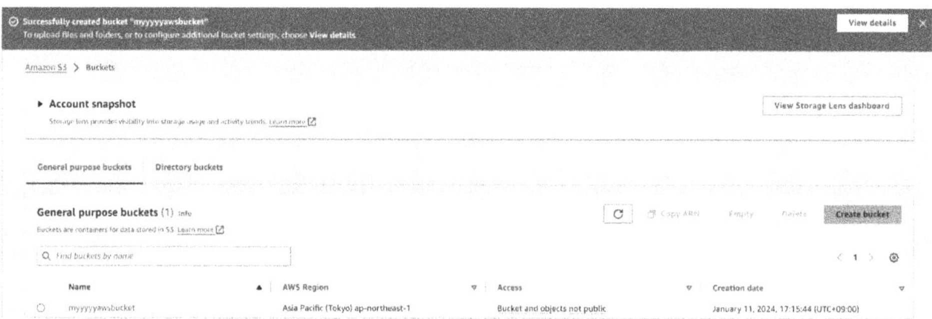

Figure 7.5　Amazon S3 bucket creation completed.

If you click on "Create bucket" in Figure 7.4, the message "Bucket creation completed" will appear as shown in Figure 7.5, and the bucket will be created.

7.1.2 *Delete S3 bucket*

To delete an S3 bucket, click the circular checkbox next to the user-created bucket name in Figure 7.6, and then click the "Delete."

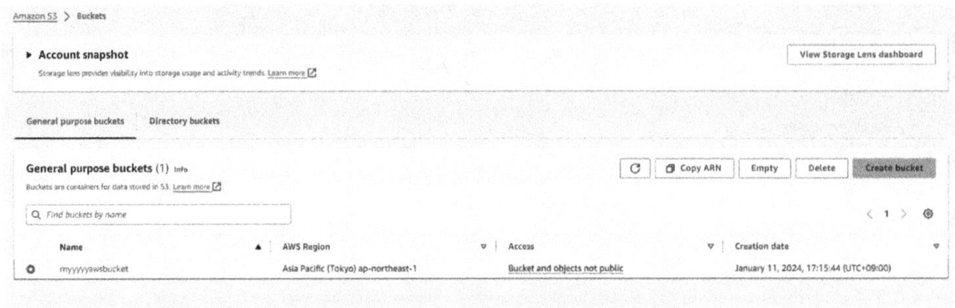

Figure 7.6 Amazon S3 bucket list.

Amazon S3 > Buckets > myyyyyawsbucket > Delete bucket

Delete bucket Info

- Deleting a bucket cannot be undone.
- Bucket names are unique. If you delete a bucket, another AWS user can use the name.
- If this bucket is used with a Multi-Region Access Point in an external account, initiate failover before deleting the bucket.
- If this bucket is used with an access point in an external account, the requests made through those access points will fail after you delete this bucket.

Learn more

Delete bucket "myyyyyawsbucket"?

To confirm deletion, enter the name of the bucket in the text input field.

myyyyyawsbucket

Cancel Delete bucket

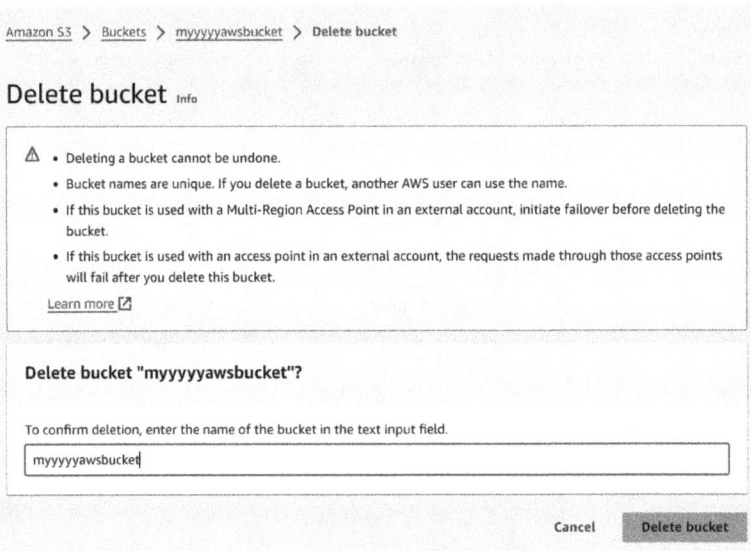

Figure 7.7 Delete Amazon S3 bucket.

If you click the “Delete,” the screen shown in Figure 7.6 will appear. Here, after entering the bucket name set by the user in the text input field below, press the” Delete bucket” button at the bottom right to delete the bucket.

Next, we will explore how to delete an S3 bucket when there are objects inside the bucket. Please refer to Figure 7.7 and click on the bucket name to check what details are inside the bucket.

When you click on the bucket name in Figure 7.8, it appears as shown in Figure 7.9. You can see that there are two objects, and to completely delete the bucket, you must delete the objects created in the bucket. Therefore, to delete an

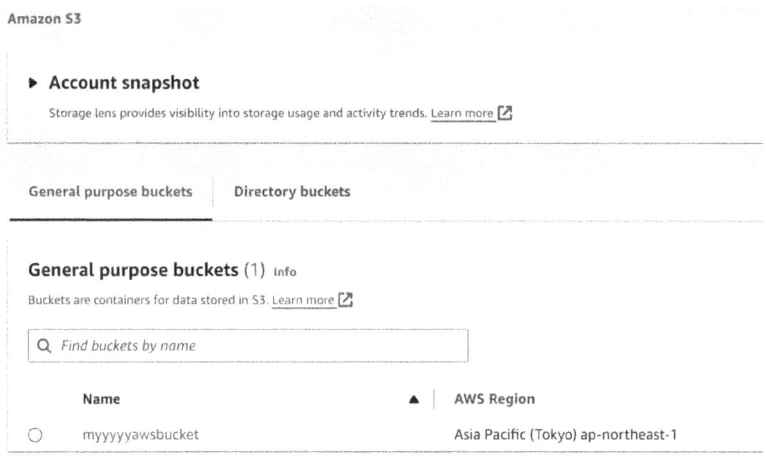

Figure 7.8 Amazon S3 bucket information.

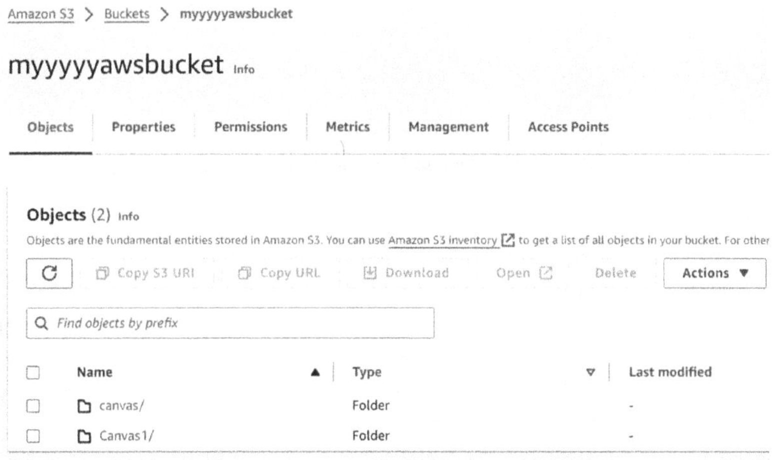

Figure 7.9 Amazon S3 bucket objects.

object, select the square checkbox next to the object name, and then click on the "Delete" to delete it.

After checking the checkbox in Figure 7.9, if you click on the "Delete", the screen will appear as shown in Figure 7.10. As shown in Figure 7.10, enter "permanently delete" in the text box and then click on the "Delete objects" button below to delete the object.

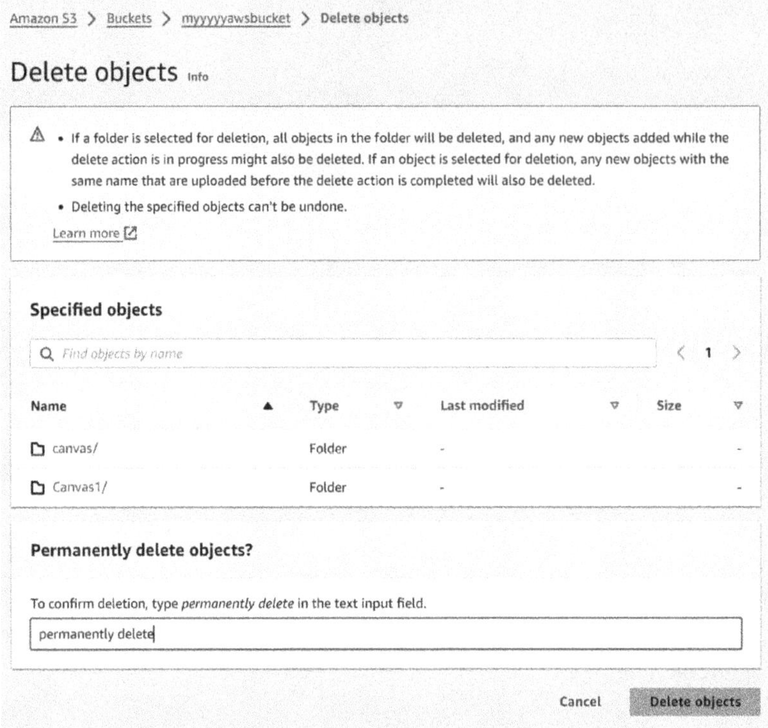

Figure 7.10 Delete Amazon S3 bucket objects.

7.2 Amazon EC2

Amazon Elastic Compute Cloud (Amazon EC2) is a major computing service of AWS that provides a cloud-based virtual server hosting platform. This web service offers computing capacity that can be scaled according to user demands in a secure manner, and allows users to easily set up and adjust the required computing capacity through an intuitive web interface. Amazon EC2 provides extensive control over server instances running on Amazon's proven computing infrastructure, while offering flexibility to quickly start or scale capacity up or down based on computing requirements. EC2 users can provision[4] and manage virtual servers (instances) as needed, allowing them to run various applications. Additionally, it provides developers and system administrators with tools for building recoverable applications and capabilities for isolation and recovery in common error situations for better management. In terms of cost, users are only billed for what they

[4]Provisioning: Allocating, arranging, and deploying system resources to meet the needs of users so that the system is ready to go when needed.

actually use, allowing for efficient cost management. The main features and concepts of Amazon EC2 are as follows:

1. **Instance:** This means a virtual server in EC2. Users can start an instance with desired size, operating system, and configuration.
2. **Instance types:** Various instance types with different performance characteristics are available. For example, there are CPU instances with high computing power, memory optimized instances, and so on.
3. **Amazon machine image (AMI):** This is an image used when starting an instance. AMIs include operating system and software configurations.
4. **Region and availability zones:** EC2 provides multiple regions and multiple availability zones within each region. This allows for high availability and fault tolerance, as well as the ability to run servers in geographically close locations.
5. **Scaling:** You can scale your application up or down using EC2. By utilizing Auto Scaling, you can automatically adjust instances according to traffic increases or decreases.
6. **Security groups and networking:** You can define firewall rules for instances and control network configurations through Amazon Virtual Private Cloud (VPC) by using security groups.
7. **Storage options:** EC2 instances offer various storage options and you can use Amazon Elastic Block Store to store data or integrate with object storage services like Amazon S3.
8. **Monitoring and logging:** You can monitor the performance of instances and applications and collect logs using Amazon CloudWatch.
9. **Security and access control:** You can manage access to resources and enhance security by using AWS Identity and Access Management (IAM).

In summary, with virtual server hosting using Amazon EC2, you can develop and deploy applications while reducing hardware costs.

7.2.1 *Getting started with EC2*

Search for EC2 in the search bar on the AWS Console home screen, as shown in Figure 7.11. After entering your search in the search bar, click on EC2 in the service category to learn more about EC2.

EC2 refers to a virtual server in the cloud. If you want detailed information about EC2, click on the instructions shown in Figure 7.12. To start EC2, click on the "Launch instance" button.

When you start an instance, the screen shown in Figure 7.13 will appear.

After assigning a name, as shown in Figure 7.13, you need to configure the application and OS image (AMI) as shown in Figure 7.14. After checking the AMI

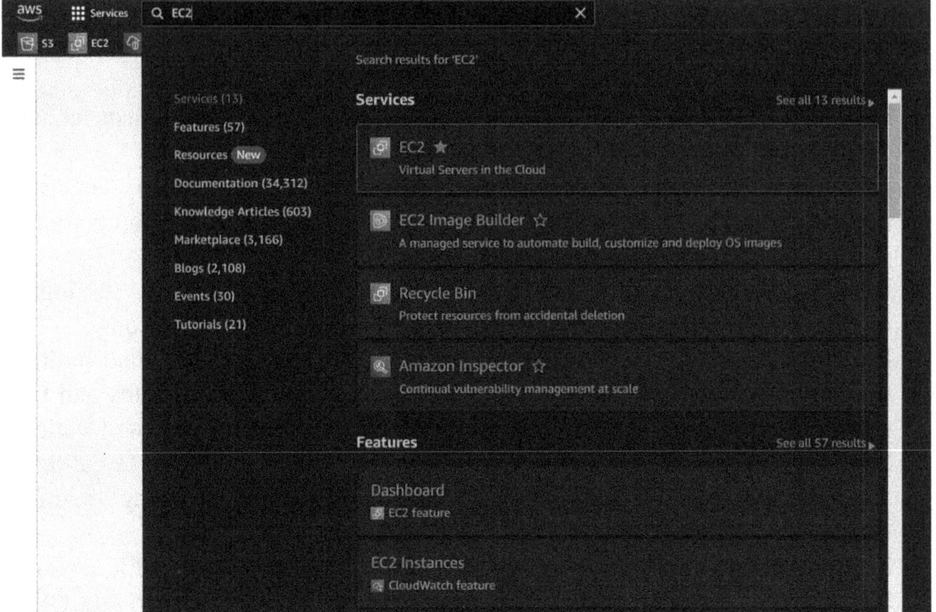

Figure 7.11 AWS console home screen.

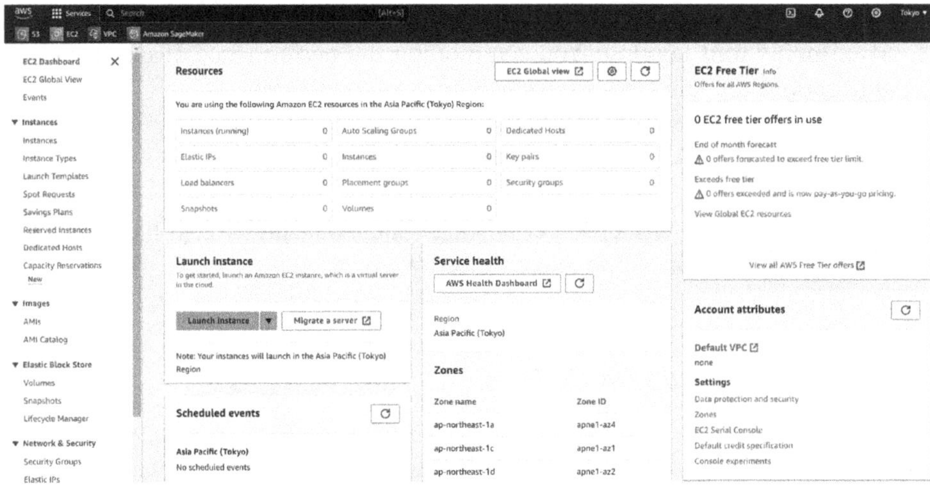

Figure 7.12 AWS EC2 main screen.

list in the Quick Start category, the user selects the appropriate AMI. If you want to find an AMI that is not shown in Figure 7.14, click on "Find more AMIs" on the right side to check for more AMIs.

In the instance type list, you can select an instance based on your desired hardware configuration. By default, t2.micro is selected, which is available in the

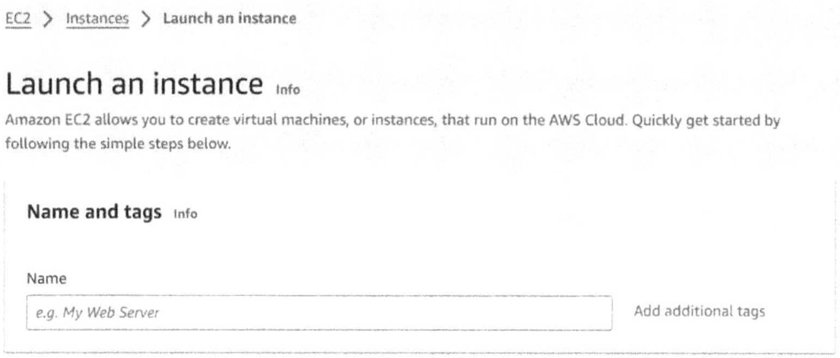

Figure 7.13 Setting EC2 instance name.

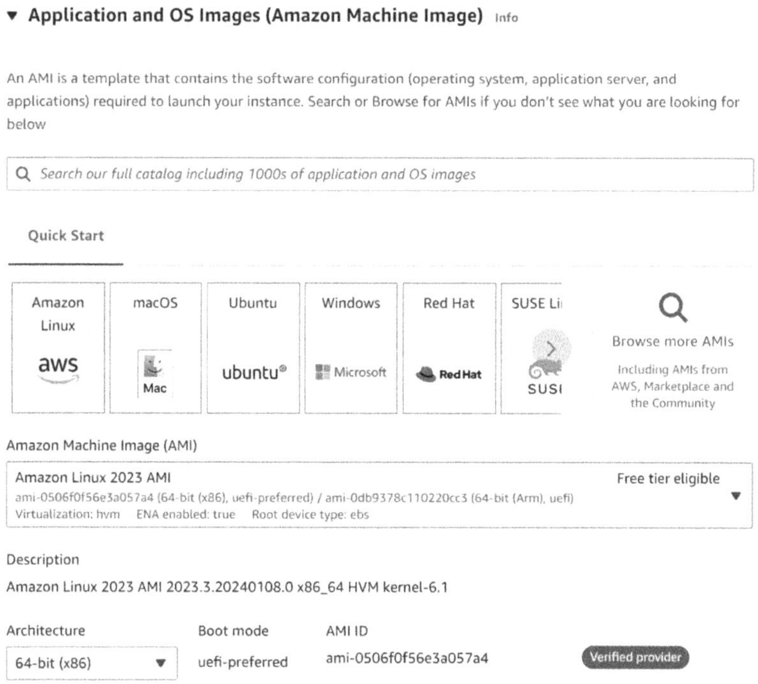

Figure 7.14 EC2 instance AMI settings.

free tier. If t2.micro is not available, you can use t3.micro instead. Through instance type settings, users can specify the required CPU and memory capacity. It is important to check the usage fee information for each instance and be aware of the service cost. After setting the instance type, you need to proceed with key pair settings. In the key pair name settings, select the key pair that you previously created. You can securely access the instance through the key pair. It is important

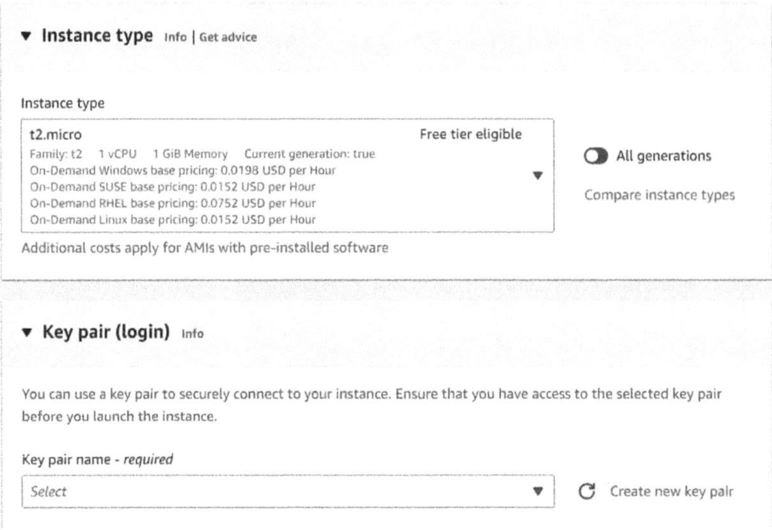

Figure 7.15 Set EC2 instance types.

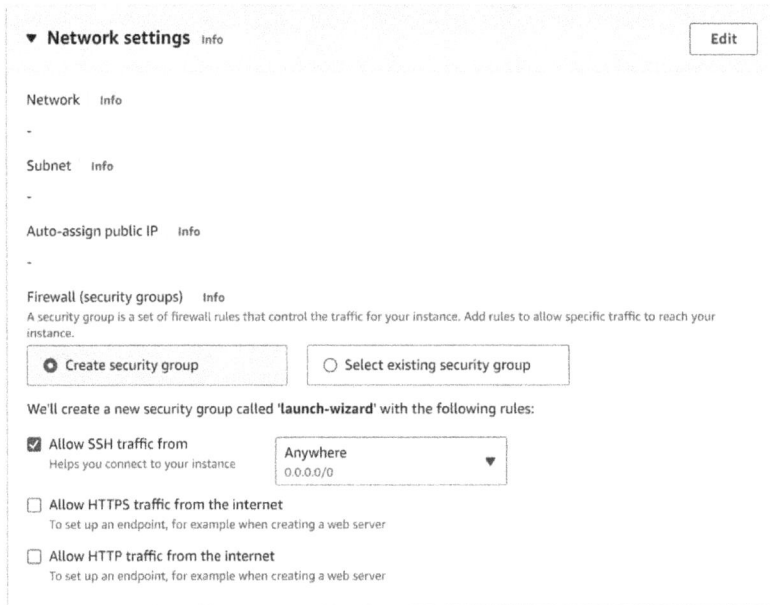

Figure 7.16 EC2 instance network settings.

to note that if you proceed without a key pair (which is not recommended), you will not be able to access the instance without the corresponding key pair.

Figure 7.16 shows the step of setting up the network to start EC2. First, select VPC and then select a subnet. Then, you can choose to enable or disable automatic

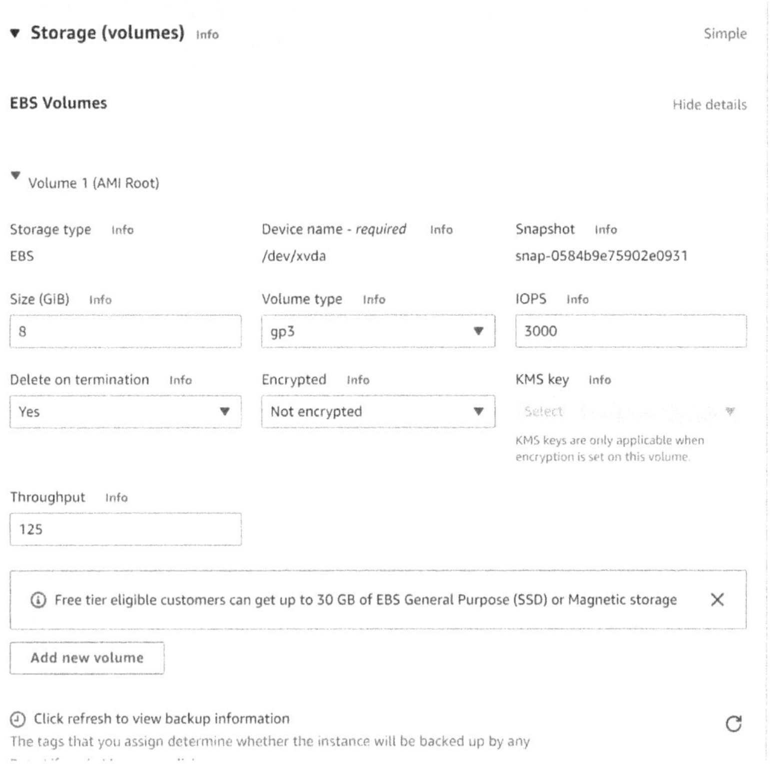

Figure 7.17 EC2 instance network configuration.

public IP assignment. In the case of a firewall (security group), it is recommended to set it up if possible, as it performs the function of controlling traffic to the instance. If there is an existing security group, select the existing security group to set up the firewall.

Figure 7.17 shows the step of setting up an EBS volume (storage space) for EC2. Amazon Elastic Block Store (Amazon EBS) provides block-level storage volumes that can be used with EC2 instances. EBS volumes behave like raw block devices with no specified format. EBS volumes attached to instances are represented as storage volumes and persist regardless of the instance's lifespan. These volumes can be used to create file systems on top of them or utilized like block devices such as hard drives.

Once the setting is completed, the summary displayed on the right side of the screen is a page that organizes the steps taken so far in an easy-to-understand manner. From here, click on the "Launch instance" button in the bottom right corner to create an EC2.

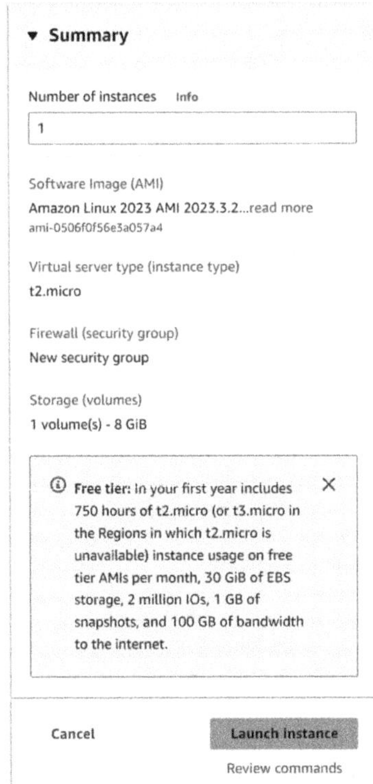

Figure 7.18 Starting EC2 instances.

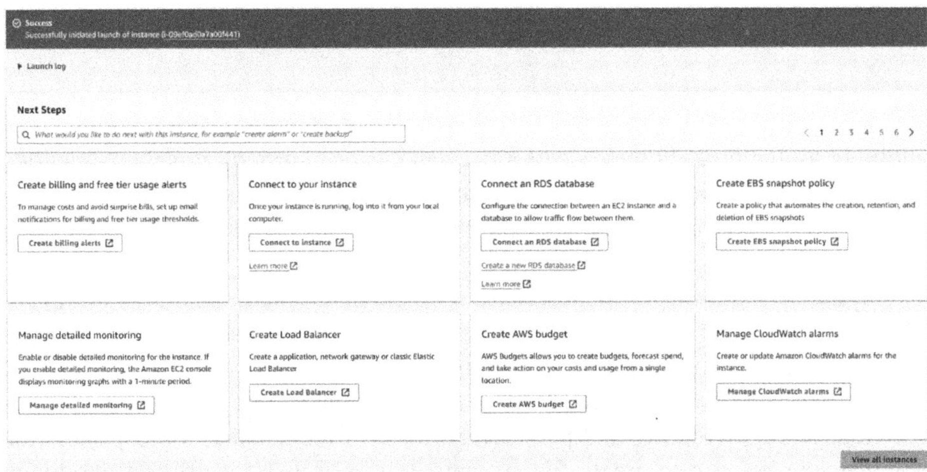

Image 7.19 EC2 instance creation complete.

When you click on the "Launch instance" button in Figure 7.18, the successful creation of the instance will be performed as shown in Figure 7.19. As a result, you can see that a virtual server has been created in the AWS cloud. Through this,

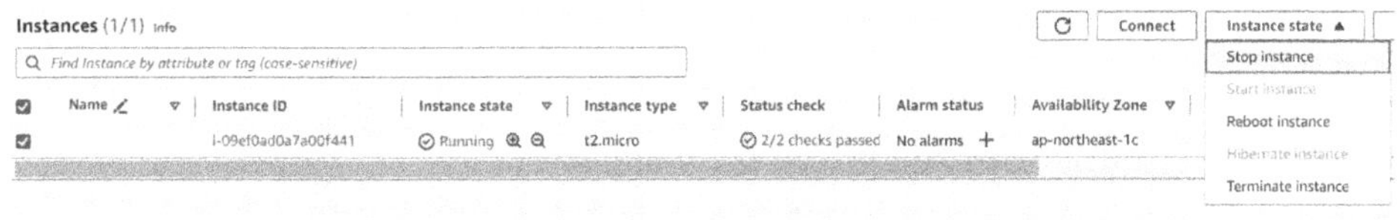

Figure 7.20 EC2 instance list.

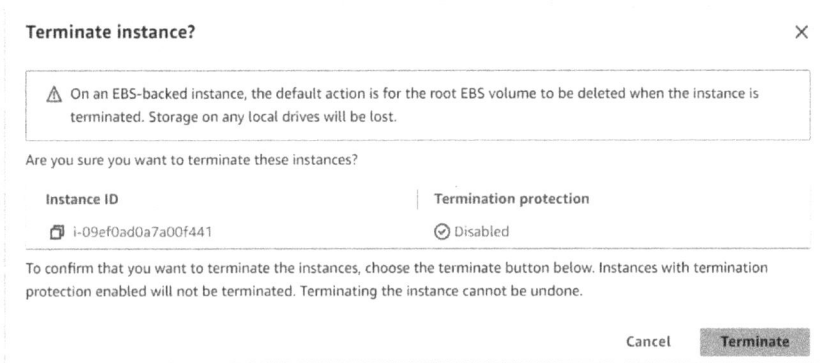

Terminate instance?

On an EBS-backed instance, the default action is for the root EBS volume to be deleted when the instance is terminated. Storage on any local drives will be lost.

Are you sure you want to terminate these instances?

Instance ID	Termination protection
i-09ef0ad0a7a00f441	Disabled

To confirm that you want to terminate the instances, choose the terminate button below. Instances with termination protection enabled will not be terminated. Terminating the instance cannot be undone.

Cancel Terminate

Figure 7.21 EC2 instance termination confirmation message.

you can set up and configure the operating system and applications running on the instance using Amazon EC2.

7.2.2 *Terminate EC2*

First, check the instance in the EC2 instance list as shown in Figure 7.20. Then, click "Instance state" and click "Terminate instance" for the instance listed.

When the confirmation message for instance termination appears as shown in Figure 7.21 select "Terminate."

As shown in Figure 7.21, the instance of Amazon EC2 is terminated by clicking on "Terminate." The instance remains displayed in the console for a while even after it has been terminated, and it is automatically deleted from the instance list afterwards. Terminated instances cannot be removed directly from the console screen.

7.3 Practice Questions

Q1. How does AWS contribute to the innovation and cost-effectiveness of cloud computing for businesses and government agencies?

Q2. Discuss the benefits of using Amazon S3 for data storage and the precautions one must take when creating and managing S3 buckets.

Q3. Explain the significance of Amazon EC2 in cloud computing and how it enables users to manage computing capacity efficiently.

Chapter 8

Amazon SageMaker

Amazon SageMaker is AWS's flagship machine learning service that helps data scientists and developers simplify the process of creating, training, and deploying machine learning models. SageMaker provides various tools and modules necessary for processing, analyzing, and deploying machine learning data. The main benefits of SageMaker are its ability to quickly build and deploy machine learning models, its features that reduce the complexity of infrastructure installation, management, and scalability, and its tools that enhance the efficiency and performance of machine learning workflows. In this section, we will delve deeply into the key components of SageMaker, its main features, utilization strategies, and configuration methods.

8.1 Technical Requirements

To effectively utilize Amazon SageMaker, certain technical requirements must be met. Firstly, an AWS account is essential to use SageMaker. Users without an account can create one on the AWS website. The account tier provided during creation, known as the Free Tier, allows users to experience various AWS services for a limited period of time for free. It is recommended to review the details of the Free Tier offerings provided by AWS at (https://aws.amazon.com/free/).

SageMaker Studio, within SageMaker, provides a browser-based integrated development environment with various tools and features necessary for the development and training of machine learning models. As SageMaker supports major machine learning frameworks such as TensorFlow, PyTorch, and MxNet, the technical requirements of these frameworks must be met. In particular, when utilizing GPU acceleration with frameworks like TensorFlow, GPU drivers and related libraries are required. These drivers and libraries can be easily configured on AWS EC2 instances using AMIs. Additionally, SageMaker is closely integrated with

Amazon S3 for data storage and processing. Therefore, configuring S3 bucket settings and granting access permissions to the bucket are necessary.

8.2 Main Functions

In this chapter, we provide documentation and practice materials on various features and utilization methods of Amazon SageMaker. Through this, you can deeply understand the key features provided by SageMaker and learn how to actually utilize them. First, we will examine the methods of data exploration and analysis using notebook instances. Next, we will explain in detail the process of model creation using the provided algorithms and model deployment through hosting. We will also cover performance analysis and management methods for completed machine learning models. This content will be very useful for those who are new to machine learning, as well as users who want to use SageMaker for model construction and analysis. Once you experience the key features of creating, analyzing, and deploying models in a managed infrastructure through this AWS service, you can focus more on machine learning algorithm research and problem-solving, freeing yourself from the complexities of infrastructure construction and management. The main features of SageMaker are as follows.

- **Fast and easy model training:** It supports various machine learning frameworks and infrastructures to enable fast and easy model training.
- **Model hosting and deployment:** Supports all tasks required to host and deploy trained models.
- **Automatic model tuning:** It automatically performs model tuning to find the optimal model.
- **Reinforcement learning:** By providing reinforcement learning algorithms and functions, you can quickly build and train reinforcement learning models.
- **Data preprocessing:** By providing built-in functions for data preprocessing and transformation, it enables fast and easy preprocessing of data.
- **Machine learning workflow management:** Provides various tools necessary for managing and monitoring machine learning workflows.
- **Continuous model improvement:** Supports continuous model improvement by allowing the use of new data to refine the model.
- **Machine learning model description:** Provides tools for explaining and interpreting machine learning models.

Through these features, SageMaker enables developers and data analysts to efficiently build, train, and deploy machine learning models. Next, we will take a closer look at SageMaker's workflow.

1. **Build stage:** The Build phase of SageMaker is the process of developing a machine learning model. In this phase, various machine learning algorithms, frameworks, and libraries are used to analyze data and create models. The Build phase of the model provides the following functionalities.
   - **Framework selection:** Supports various machine learning frameworks such as TensorFlow, PyTorch, and Apache MxNet. Users can choose their preferred framework to develop machine learning models.
   - **Model training:** Users can train machine learning models on various types of instances. In addition, SageMaker manages tasks such as uploading datasets, model configuration, hyperparameter tuning, monitoring jobs, and evaluating trained results.
   - **Auto model tuning:** Provides a function to automatically tune the hyperparameters of the model. This feature tries various values of hyperparameters to find the optimal model configuration.
   - **Distributed learning:** Efficiently trained datasets and models using distributed learning. Distributed learning is the process of performing model training simultaneously across multiple services.
   - **Model validation and evaluation:** Provides tools and functions to evaluate the accuracy of the model and improve prediction performance. To do this, it validates the data, evaluates the model's performance, and tunes the model.
   - **Model deployment:** Supports saving machine learning models to Amazon S3 and deploying them on various services such as AWS Lambda, Amazon EC2, and more. Additionally, SageMaker provides tools for monitoring and maintaining models after deployment.

In SageMaker, various notebook instances are provided, including Jupyter and Anaconda as examples. Figure 8.1 shows the features of the SageMaker notebook instance environment.

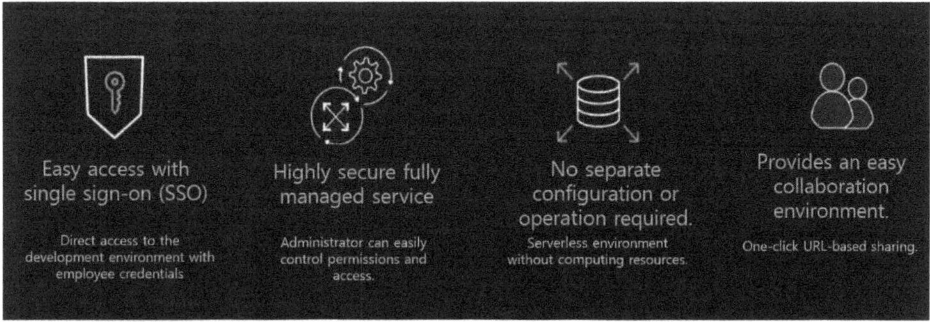

Figure 8.1 Characteristics of SageMaker notebook instances.

2. **Training stage:** The training phase is about training the data. SageMaker prepares the necessary environment for users to easily train by allowing them to select machine learning algorithms, process data, and configure models. This environment includes everything from machine learning algorithms to hardware and network settings. In the training process of SageMaker, various options are provided. You can use frameworks like TensorFlow, PyTorch, and MXNet to train deep learning models, and these frameworks are already installed in SageMaker, so there is no need to install them separately. Through the SageMaker Python SDK, you can use these frameworks and also train with various types of compute instances. These instances have various sizes of CPU, GPU, and memory options, making it easy to handle large-scale data. Once the model training is complete, SageMaker saves the results. This saved model can be used in SageMaker endpoints to make predictions on data using the trained model.
3. **Deploying:** Similar to the training phase, Amazon SageMaker also manages deployment infrastructure and provides the following additional features.
   - **A real-time endpoint:** Through this, a HTTPS API is generated from the model to provide predictions. This means automatic scaling is possible.
   - **Batch conversion:** Multiple data points or data sets can be entered into a model at once and processed simultaneously, allowing you to get results more efficiently and quickly than if you processed the data individually.
   - **Infrastructure monitoring using Amazon CloudWatch:** Through this, you can view metrics in real-time and check logs of infrastructure performance.
   - **Amazon SageMaker monitor:** Captures the data sent to the endpoint, identifies issues in the data by comparing it to the baseline, and sends notifications.
   - **Amazon SageMaker Neo:** Compiles models for specific hardware architectures and deploys optimized versions using lightweight execution environments.
   - **Amazon elastic inference:** This adds GPU acceleration to CPU-based instances to find the cost and performance ratio of the prediction infrastructure (Figure 8.2).

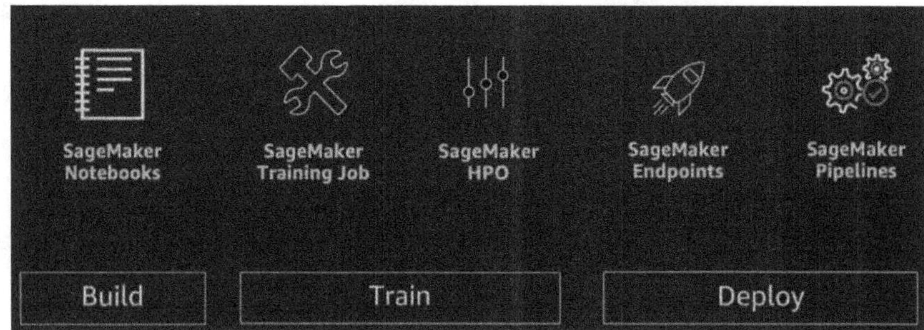

Figure 8.2 Main capabilities of Amazon SageMaker.

8.3 Major Algorithms

Amazon SageMaker offers various machine learning algorithms. The algorithms provided by SageMaker can be used in conjunction with various types of learning, such as supervised learning, unsupervised learning, and reinforcement learning. Figure 8.3 shows a list of algorithms commonly used in the Amazon SageMaker Studio environment, as well as various other machine learning algorithms.

8.4 Getting Started with SageMaker

Let's go to the AWS homepage to get started with Amazon SageMaker. You can access the AWS homepage at the following address: https://aws.amazon.com/ko/.

8.4.1 *Account creation*

Before examining the content of Amazon SageMaker, you need to create an AWS account. Therefore, in this chapter, we will learn how to create an AWS account. First, let's go from the AWS homepage to the Amazon SageMaker homepage. There are three ways to go from AWS to the Amazon SageMaker homepage. The first way is shown in Figure 8.4, where you can click on the magnifying glass tab and search for SageMaker to go to the SageMaker homepage. The second way is for users to directly access the AWS SageMaker homepage address (https://aws.amazon.com/ko/sagemaker/). Lastly, as shown in Figure 8.4, you can click on the "Products" tab and then click on the "Main services" tab to navigate to Amazon SageMaker within the category. Among these three methods, choose the one that

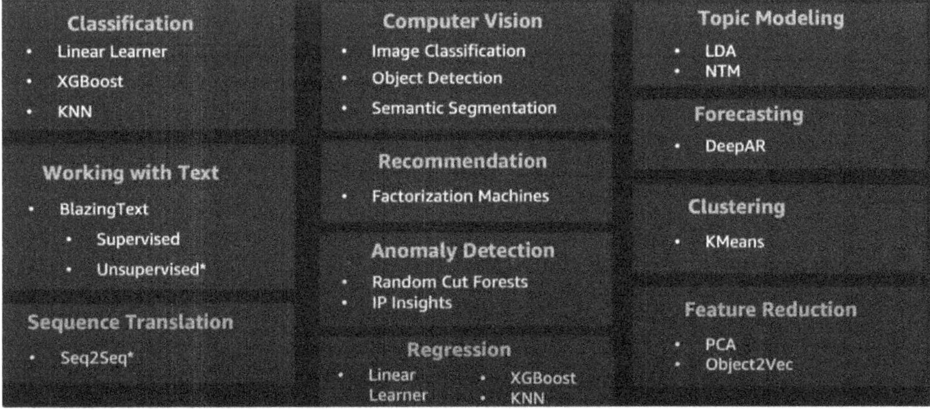

Figure 8.3 Built-in algorithms provided by SageMaker.

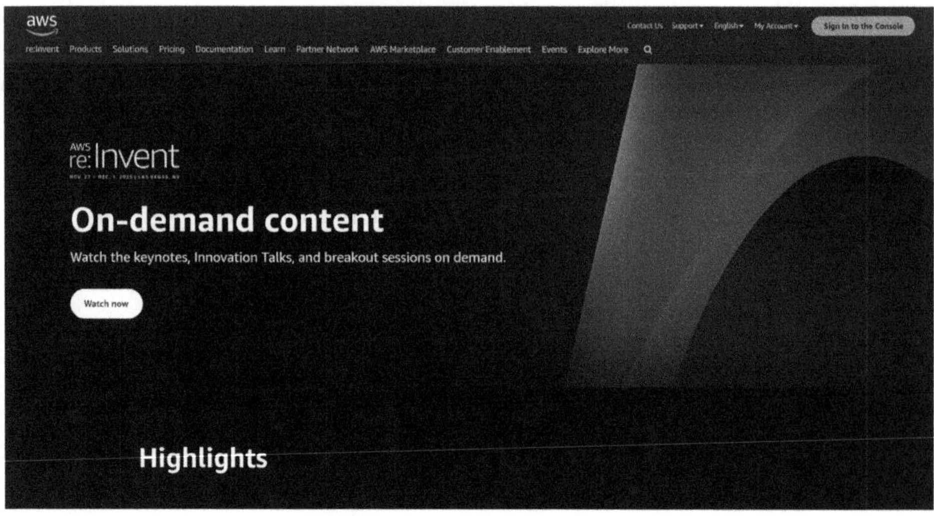

Figure 8.4 AWS home screen.

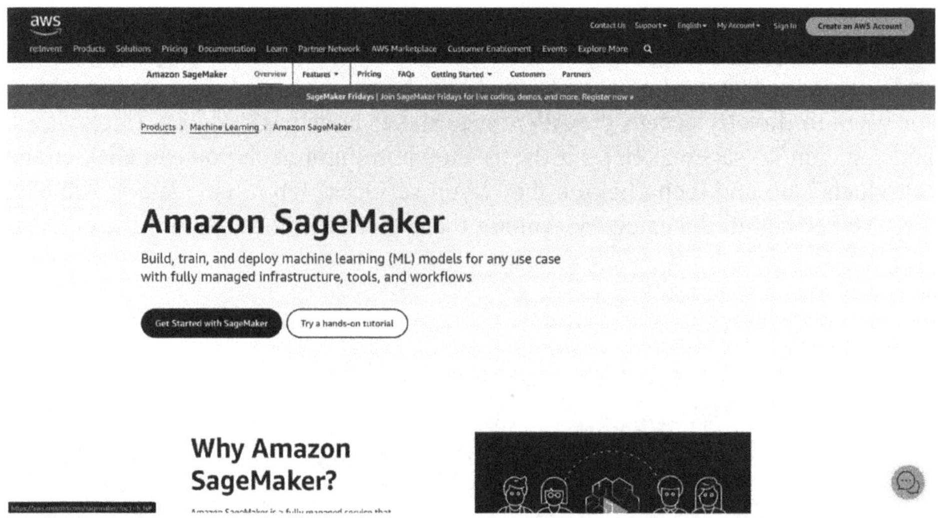

Figure 8.5 Amazon SageMaker home screen (new).

is most convenient for you to go to the AWS SageMaker homepage. Figure 8.5 shows the screen that appears when the user has never visited the AWS homepage before. Figure 8.6 shows the screen that appears when the user has previously visited the AWS homepage. Therefore, the screen of the AWS SageMaker homepage appears in two cases, as shown in Figures 8.5 and 8.6.

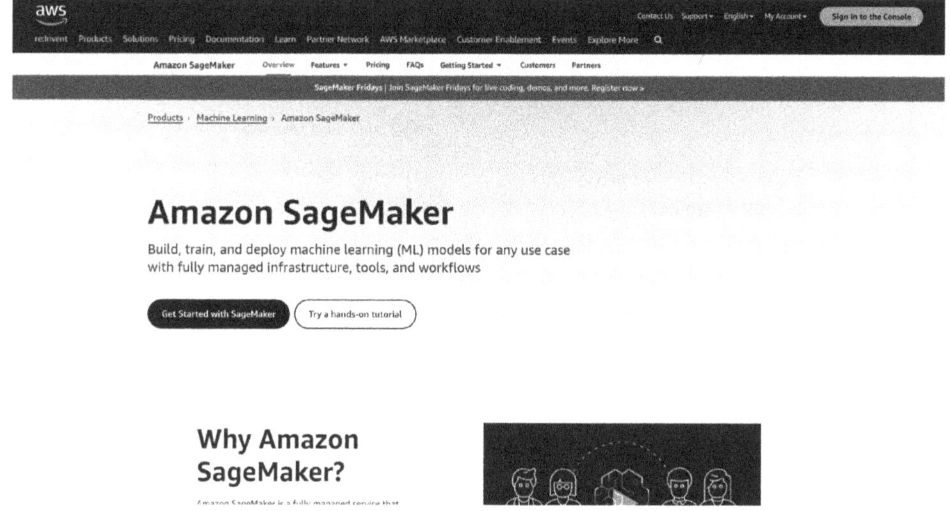

Figure 8.6 Amazon SageMaker home screen (member).

8.4.2 *How to create an account*

8.4.2.1 *New user*

As the first method of creating an AWS account, if the screen appears as shown in Figure 8.5, you can click on the "Create AWS Account" tab in the top right corner to go to the screen for creating an AWS account. The screen shown in Figure 8.7 will appear when you click the "Create AWS Account" tab. When this screen appears, you should enter the email address and account name that you will use for AWS. When setting your email address, make sure to enter it in the correct email address format, and your account name should be in English or numbers. This can be changed in your account settings after signing up. If you do not adhere to the above instructions, an error message like that shown in Figure 8.8 will appear. If you have completed the above steps without any errors, click on the "Verify Email Address" tab to proceed to the next step.

Once email address verification is completed in Figure 8.7, a screen similar to that shown in Figure 8.9 will appear. As shown in Figure 8.9, you can enter the verification code after confirming the email containing the code. The verification code will have been sent to the email address set by the user in Figure 8.7. After completing the verification code entry, a screen for setting the user password will appear. The following will explain how to set the user password. In Figure 8.9(a), after entering the verification code and pressing the confirm button, a screen like Figure 8.9(b) will appear. As shown in Figure 8.8, the password must be at least eight characters long and must include either letters or numbers, or non-letter and

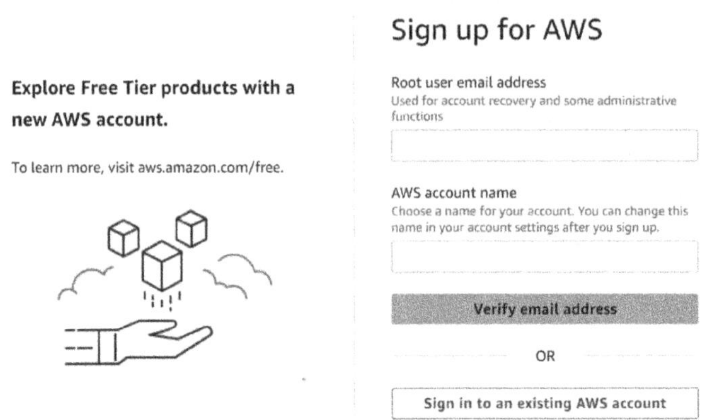

Figure 8.7 AWS membership registration.

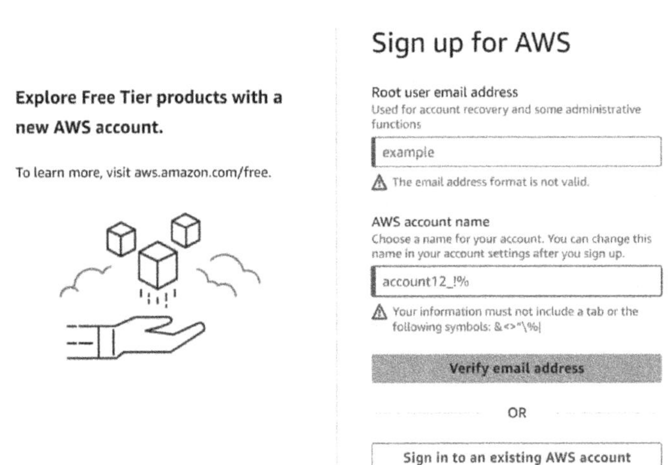

Figure 8.8 AWS membership sign-up error screen.

non-numeric characters. In Figure 8.9(b), there are conditions to set the user password. As seen in Figure 8.8, the password must include at least three of the four conditions. Once the setup is complete, click the "Continue" button below to proceed to the next step.

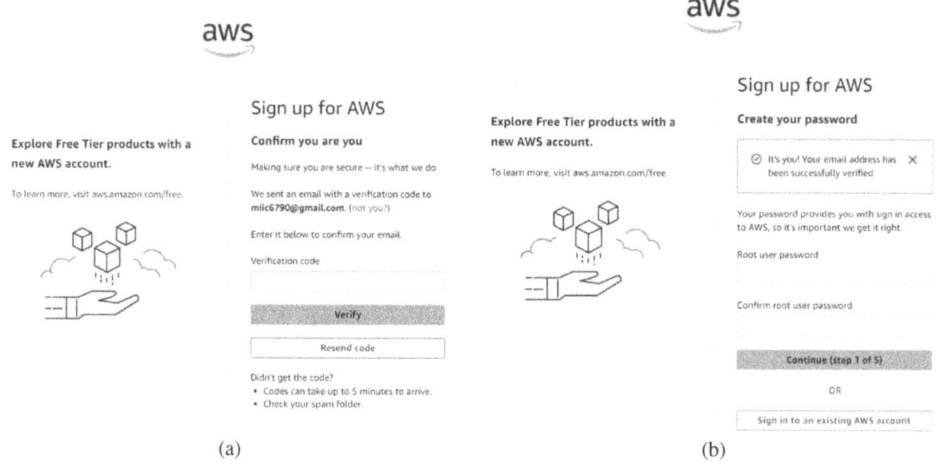

Figure 8.9 AWS membership registration process. (a) Membership registration step. (b) User password input.

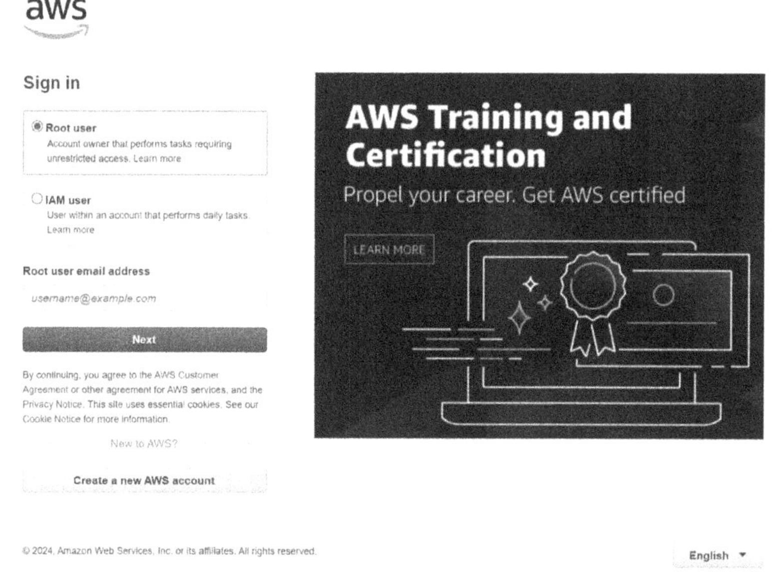

Figure 8.10 AWS login.

Top right to go to the login screen for AWS SageMaker. When you click on the tab, the screen will appear as shown in Figure 8.7. As shown in Figure 8.7, there is a screen to log in to the AWS SageMaker Console and a tab at the bottom of the screen that says “Create a new AWS account.” To create an AWS account, click on the “Create a new AWS account” tab at the bottom of the screen as shown in Figure 8.10, and the screen in Figure 8.9 will appear. Once Figure 8.9 appears,

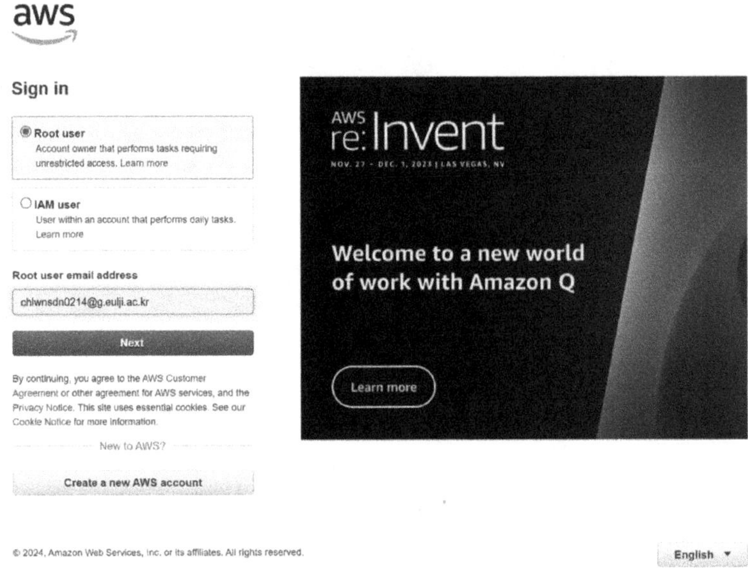

Figure 8.11 AWS login.

follow the instructions from Figure 8.7 onwards as described in the "Account Creation Method."

Figure 8.11 shows the screen for entering the user's address, where the user's name, phone number, and address are inputted. For practice, select "Individual" for the AWS usage plan category, and enter the user's name in the "Full Name" field. After selecting the country code in the "Phone Number" field, input the user's cellphone number. In the "Country or Region Settings" field, the user sets their current country by clicking the downward, you will need to enter the user email address you set during registration in the text box and proceed to the next step by clicking the "Next" button.

After entering the user's email address and clicking the "Next" button in Figure 8.11, the screen shown in Figure 8.12 will appears. Once this screen appears, the user can log in by entering the password they set when signing up.

After entering the password, press the login button, and the AWS Console screen shown in Figure 8.13 will appear.

8.4.3 *Regional settings*

Next, let's find out how to configure the region in SageMaker. As you can see in Figure 8.14, the region configuration can be selected in the top right tab according to the user's preference. This screen shows an example of a user selecting and configuring the US East (Northern Virginia) region.

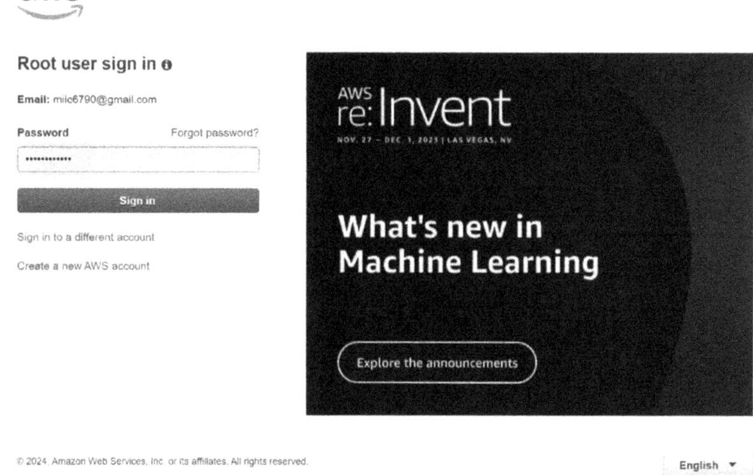

Figure 8.12 AWS password input window.

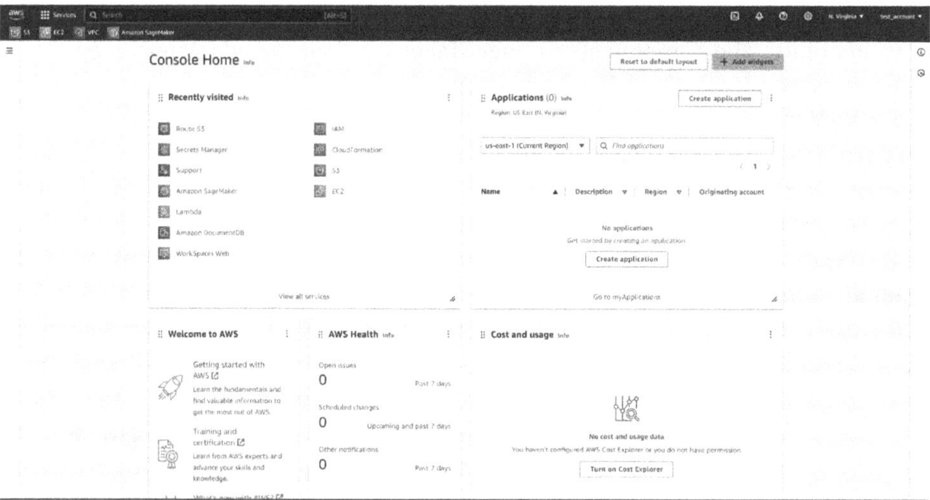

Figure 8.13 AWS console home screen.

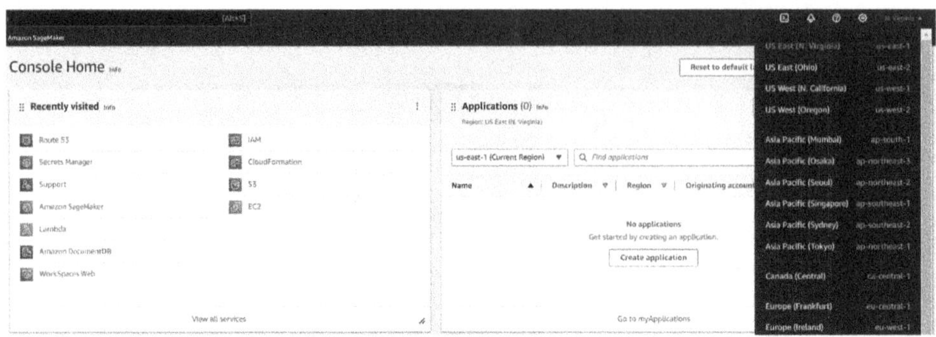

Figure 8.14 AWS SageMaker region screen.

8.5 SageMaker Domain

SageMaker Domain is a core feature provided by SageMaker Studio in AWS, which simplifies the construction, training, and deployment processes of machine learning models for users via the necessary infrastructure. This domain supports integration with AWS's major data services, such as Amazon S3, Amazon Athena, Amazon QuickSight, and Amazon Redshift, allowing users to process and analyze data smoothly. The components of SageMaker Domain include Amazon EFS volumes, an authenticated user list, various security settings, applications, policies, and Amazon VPC. Within the domain, users can share notebook files and other necessary artifacts with each other, and they can create and manage multiple SageMaker domains within a single AWS account. Next, let's find out how to set up a SageMaker domain. Firstly, when the user logs in, the AWS Console main screen will appear. There are a few methods to navigate to the SageMaker Console screen from this screen. First, there is a search bar next to the magnifying glass where you can search for AWS service, as shown in Figure 8.15. Here, you can search for SageMaker and navigate to the SageMaker homepage. Searching for SageMaker will bring up the screen shown in Figure 8.16. From here, click on AWS SageMaker to proceed. The AWS SageMaker Console main screen is shown in Figure 8.17.

When you search for SageMaker in the "Search" tab in Figure 8.16, the screen shown in Figure 8.17 will appear.

8.5.1 *Domain creation*

To start SageMaker, you need to create a domain. To create a domain, click on the "Get Started" tab at the top left corner, as shown in Figure 8.17. Once you click this tab, the SageMaker start page will appear, as shown in Figure 8.18. Once this page appears, click on "SageMaker domain settings" at the bottom of the screen.

There are two ways to set up a domain, as shown in Figure 8.19. In this chapter, let's explore a quick setup method to briefly explain the process of domain

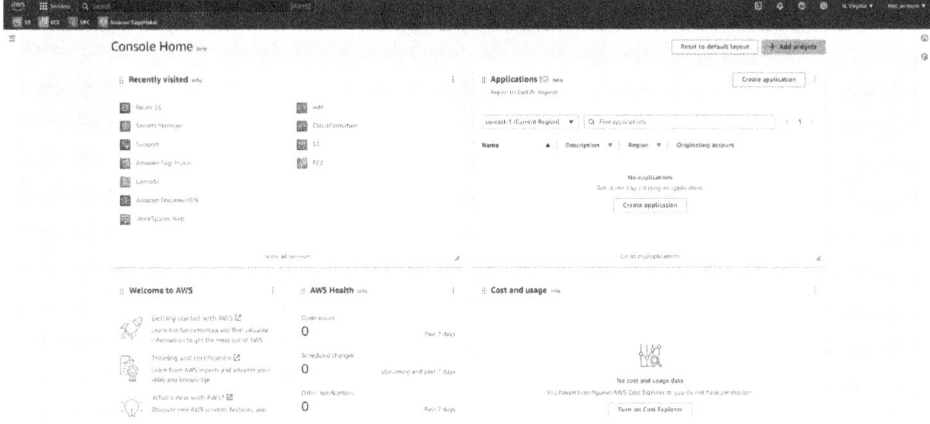

Figure 8.15 AWS console home screen.

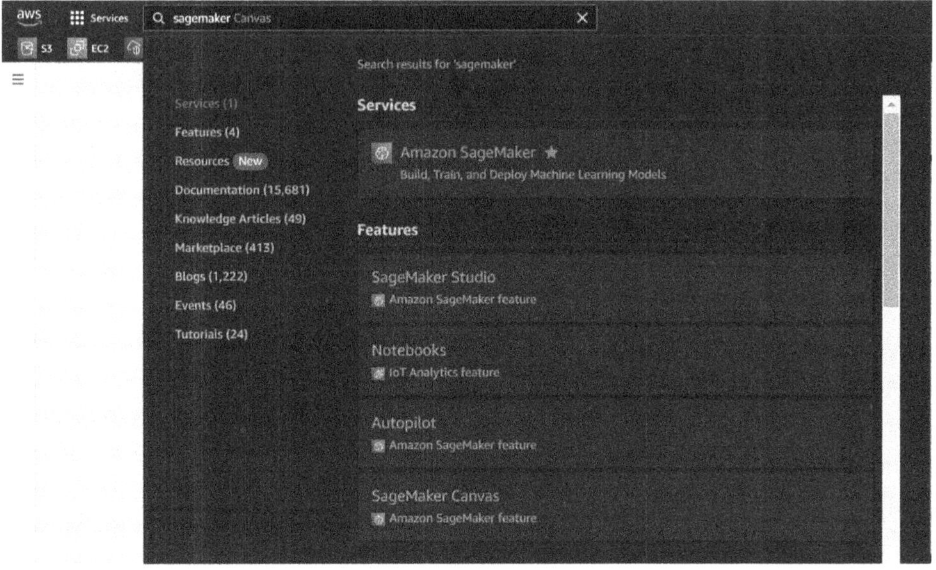

Figure 8.16 AWS search screen.

creation. After clicking on the "Quick setup" as shown in Figure 8.19, click on the "Settings" tab below to create the domain.

Once the domain creation is complete, as shown in Figure 8.20, the domain name and the user name are automatically set, and the creation status is displayed as "Pending."

After about 5–10 minutes, when Figure 8.21 is in the InService state, the domain creation is completed.

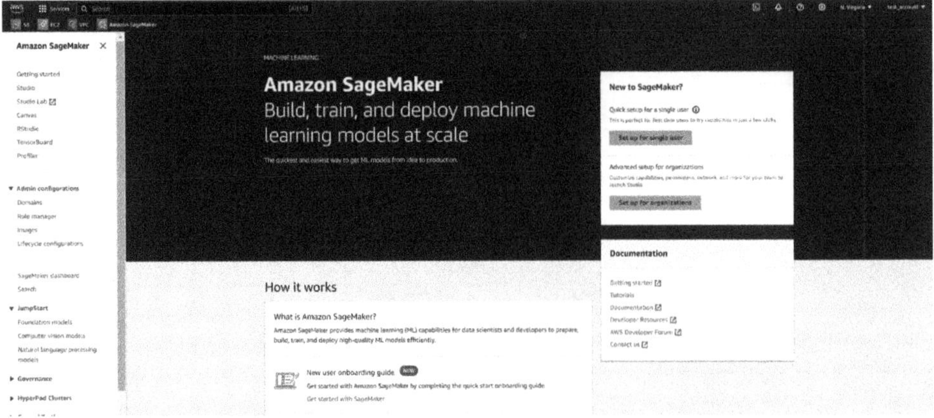

Figure 8.17 AWS SageMaker console home screen.

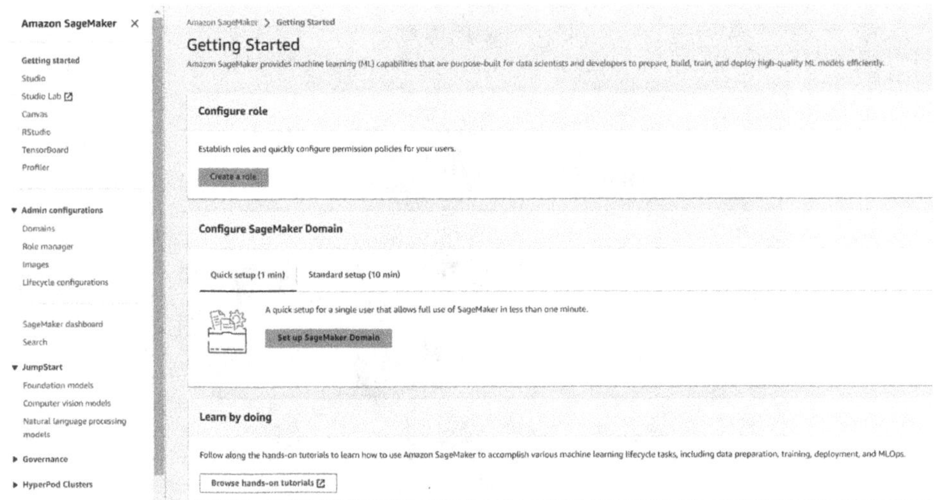

Figure 8.18 SageMaker start screen.

8.5.2 *Domain deletion*

To delete a domain, you must meet the following requirements.

- To delete a domain, you need to have administrator privileges.
- Only apps with the following conditions can be deleted. If you try to delete a domain where a user profile or app has been created, an error will occur.
- If you wish to delete a domain, it cannot include user profiles or shared spaces. To delete a user profile or shared space, it cannot include any apps without errors occurring in the user profile or space.

Amazon SageMaker > Set up SageMaker Domain

Set up SageMaker Domain

Set up for single user (Quick setup)

Set up for organizations

Set up

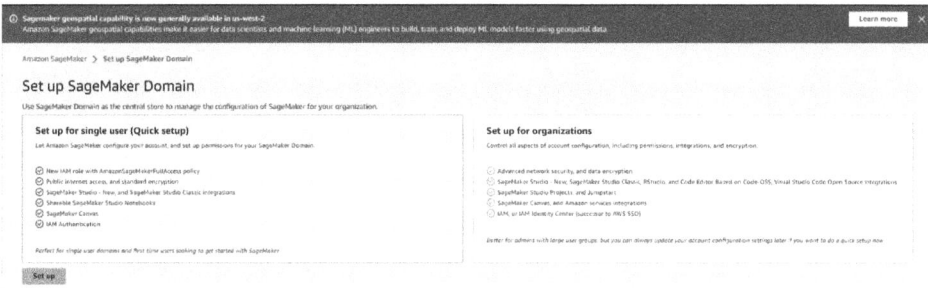

Figure 8.19 Setting up SageMaker domain name.

Amazon SageMaker > Domains

Domains Info

A domain includes an associated Amazon Elastic File System (EFS) volume; a list of authorized users; and a variety of security, application, policy, and Amazon Virtual Pri
the EFS for notebooks, Git repositories, and data files.

Domains (1) Info

Find domain name

Name	Id	Status
QuickSetupDomain-20240122T010853	d-mh8rtcu3crjz	Pending

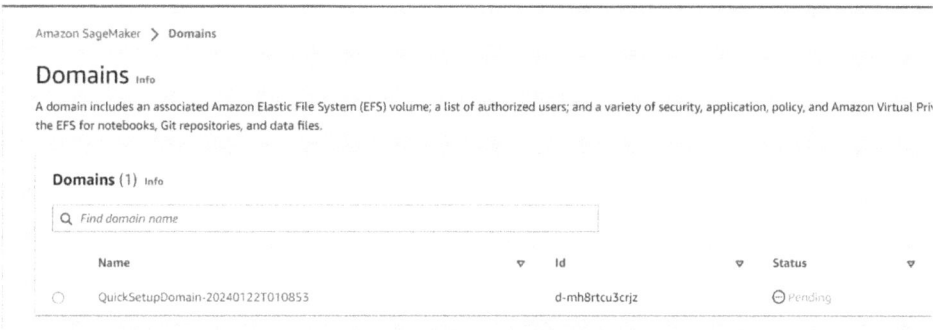

Figure 8.20 SageMaker automatically generates domains.

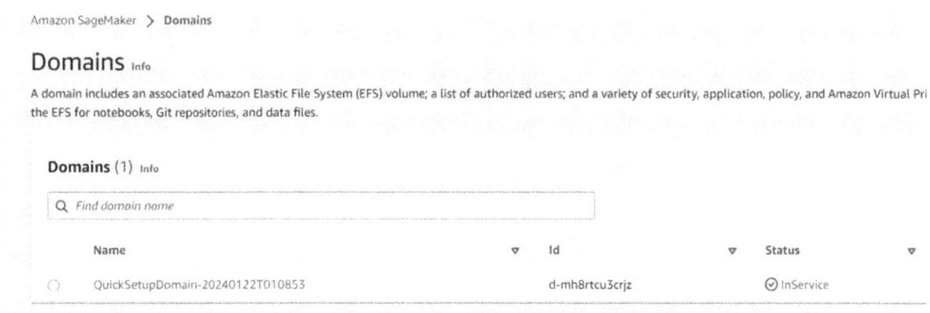

Figure 8.21 SageMaker domain creation completed.

When you delete a domain, the following state occurs.

- For apps, the data stored in the user's home directory is saved, and any unsaved data on the laptop will be lost.
- For user profiles, users will no longer be able to access the corresponding domain and will be unable to access the home directory. However, the data will not be deleted.

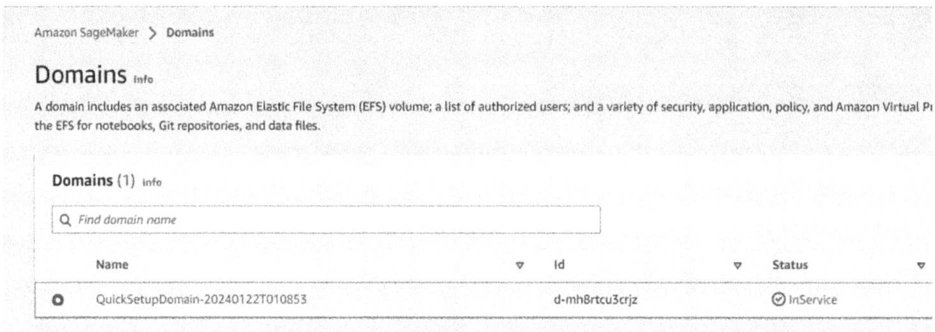

Figure 8.22 SageMaker domain interface.

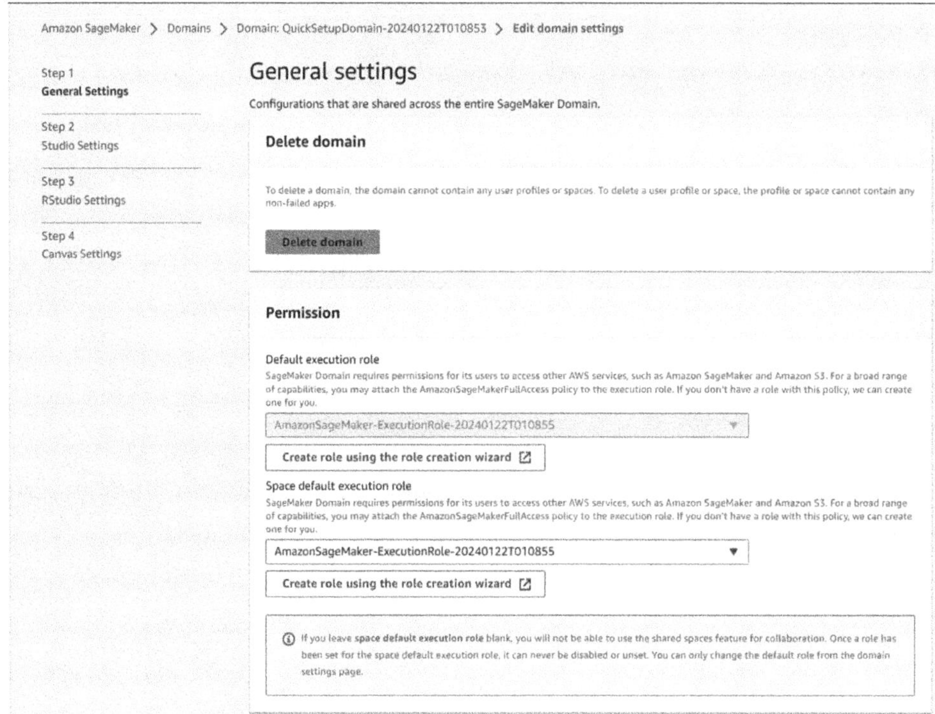

Figure 8.23 Editing SageMaker domain settings.

Before deleting a domain, there must be no information about user profiles in the domain details, as shown in Figure 8.22. If there is user profile information in the domain details, the user profile must be deleted before deleting the domain. Click the circular button next to the domain name in Figure 8.22 and then click the "Edit" tab to start the process of deleting a domain.

As you can see in Figure 8.23, there is a "Delete Domain" tab that allows users to delete a domain. You can delete the domain by clicking on this tab.

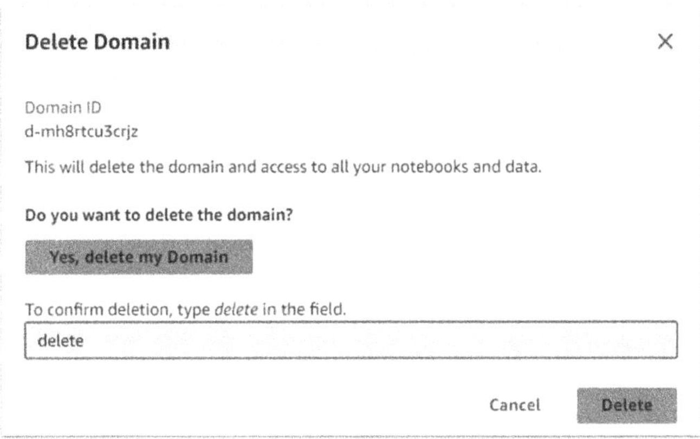
Delete Domain
Domain ID
d-mh8rtcu3crjz
This will delete the domain and access to all your notebooks and data.
Do you want to delete the domain?
Yes, delete my Domain
To confirm deletion, type *delete* in the field.
delete
Cancel
Delete

Figure 8.24 SageMaker domain deletion message.

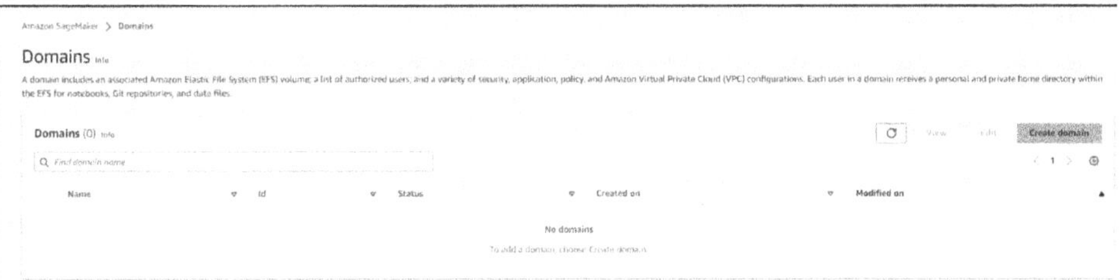
Domains
Domains (0)
No domains

Figure 8.25 SageMaker domain deletion completed.

When you click on the "Delete Domain" tab in Figure 8.23, the screen shown in Figure 8.24 will appear.

When the domain deletion message appears, click the "Yes, delete domain" button, then enter "delete" in the text field below to proceed with domain deletion. If you have clicked the "Delete" button, it may take a few minutes for SageMaker to delete the domain internally. Once the domain deletion process is complete, you can confirm that the information for the domain has been deleted by referring to Figure 8.25, which shows that there is no domain on the screen.

8.6 User Profile Settings

The user profile represents a single user within the domain. It is created when the user registers with AWS and serves as a form of reference for sharing, reporting, and other user-centric functionalities. User profile settings can be amended on the domain details screen. Refer to Section 8.5 for instructions on the domain details screen. If there is no domain, create a domain. After creating the domain, clicking

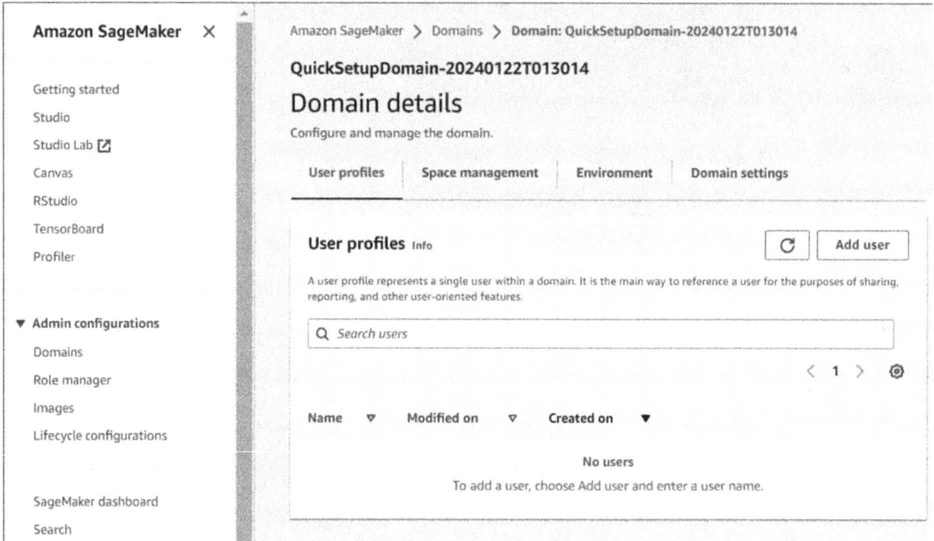

Figure 8.26 SageMaker domain details.

on the domain name will display the domain details screen, as shown in Figure 8.26. We will explore the method of setting up user profiles here.

8.6.1 *User profile creation*

To create a user profile, first click on the "Add user" tab as shown in Figure 8.26. As you can see in Figure 8.27, the user profile name and execution role are automatically set. In this step, you can set the user profile name to any desired name following the format below, up to a maximum of 63 characters. The use of existing execution roles is possible, and if there is no execution role, you can create a new one by clicking on the arrow direction below Figure 8.27 and selecting the "Create a new role" tab. Once all the settings are completed, click on the "Next" tab to proceed to the next step.

As shown in Figure 8.28, the next step is to set the Jupyter Lab version. The existing Jupyter server is automatically executed with the version selected by default for all users. The Jupyter Lab version should be set to 3.0, which is automatically set by AWS SageMaker. After configuring, click the "Next" button located at the bottom right to proceed to the next step.

In step 3, you need to configure the RStudio settings. Here, the step of configuring the RStudio IDE for the organization is done automatically. Therefore, press the "Next" button at the bottom right to proceed to the next step (Figure 8.29).

In the Canvas setup phase, you can configure settings for time series forecasting, as shown in Figure 8.30. When activating time series forecasting, you need to specify the Amazon Forecast role. If you want to create an Amazon Forecast role,

Amazon SageMaker > Domains > Domain: QuickSetupDomain-20240122T013014 > **Add user profile**

Add user profile

Step 1
General settings
Step 2
Studio settings
Configure Studio IDE and Notebooks for your organization.
The Jupyter Server runs with the selected version by default for the user. Permissions to run Jupyter Lab versions are defined by an IAM policy. You must restart the Jupyter Server app to make the version changes effective.
Step 3
RStudio settings
Step 4
Canvas settings

General settings
User profile and details.

User profile

Name
default-1705854969940
The name can have up to 63 characters. Valid characters: A-Z, a-z, 0-9, and - (hyphen)

Execution role
The default execution role for both users and spaces in the domain. The execution role must have the AmazonSageMakerFullAccess policy attached.
AmazonSageMaker-ExecutionRole-20240122T013015
Create role using the role creation wizard

Tags - *optional*

Add tag
You can attach up to 50 tags

Cancel Next

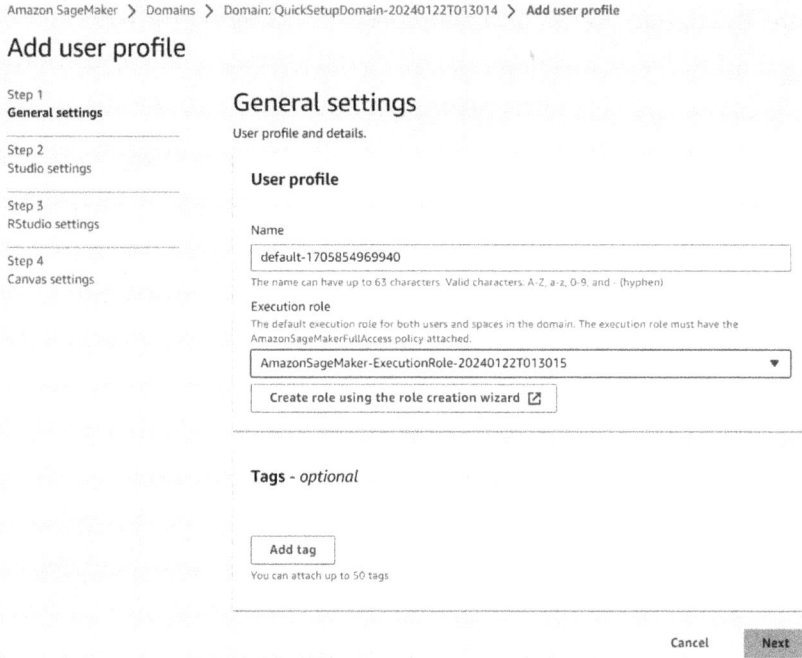

Figure 8.27 General settings for the SageMaker user profile.

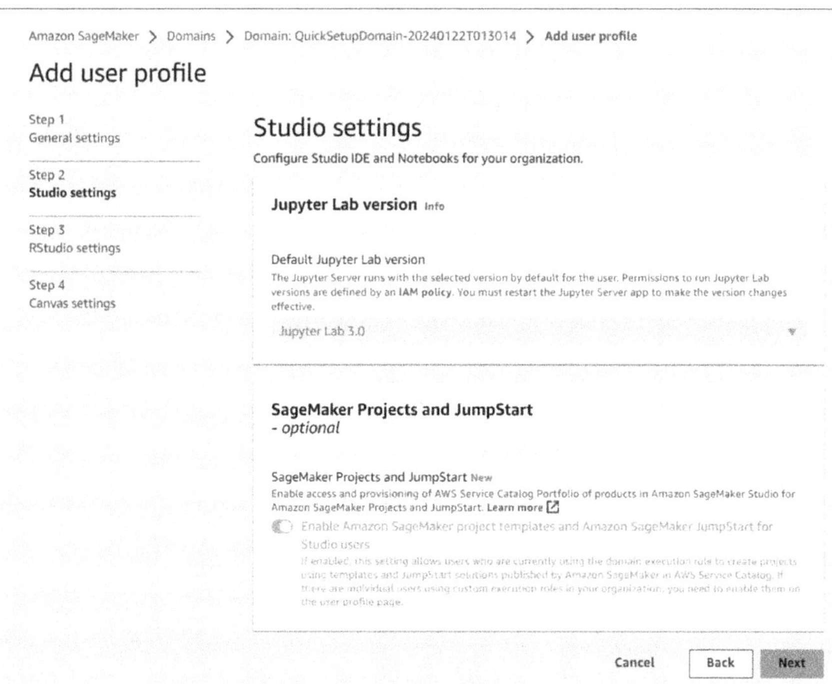

SageMaker Projects and JumpStart

SageMaker Projects and JumpStart New
Enable access and provisioning of AWS Service Catalog Portfolio of products in Amazon SageMaker Studio for Amazon SageMaker Projects and JumpStart. Learn more

Figure 8.28 SageMaker user profile studio settings.

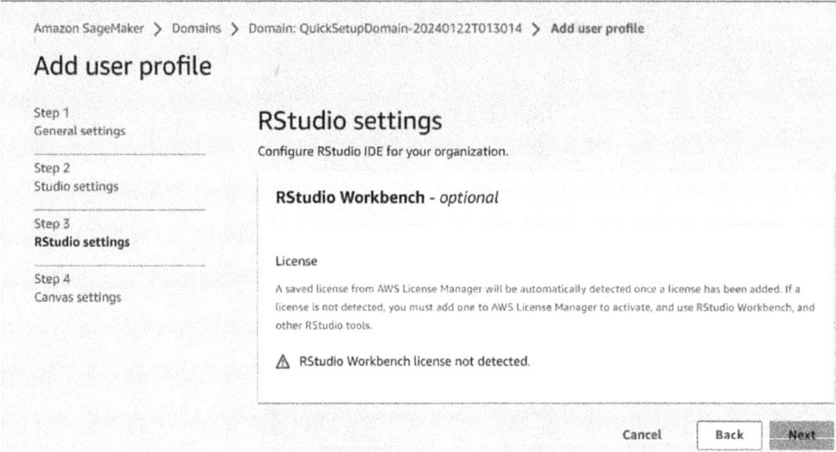

Figure 8.29 SageMaker user profile RStudio settings.

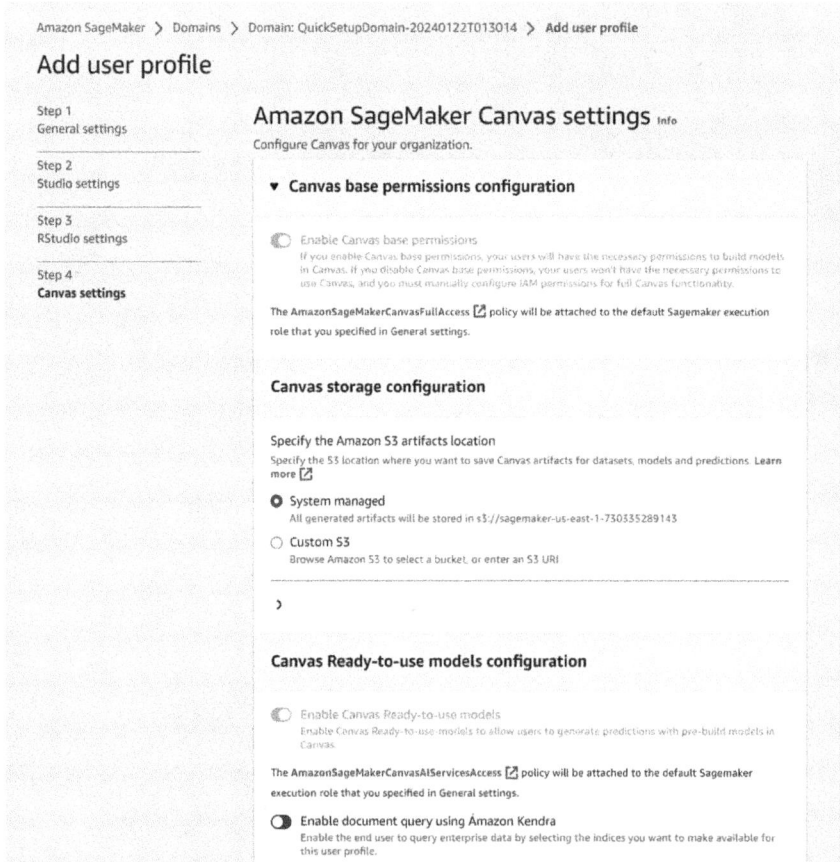

Figure 8.30 SageMaker user profile Canvas settings.

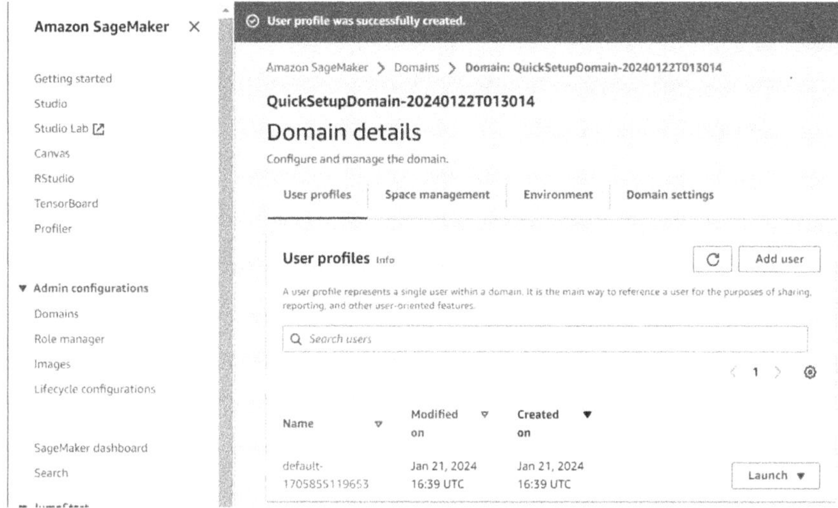

Figure 8.31 List of SageMaker user profiles.

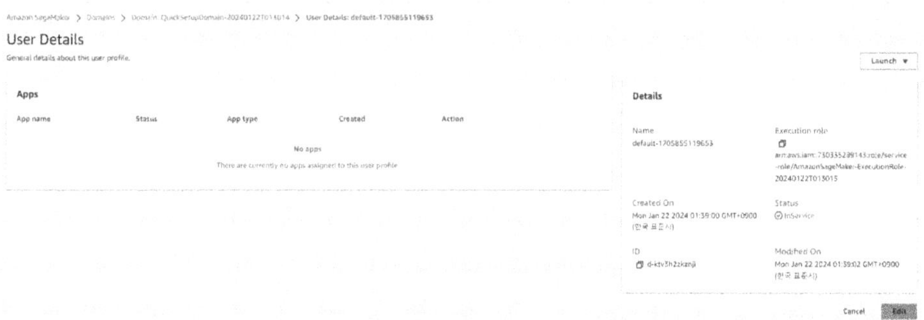

Figure 8.32 SageMaker user details screen.

click the "Create and use a new execution role" button, and then click the "Submit" button at the bottom right. If you want to use an existing execution role, click the "Use and existing execution role" button, and then click the "Submit" button to add it to your user profile.

Once you add a user profile and the process is completed, it will appear on the screen, as shown in Figure 8.31. After creating an additional user profile here, you can then use the features of SageMaker.

8.6.2 *User deletion*

Let's find out how to delete a user profile. Click on the user profile name, as shown in Figure 8.31. After clicking the screen shown in Figure 8.32 will appear, and there is an edit button at the bottom right of the user details screen.

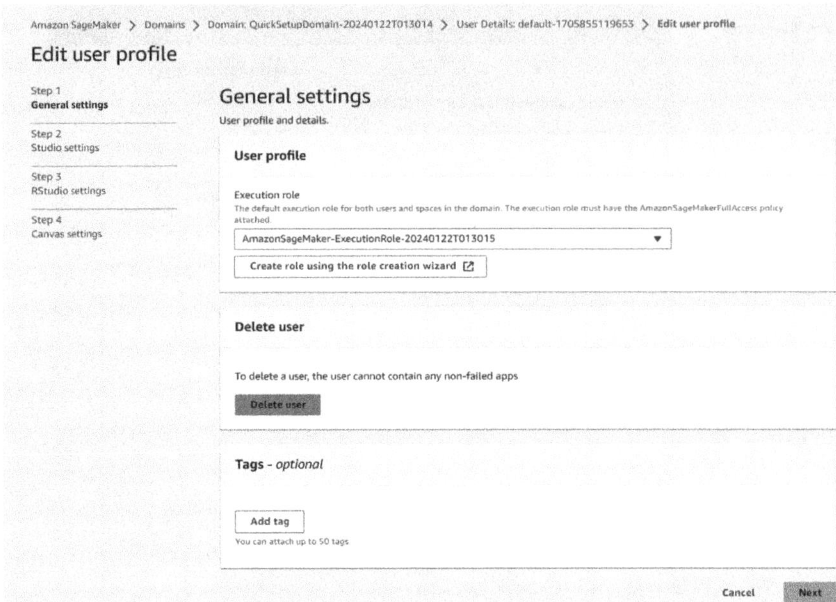

Figure 8.33 SageMaker user profile editing screen.

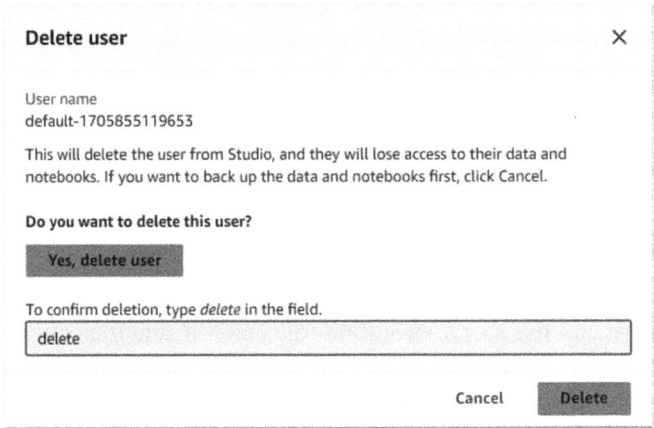

Figure 8.34 SageMaker user profile deletion message.

As you can see in Figure 8.33, there is a “Delete user” tab in the middle of the screen. Click on this tab to delete the user.

To delete a user, click the phrase “Yes, delete user.” and then enter “delete” in the text field below to delete the user profile (Figure 8.34).

As shown in Figure 8.35, if the deletion of a user profile on the SageMaker domain details page has been completed, you can see that the user profile has been deleted on the user profile screen.

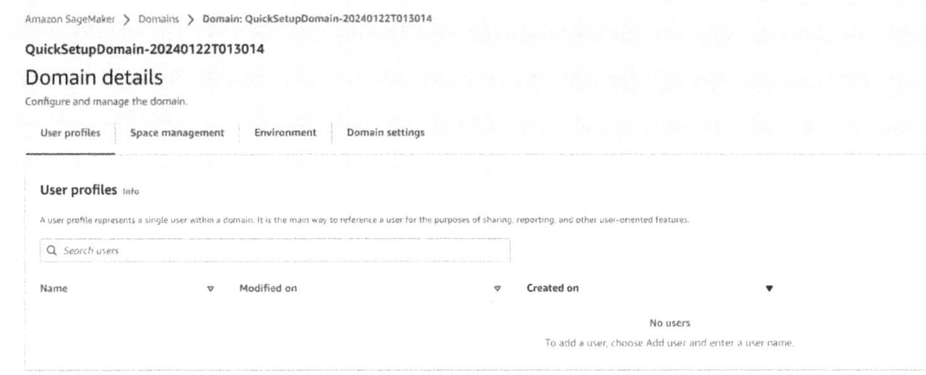

Figure 8.35 SageMaker user profile deletion complete.

8.7 Practice Questions

Q1. What are the primary benefits of Amazon SageMaker for developers and data scientists in machine learning model development?

Q2. Describe the process of model training and deployment in Amazon SageMaker.

Q3. How does Amazon SageMaker support continuous improvement of machine learning models?

Chapter 9

SageMaker Canvas

Amazon SageMaker Canvas is a visual interface for Amazon SageMaker that allows users to create machine learning models and perform predictions without coding. Canvas provides an intuitive environment for data connection, access, and preparation, and uses automated machine learning techniques to train and build models based on datasets. This tool is integrated with SageMaker Studio, making it easy for developers and general users to share the generated models. Additionally, Canvas supports various algorithm models such as binary classification, multi-class classification, numerical regression, and time series prediction, allowing users to utilize various problem-solving functionalities such as error detection and optimization without coding.

9.1 SageMaker Canvas Tutorial

If you access the AWS SageMaker homepage URL, you will see the screen shown in Figure 9.1 (https://aws.amazon.com/ko/sagemaker/). Click on the "Get Started with SageMaker" tab to go to the SageMaker Console page. If you haven't logged in, log in and then you can run SageMaker.

When you click on the "Get Started with SageMaker" tab in Figure 9.1, the screen shown in Figure 9.2 will appear. On the left side of Figure 9.2, you can see the "Canvas" tab, so click on it to run the Canvas. If the user does not have a domain, they need to create a domain and a user profile.

After the user logs in, there should be a domain and a user profile. If there is no domain or user profile, it should be created. As shown in Figure 9.3, select the user profile and click on the "Open Canvas" tab to run the Canvas.

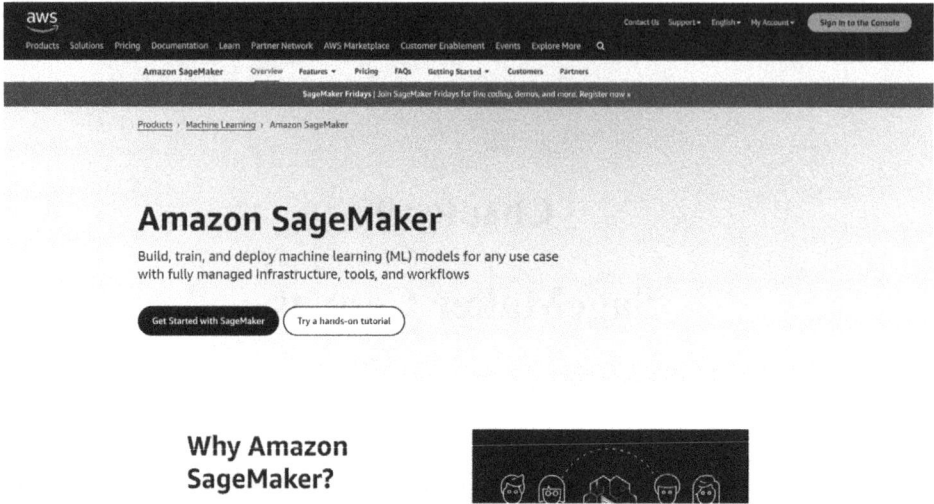

Figure 9.1 AWS SageMaker homepage.

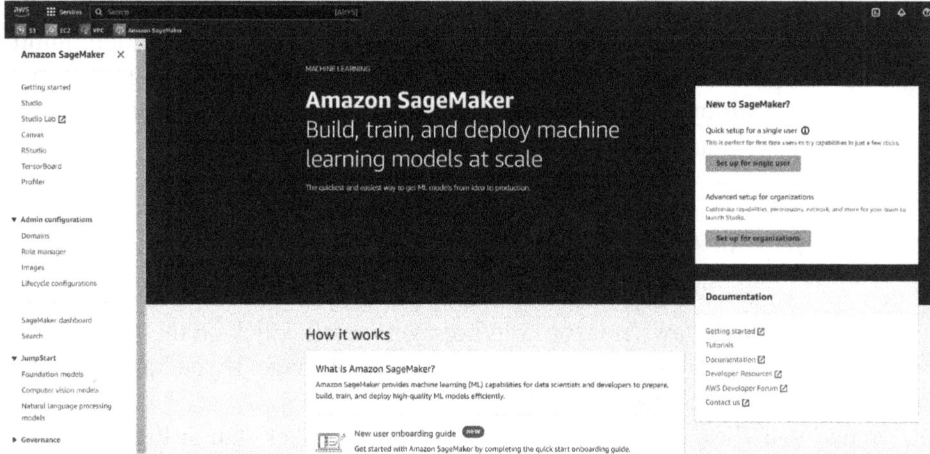

Figure 9.2 SageMaker console home screen.

9.1.1 *Importing data*

When you run AWS SageMaker Canvas, the screen shown in Figure 9.4 will appear after about 3–5 minutes. On the Canvas startup screen, you can learn how to use it. First, click on "Learn how to import datasets" to learn how to import a dataset.

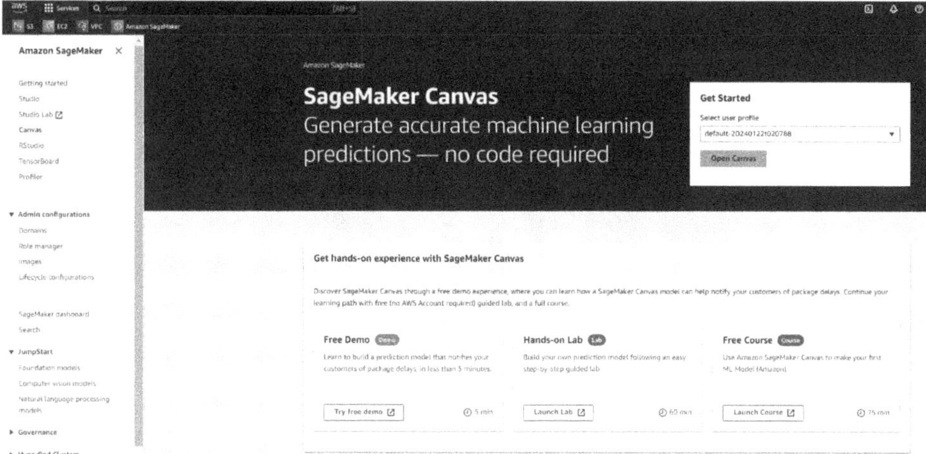

Figure 9.3 SageMaker Canvas.

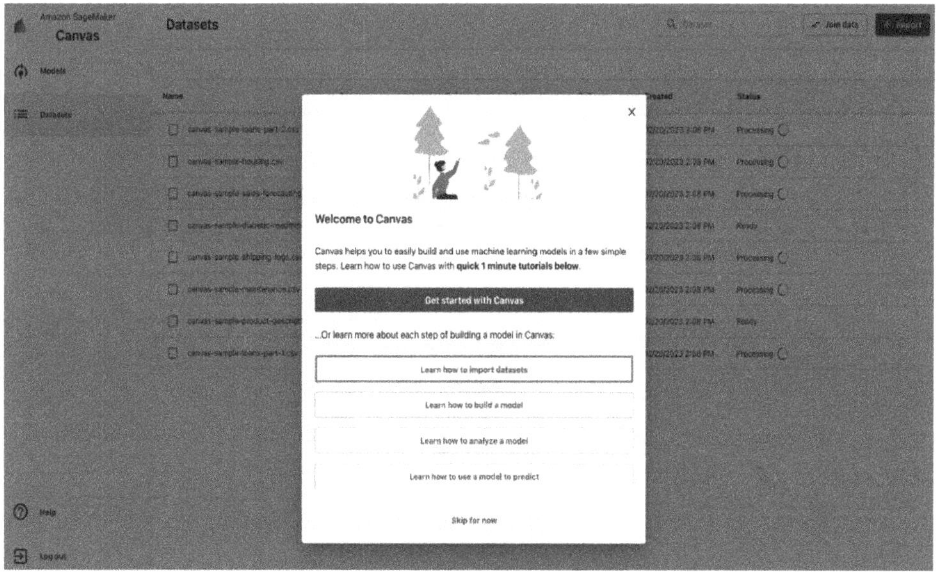

Figure 9.4 Canvas start screen.

In Figure 9.4, click the "Learn How to import datasets" tab on the screen.

To import a dataset, you first need to click on the "Datasets" tab on the left as shown in Figure 9.5.

Then, click on the "Import" tab in the upper right corner, as shown in Figure 9.6. Clicking on the "Import" tab will initiate the process of importing a new dataset. You can import CSV files with up to 1,000 columns from various data

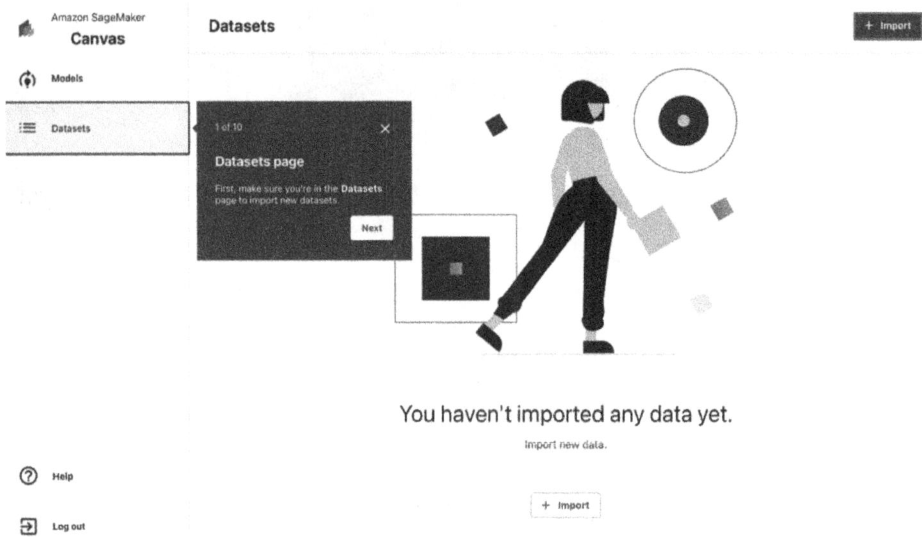

Figure 9.5 Canvas datasets screen.

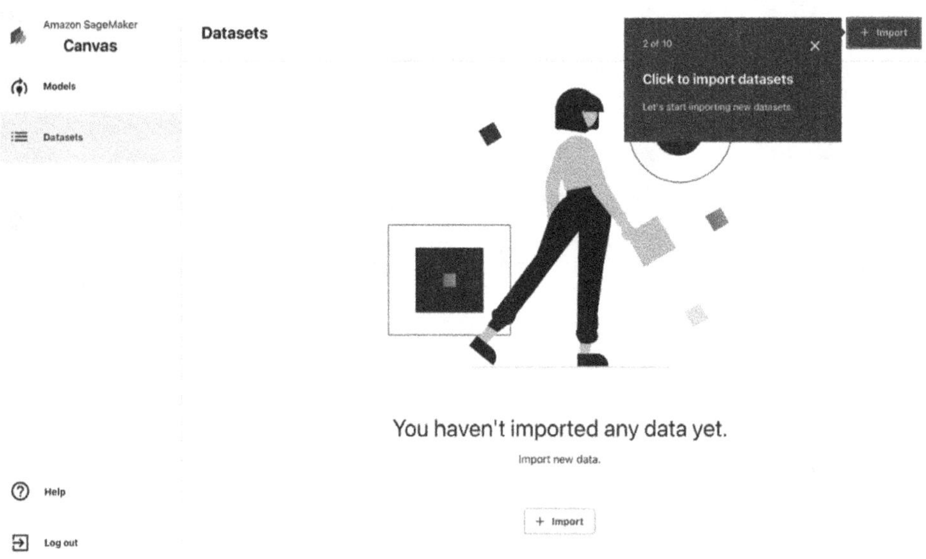

Figure 9.6 Canvas datasets import.

sources such as local uploads, S3, Snowflake, and Redshift. Clicking on the "See More" button will provide more details about dataset eligibility and information about the data source (Figure 9.7).

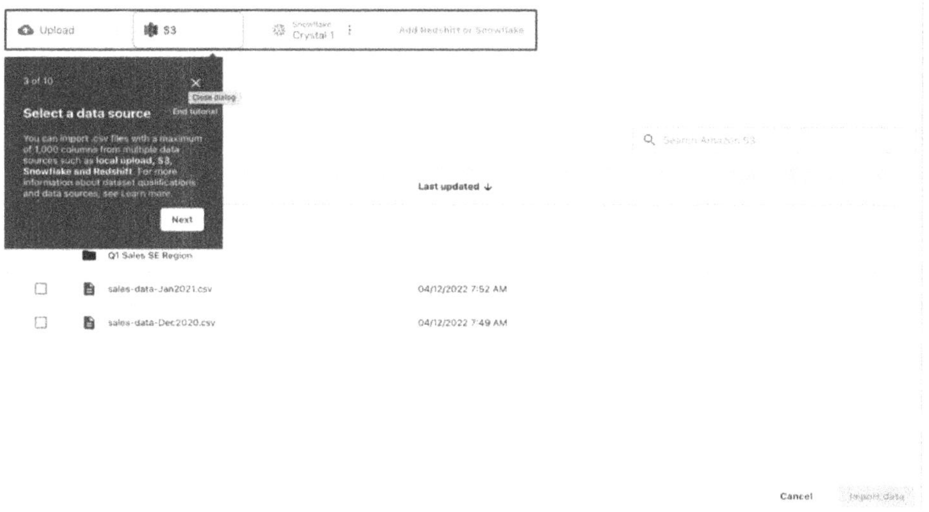

Figure 9.7 Canvas select data source.

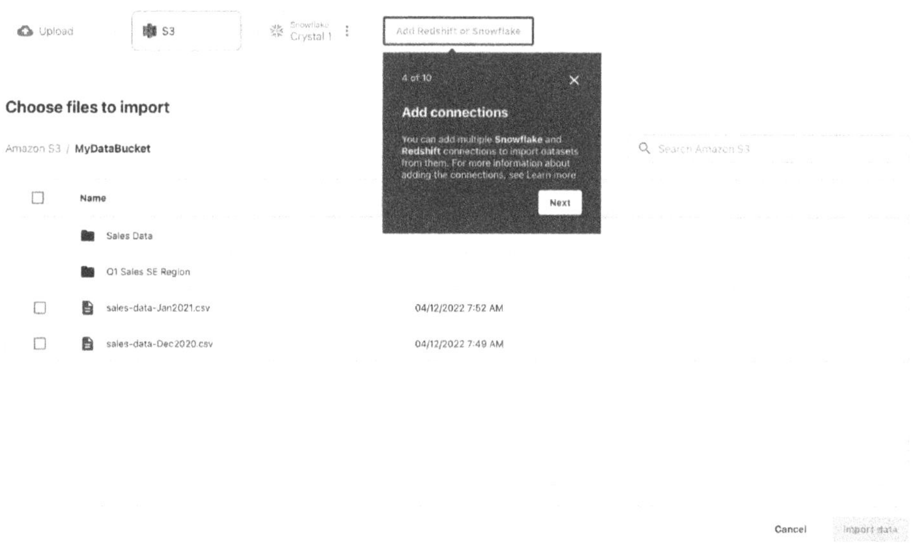

Figure 9.8 Adding connections to Canvas.

You can also import datasets from the SnowFlake and Redshift connections by adding the connections. By clicking on "See more," you can find detailed information about adding connections (Figure 9.8).

Click the checkboxes to select a dataset. You can choose a dataset to retrieve from a previously selected data source. Additionally, you can import multiple datasets at once (Figure 9.9).

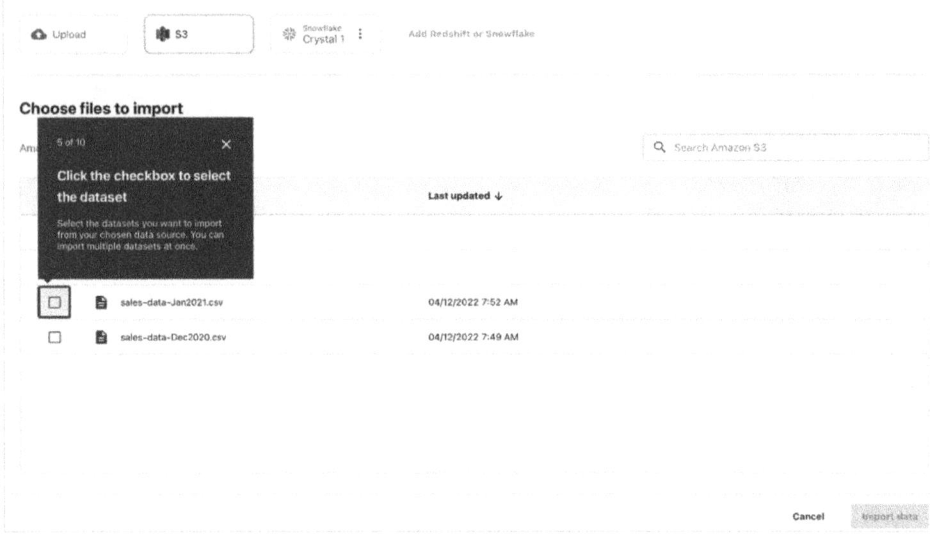

Figure 9.9 Canvas import file.

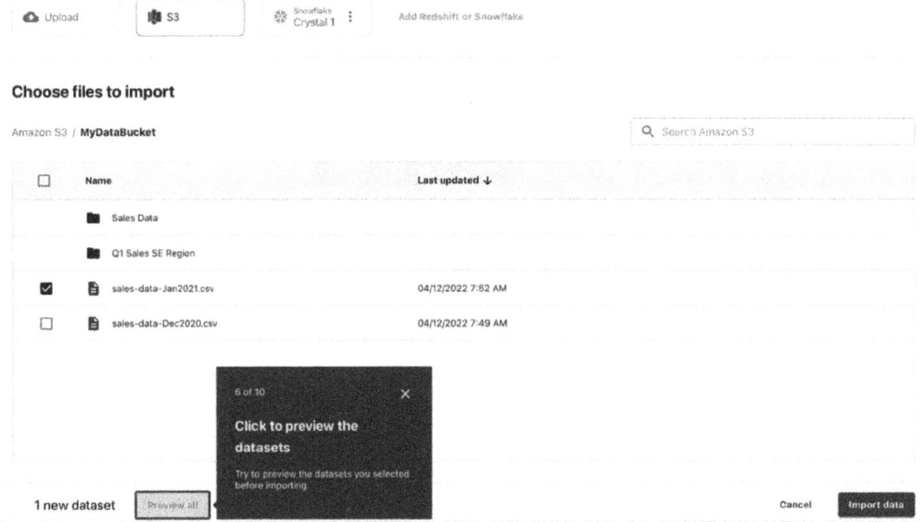

Figure 9.10 Canvas data file.

To preview the dataset, click on it. Therefore, before importing, confirm the selected dataset (Figure 9.10).

100 rows are displayed in the preview of the dataset. Scroll down to the preview section to review the preview of each dataset and make sure you are importing the correct dataset.

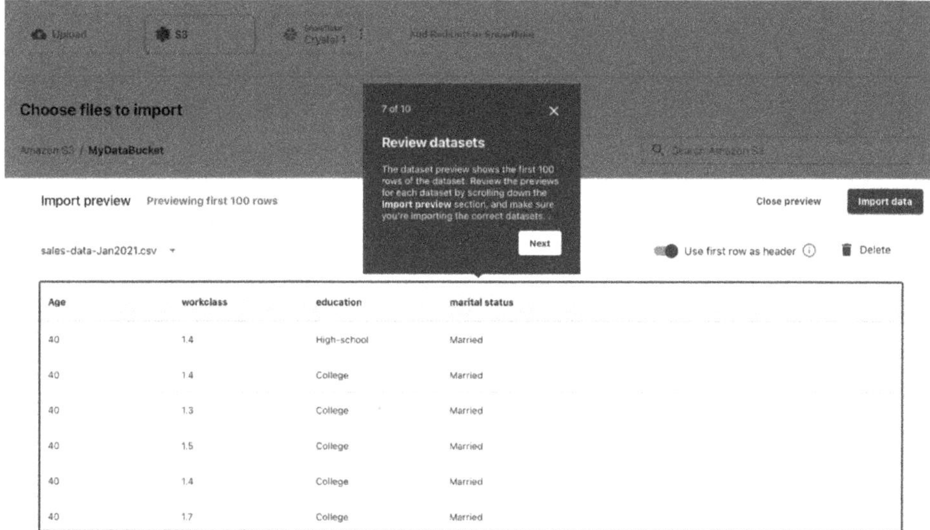

Figure 9.11 Canvas preview data.

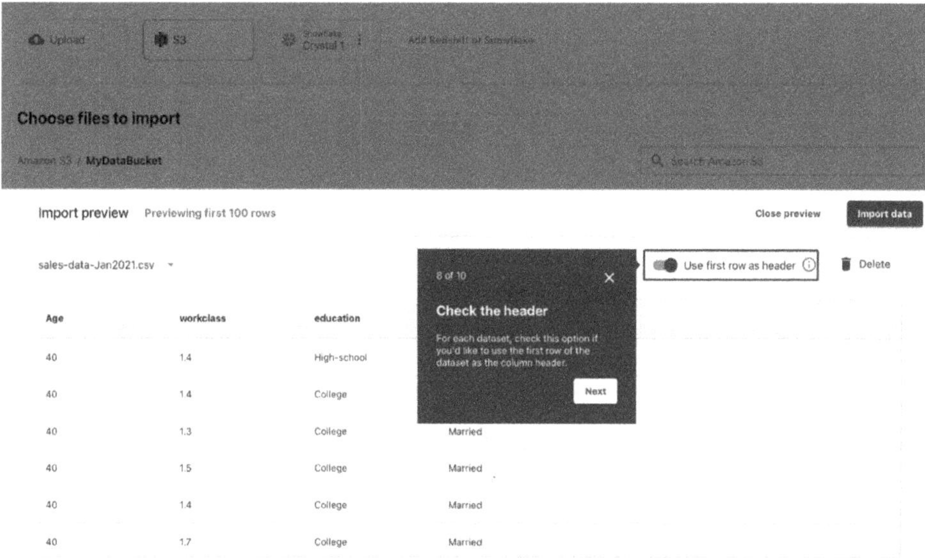

Figure 9.12 Canvas check header.

If you want to use the first row of each data set as the column headers for the dataset, select this option as shown in Figure 9.11.

As shown in Figure 9.12, users can import the dataset by clicking the "Import data" button after completing the review of the dataset.

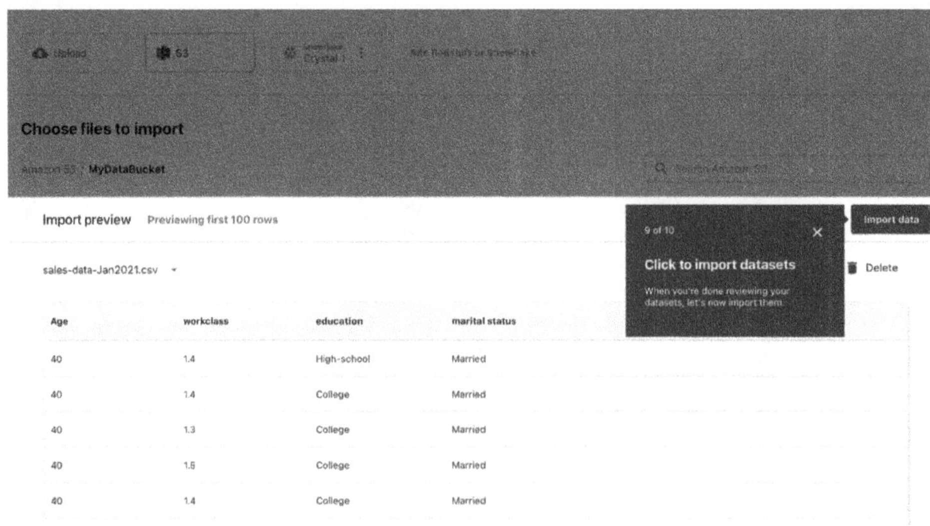

Figure 9.13 Canvas import datasets.

Figure 9.14 Canvas import datasets result.

Figure 9.13 shows a screen displaying the imported dataset. Now, on the dataset page, you can see the newly imported dataset on the Canvas (Figure 9.14).

9.1.2 *Creating a model*

Next, click on the “Learn how to build a model” tab, as shown in Figure 9.4, to rerun the tutorial. Search for the dataset to create a model on the dataset page. If

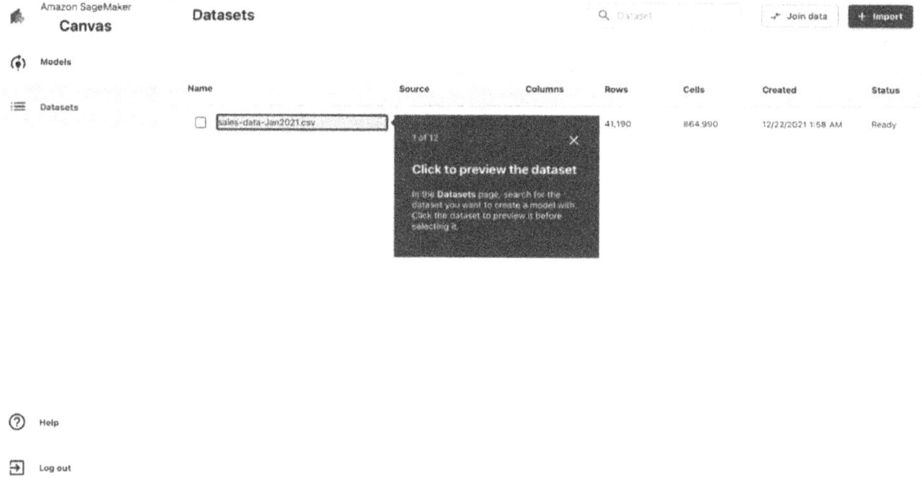

Figure 9.15 Canvas sample datasets.

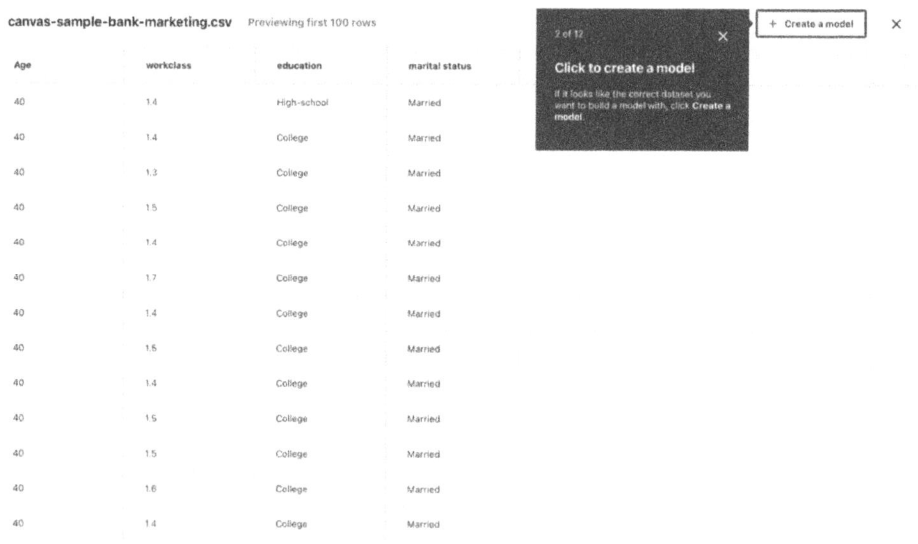

Figure 9.16 Creating a Canvas model.

you want to preview it before selecting, click on the dataset as shown in Figure 9.15.

If there are no issues with the dataset, click on “Create a model” to create the model (Figure 9.16).

Before creating a model, review the column statistics of the selected dataset, such as data types, missing values, inconsistent values, and the number of unique

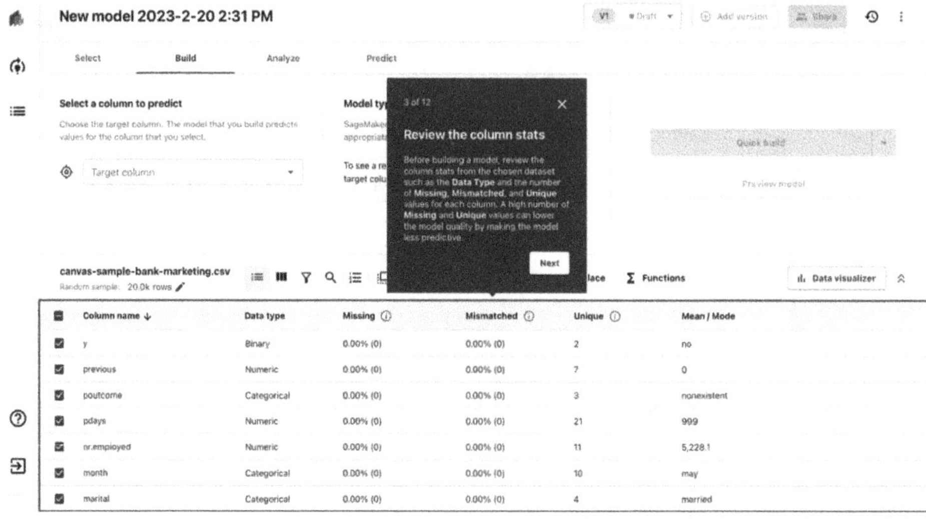

Figure 9.17 Reviewing the datasets.

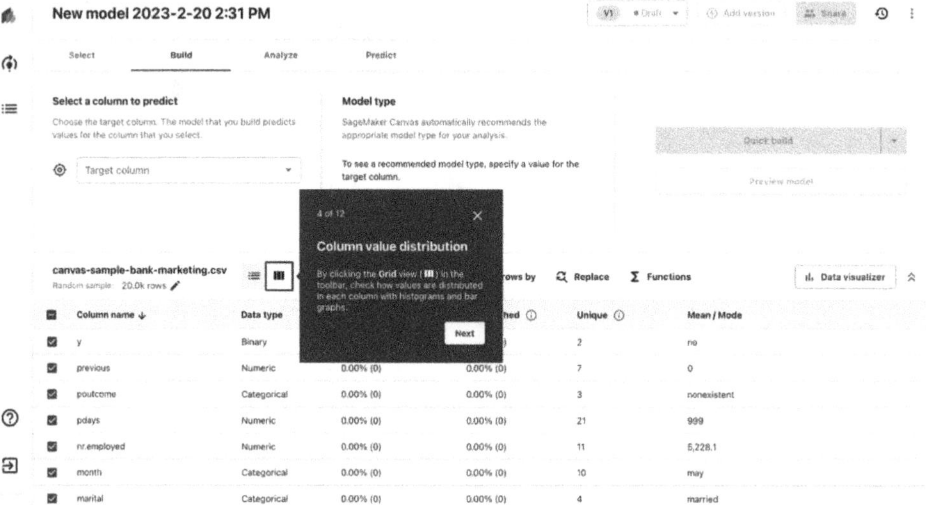

Figure 9.18 Canvas column value distribution.

values. If there are many missing or unique values, the predictive power of the model may decrease, leading to a decrease in model quality (Figure 9.17).

In the tool collection, click on "Grid View" to use histograms and bar graphs to see how values are distributed in each column (Figure 9.18).

In Figure 9.19, the y column indicates the possibility of enrolling in a new certificate of deposits using the two different values entered.

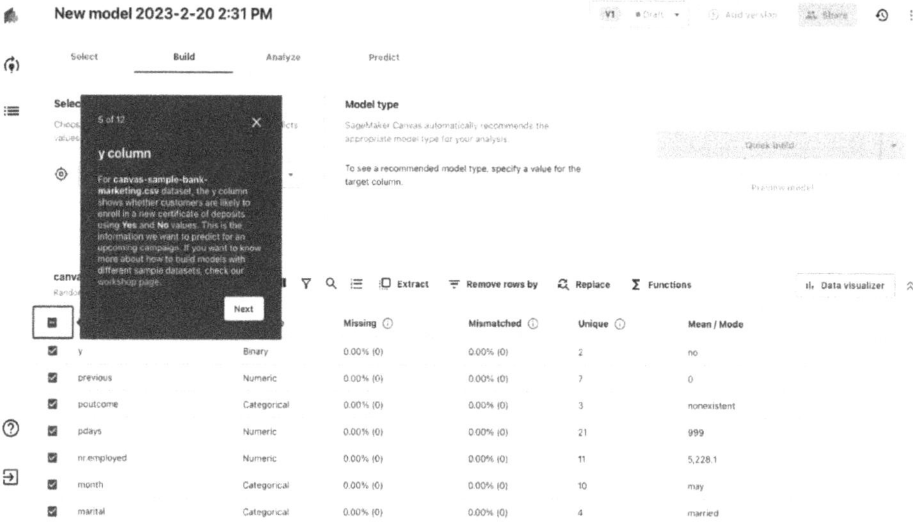

Figure 9.19 Canvas y column.

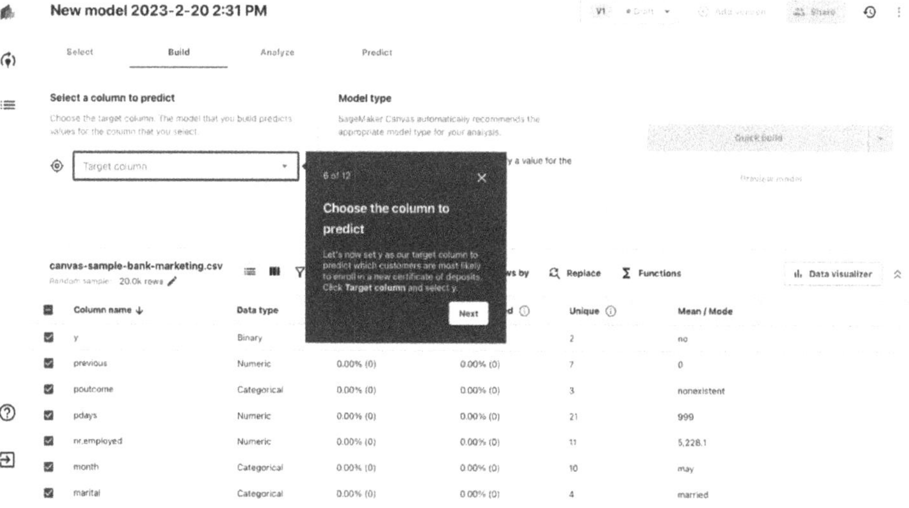

Figure 9.20 Canvas target column.

Now let's set the target column to predict which customer is most likely to enroll in a new certificate of deposits. Click on the Target column and select the y column (Figure 9.20).

As shown in Figure 9.21, Canvas selects the appropriate model type for prediction.

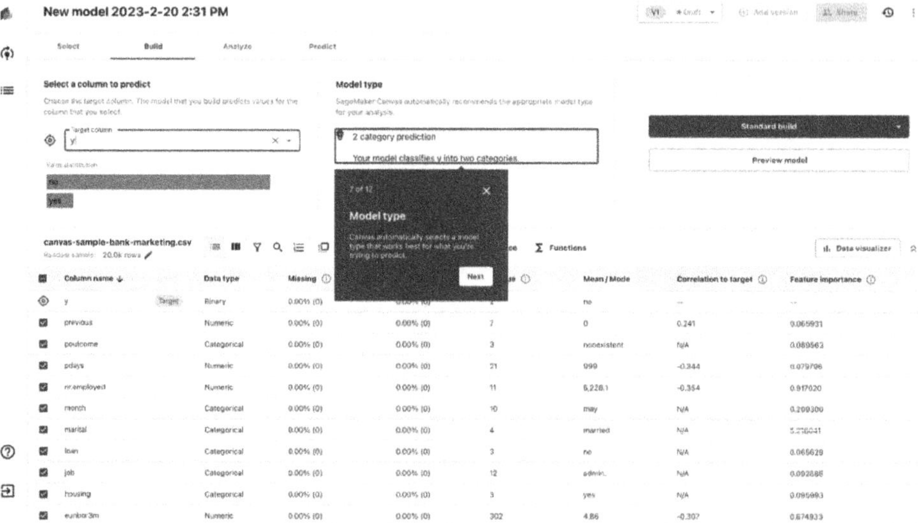

Figure 9.21 Canvas choose model type.

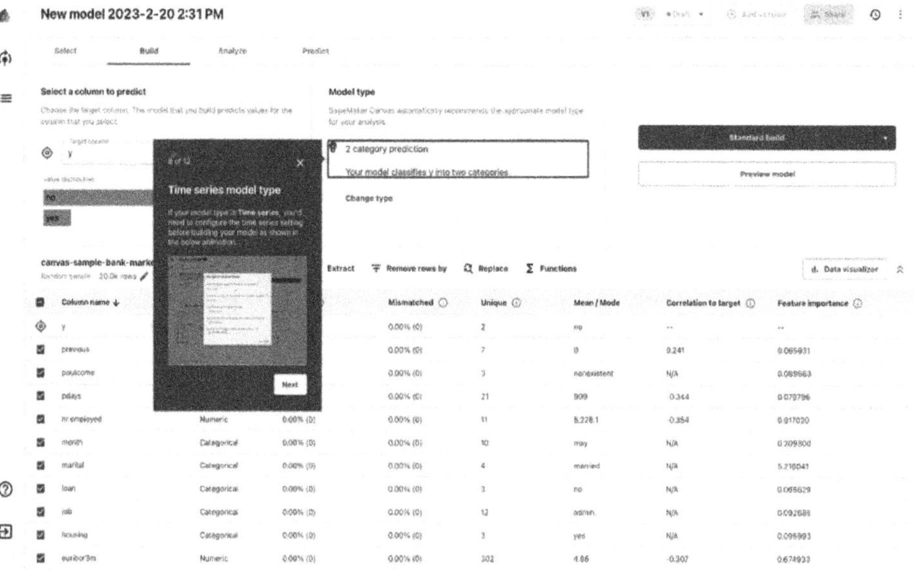

Figure 9.22 Canvas time series model type.

If the model type is Time series, you need to configure the time series settings before creating the model, as shown in Figure 9.22.

As shown in Figure 9.23, preview model is a feature that allows you to preview the results of a model. Time series model type does not support preview model.

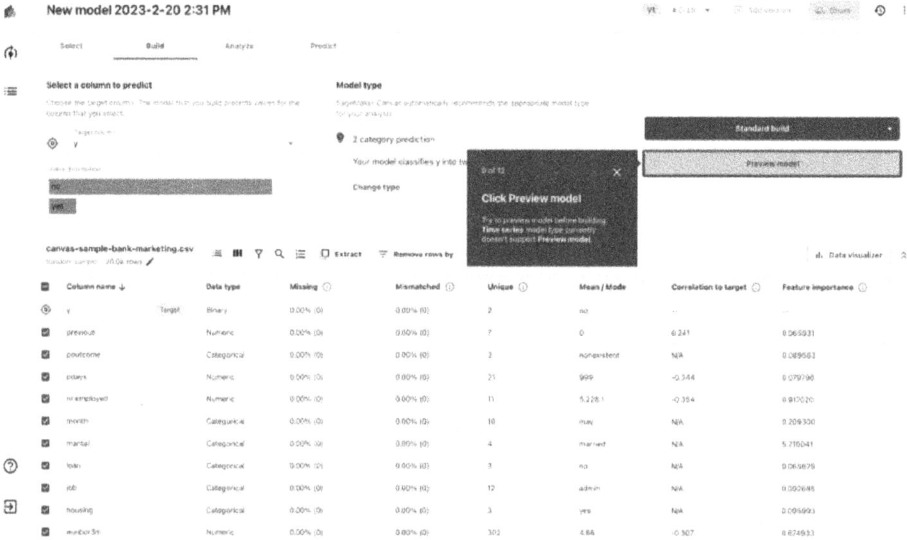

Figure 9.23 Canvas preview model.

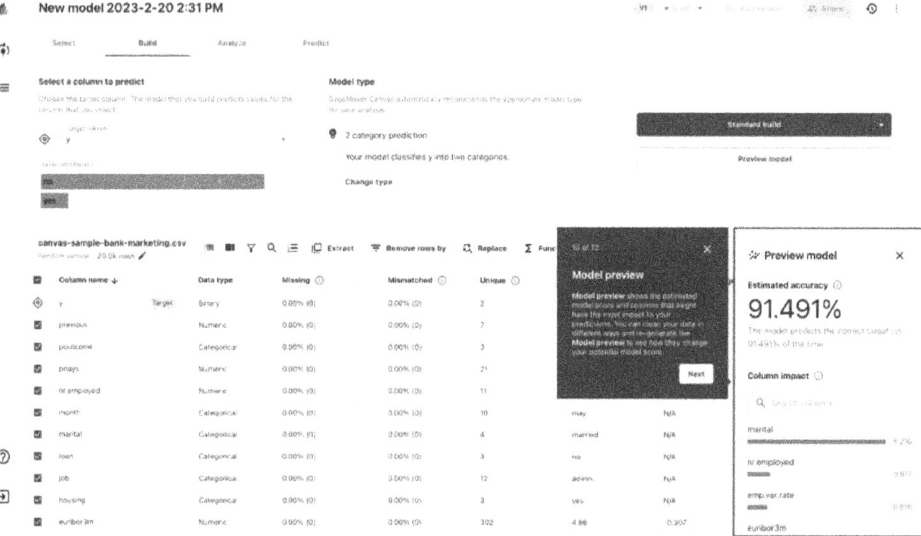

Figure 9.24 Canvas preview model result.

Model preview displays estimates of model scores and columns which may have the biggest impact on your predictions. By organizing data in various ways and regenerating the model preview, you can see how the model scores change (Figure 9.24).

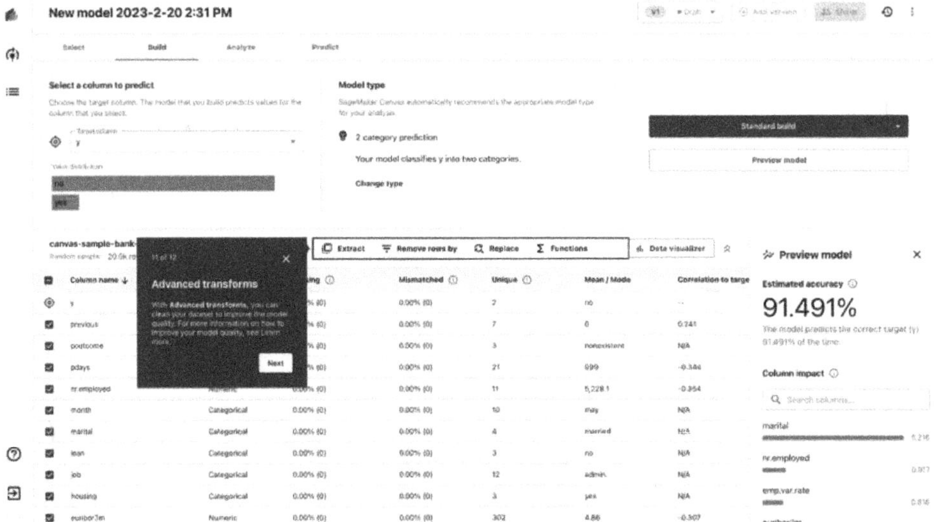

Figure 9.25 Canvas advanced transforms.

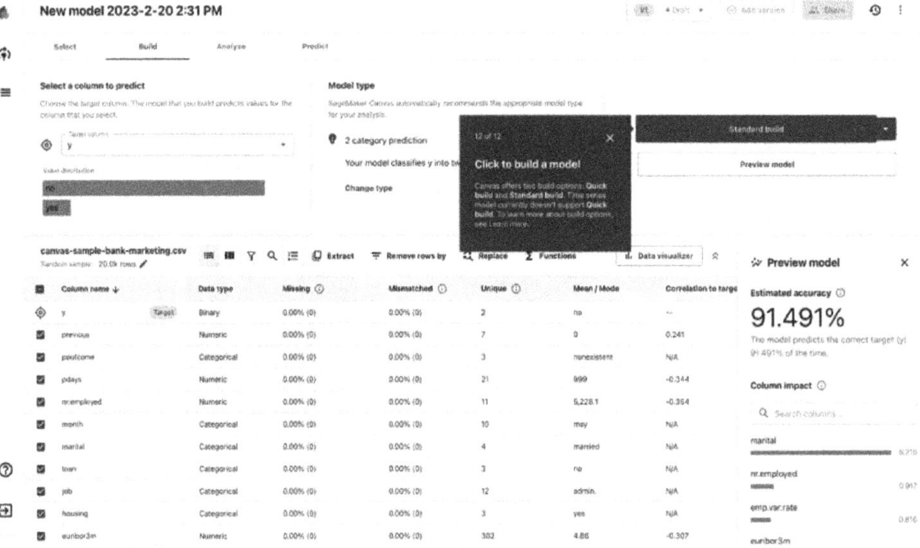

Figure 9.26 Click Canvas to build a model.

Using advanced transforms can help organize the dataset and improve the quality of the model. By clicking on "Learn more," you can find detailed information on how to enhance model quality (Figure 9.25).

Canvas provides two build options: Quick build and Standard build. In addition, the time series model does not currently support Quick build. Click on "Learn more" for more detailed information on the build options (Figure 9.26).

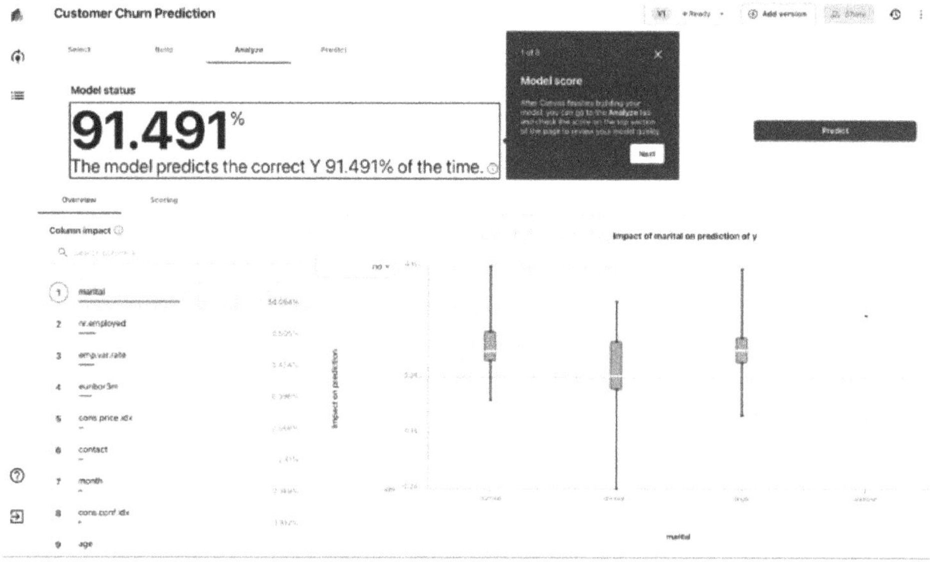

Figure 9.27 Canvas model score.

9.1.3 *Analyzing the model*

To learn how to analyze a model, click on the "Learn how to analyze a model" tab shown in Figure 9.4. After completing the model creation, navigate to the "Analyze" tab to see the analysis results for the model at the top of the page (Figure 9.27).

Next, we can identify the columns that have a significant impact on the prediction. Ultimately, by looking at the canvas-sample-bank-marketing.csv dataset, we can see that the marital column has a greater impact compared to other factors (Figure 9.28).

To have a more detailed understanding of the values of the generated model, click on the "Scoring" tab (Figure 9.29).

Using score visualization, you can examine the accuracy of the predictions by checking the number of values predicted accurately or inaccurately. By clicking on "Learn more," you can find detailed information about score details for various model types (Figure 9.30).

In Scoring insights, you can check the frequency at which the model predicts a specific value and the frequency at which the model accurately predicts that value. In this example, you can also identify the frequency at which the model predicts the value it desires to predict (Figure 9.31).

To learn more about the model quality, click on "Advanced metrics", as shown in Figure 9.32, to view the detailed report.

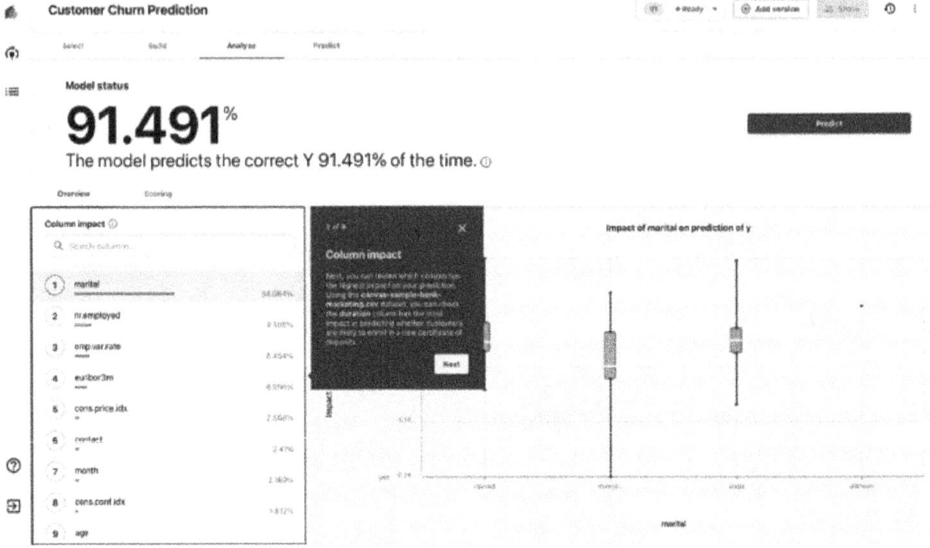

Figure 9.28 Canvas data column impact.

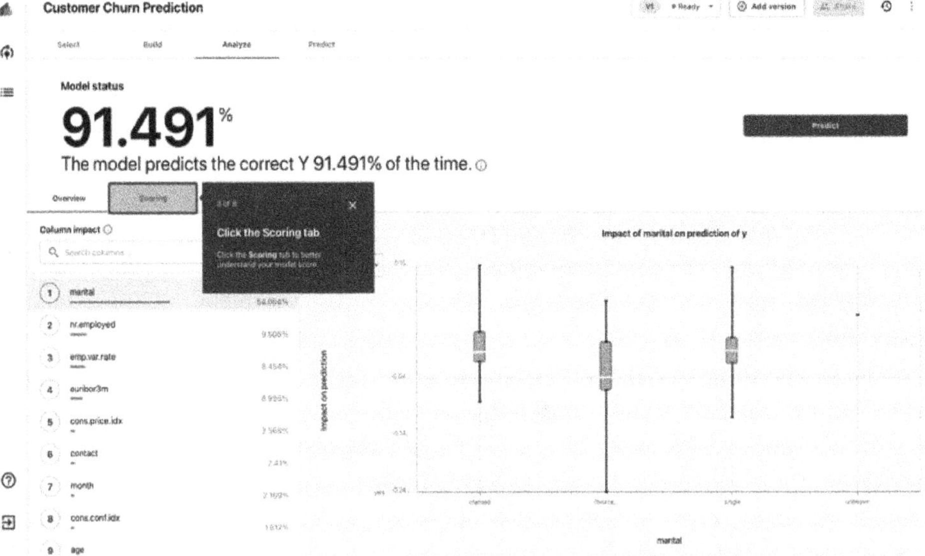

Figure 9.29 Canvas model scoring.

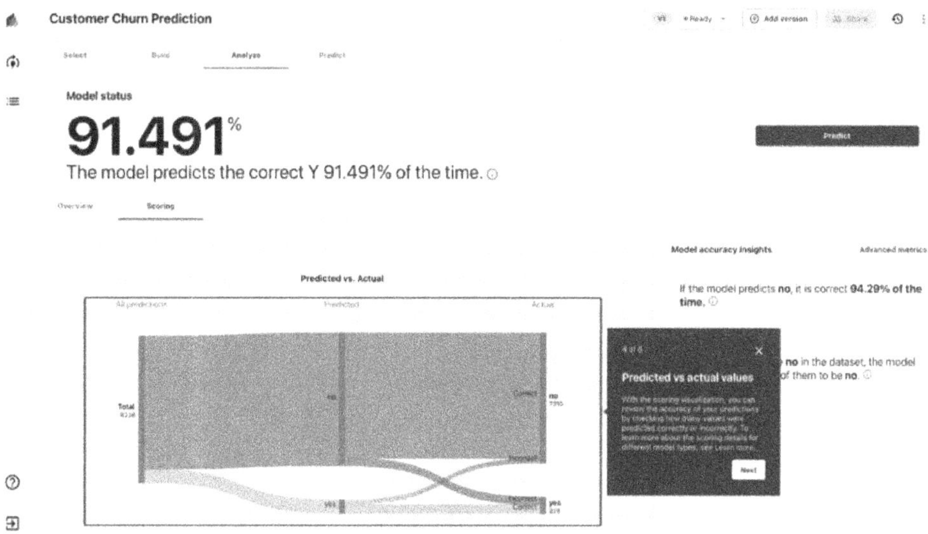

Figure 9.30 Canvas predicted versus actual values.

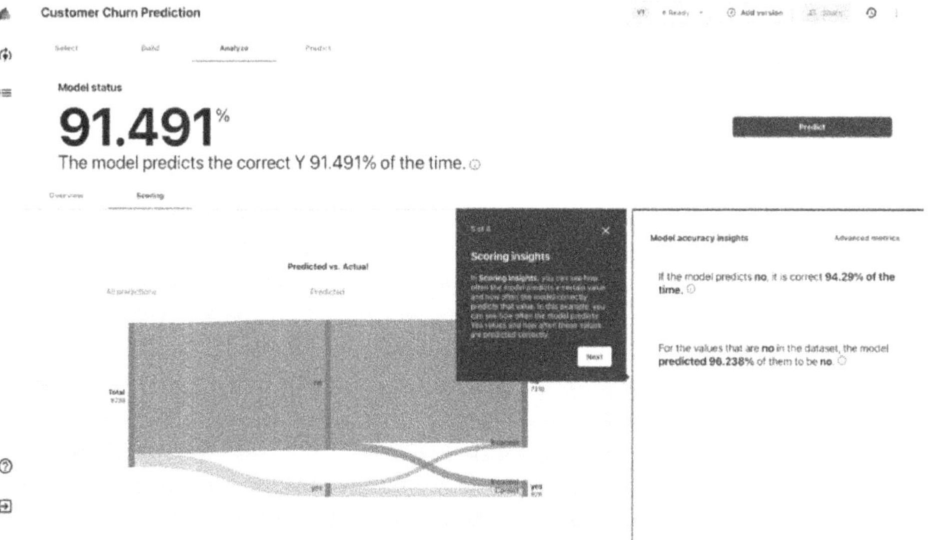

Figure 9.31 Canvas scoring insights.

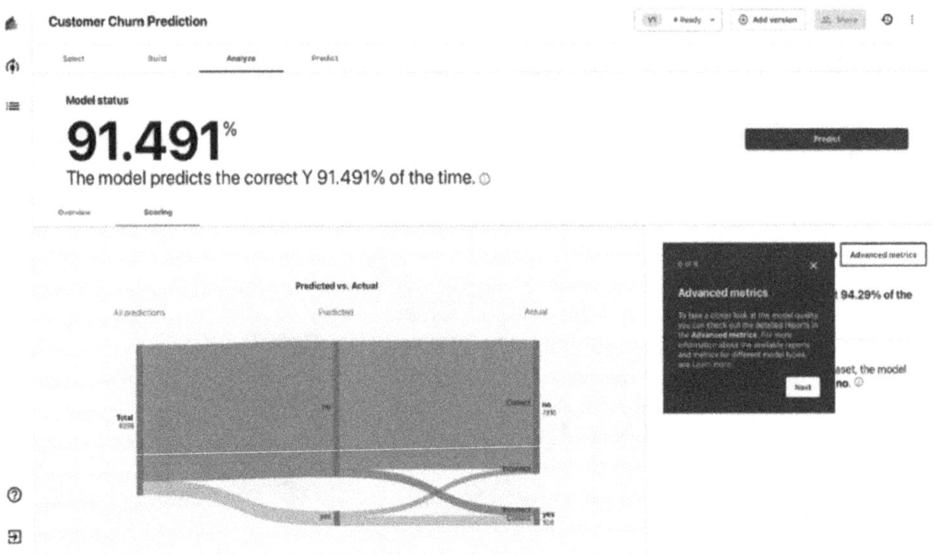

Figure 9.32 Canvas advanced metrics.

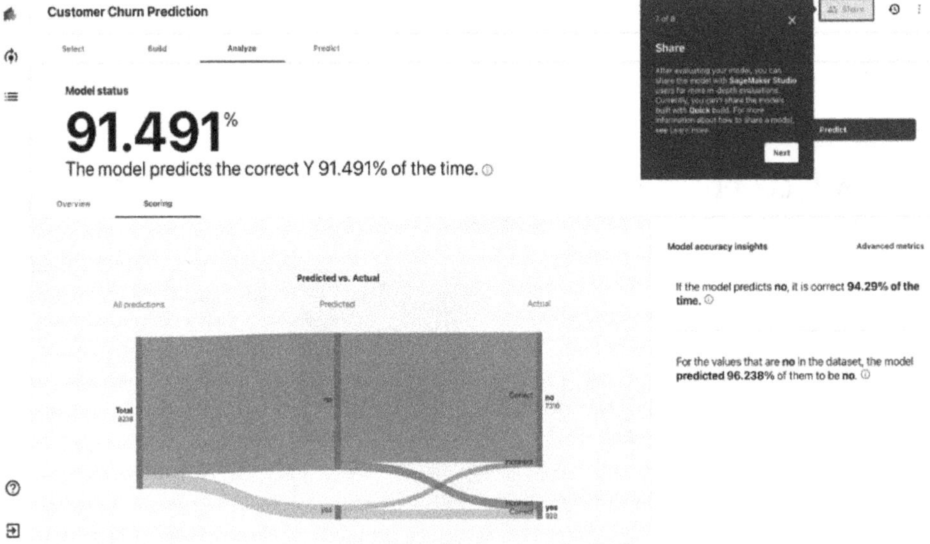

Figure 9.33 Canvas data share.

After evaluating the model, you can share it with SageMaker Studio users to conduct more in-depth evaluations. Currently, models built using Quick build cannot be shared (Figure 9.33).

You can make predictions using the model you wrote using the Predict option shown on the right side of Figure 9.34.

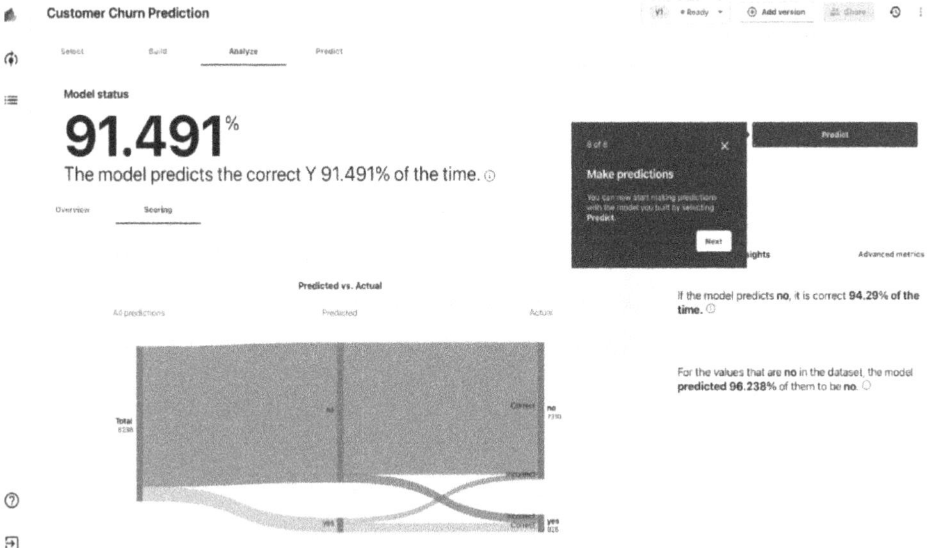

Figure 9.34 Canvas make predictions.

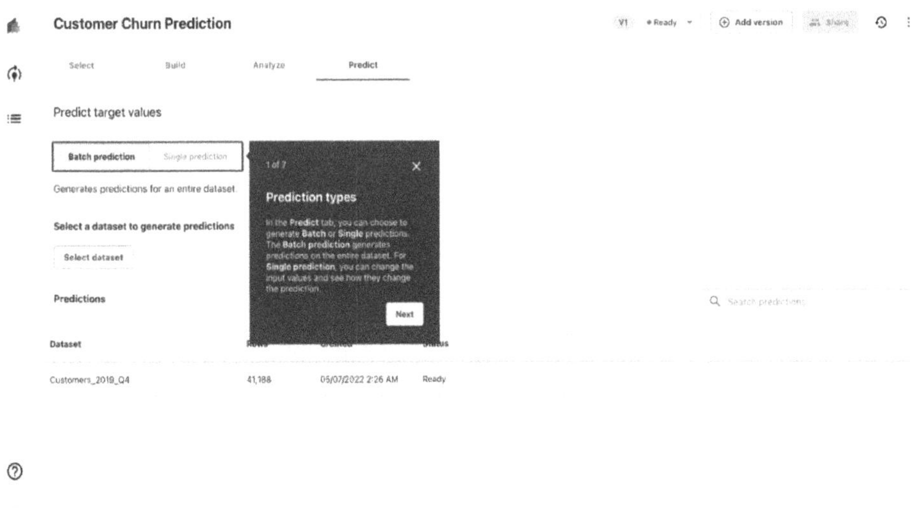

Figure 9.35 Canvas prediction type.

9.1.4 *Using models for predictions*

Next, click on the [Learn how to use a model to predict] tab shown in Figure 9.4 to run the tutorial. In this step, you can choose to generate either batch or single prediction. Batch prediction generates predictions for an entire dataset, such as CSV. Single prediction generates predictions for values directly input as single data (Figure 9.35).

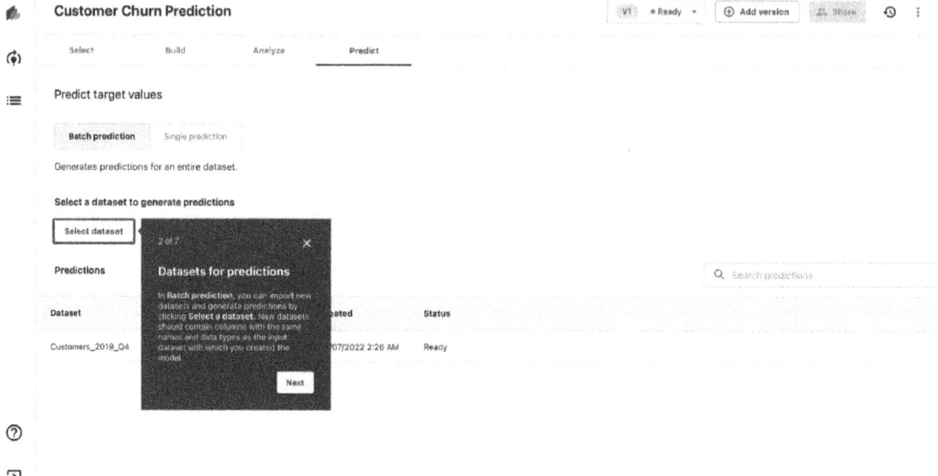

Figure 9.36 Canvas datasets for predictions.

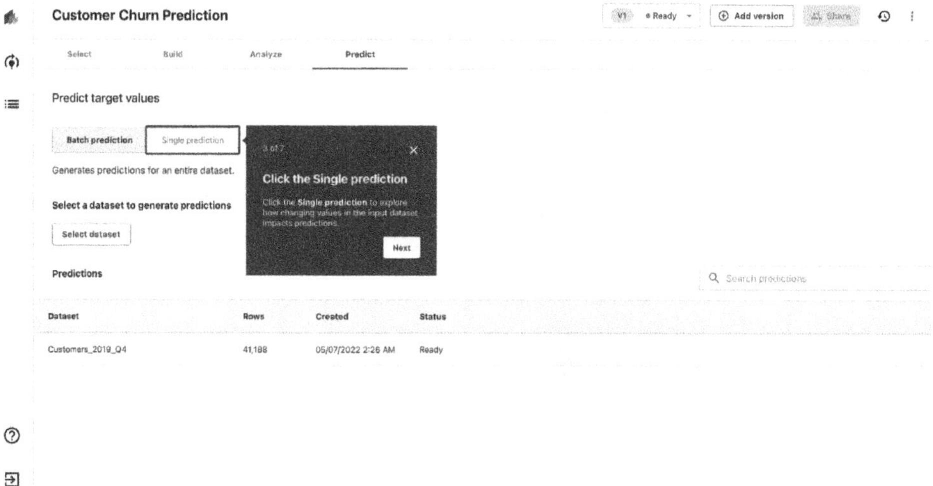

Figure 9.37 Canvas learn model predict.

In predictive modeling, click on "Select data set" in order to import a new data set and generate predictions. The new data set should include columns with the same names and data types as the input data set used for model generation (Figure 9.36).

Let's examine how changing the values of an input dataset affects a prediction by clicking on a single prediction (Figure 9.37).

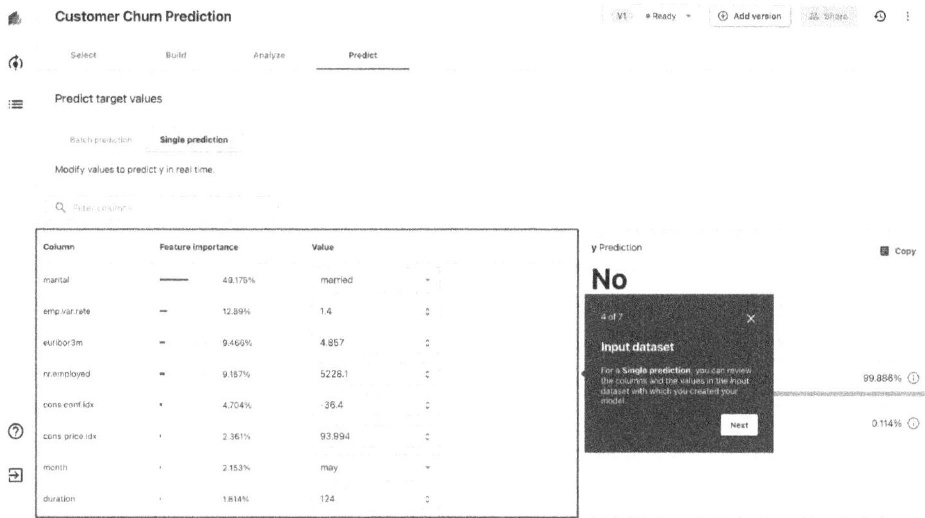

Figure 9.38 Canvas input dataset.

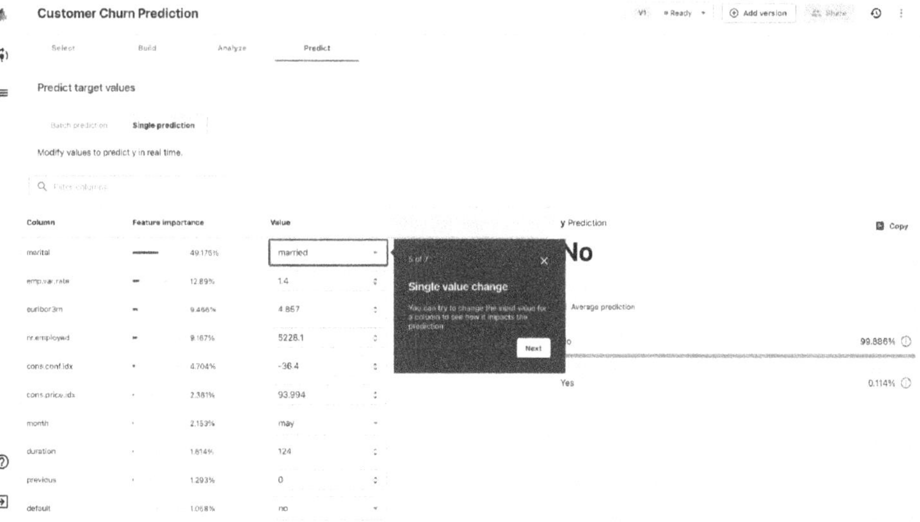

Figure 9.39 Canvas single value change.

In the case of single prediction, you can review the columns and values of the input dataset used to generate the model (Figure 9.38).

You can confirm how changing the values of variables affects predictions.

Change the single value as shown in Figure 9.39.

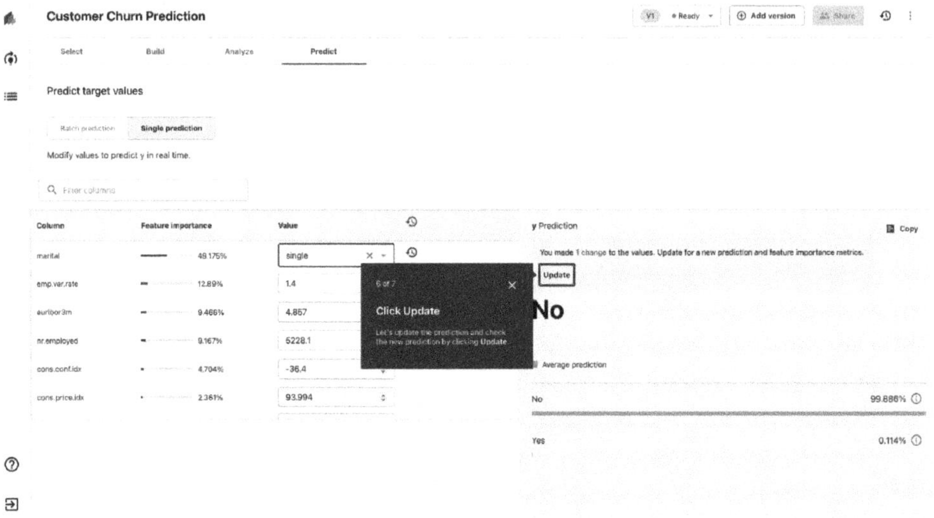

Figure 9.40 Canvas click update.

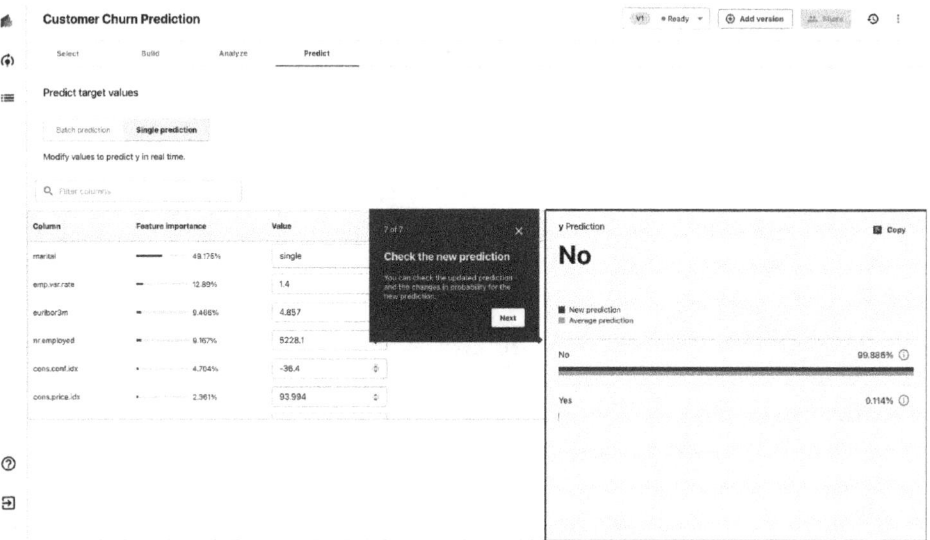

Figure 9.41 Canvas — check the new prediction.

In Figure 9.40, you can see the probability change of updated predictions and new predictions (Figure 9.41).

9.2 Practice Questions

Q1. What is Amazon SageMaker Canvas and its primary purpose?

Q2. How does Amazon SageMaker Canvas facilitate model training and deployment for non-coders?

Q3. Describe the types of problems Amazon SageMaker Canvas can solve and the functionalities it provides.

Chapter 10

SageMaker Canvas Practice

10.1 Prediction of Survival Probability for Titanic Passengers

This practical exercise demonstrates how to build a predictive model with AWS SageMaker Canvas using the Titanic survivor dataset from Kaggle. We use passenger data, using details such as name, age, gender, and social-economic class, to answer the question, "Which types of people had a higher chance of survival?" You can check the data on Titanic passengers by accessing the Kaggle website (https://www.kaggle.com/c/titanic/data) (Figure 10.1).

As shown in Figure 10.2, the data regarding the Titanic can be downloaded by clicking "Download All."

The Titanic data utilizes both the Train and Test datasets, with each having 891 and 418 rows respectively, and 11 attributes including the label. The attribute names and detailed descriptions can be found in Table 10.1.

When you run Canvas, the screen shown in Figure 10.3 will appear. Click "Skip for now," to skip the tutorial.

If you skip the tutorial, the screen shown in Figure 10.4 will appear. From here, we will upload the csv file that was obtained from the Kaggle site. Therefore, as shown in Figure 10.4, click on the "Import" tab in the top right corner to import the data.

You can import data by dragging Train.csv and Test.csv data to the screen.

After the data is uploaded, you can import it to Canvas using the "Import data" button located at the bottom right corner (Figures 10.5 and 10.6).

You can check the imported data in the dataset (Figure 10.7).

To make predictions using the Titanic data on Canvas, it is necessary to create a model. Therefore, check the train.csv file, which is the data to be trained (Figure 10.8).

Before building a model on Canvas, click on train.csv to preview the Titanic training data in advance. As you do so, you can identify the data attributes and values, as well as any empty cells in the data (Figure 10.9).

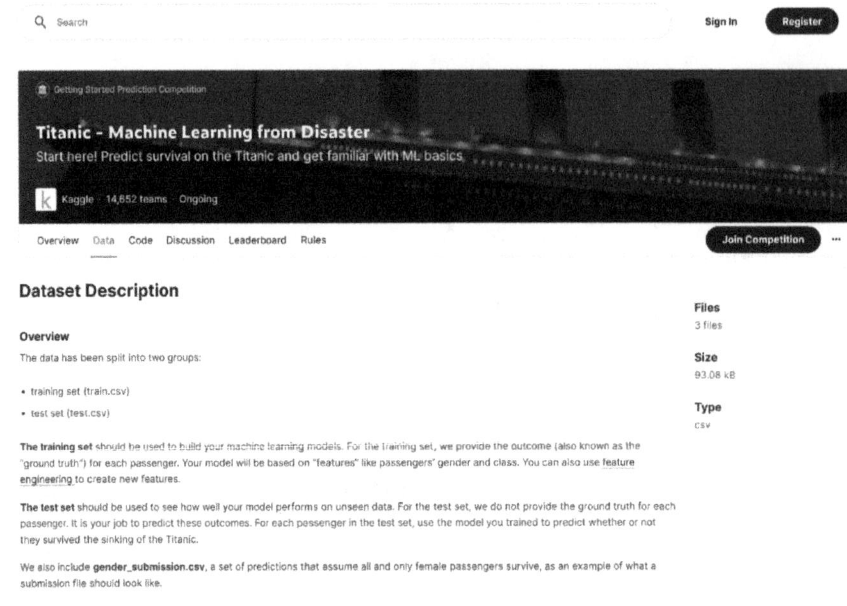

Figure 10.1 Kaggle Titanic data screen.

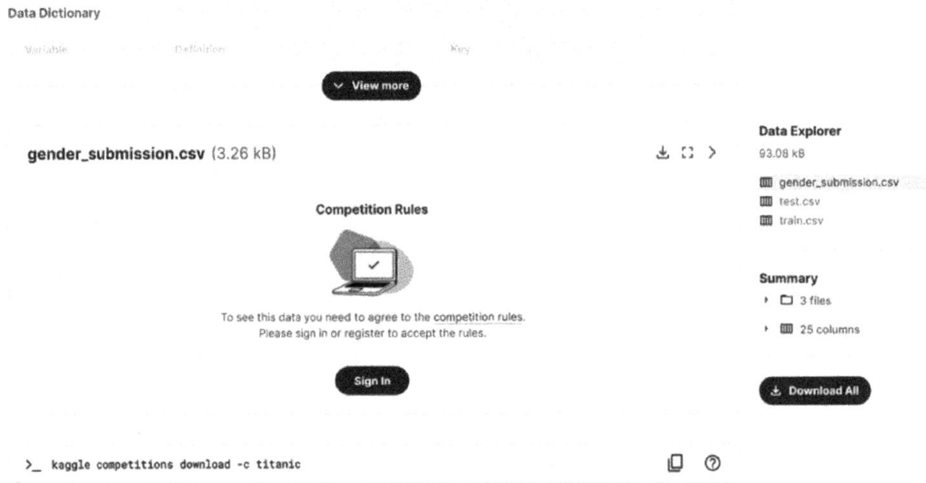

Figure 10.2 Kaggle Titanic dataset.

As shown in Figure 10.10, when the "Create new model" window appears, set the model name and then click the "Create" button to generate the model.

After generating the model shown in Figure 10.10, the screen shown in Figure 10.11 will appear. When examining the Titanic training data, a variety of

Table 10.1 Titanic dataset description.

Variable	Definition	Detailed Explanation
Survival	Survival status	0 = death 1 = survival
Pclass	Ticket class (economy)	1 = First class, 2 = Second class, 3 = Third class
Sex	Gender	Passenger gender
Age	Age	Passenger name
Sibsp	The number of siblings/couples who boarded the Titanic	The number of siblings or spouses who boarded together
Parch	The number of parents/children who boarded the Titanic	The number of parents or children who are riding together
Ticket	Ticket	Ticket
Fare	Room rate	Fee
Cabin	Room number	Cabin number
Embarked	Port of embarkation	C = Cherbourg, Q = Queenstown, S = Southampton

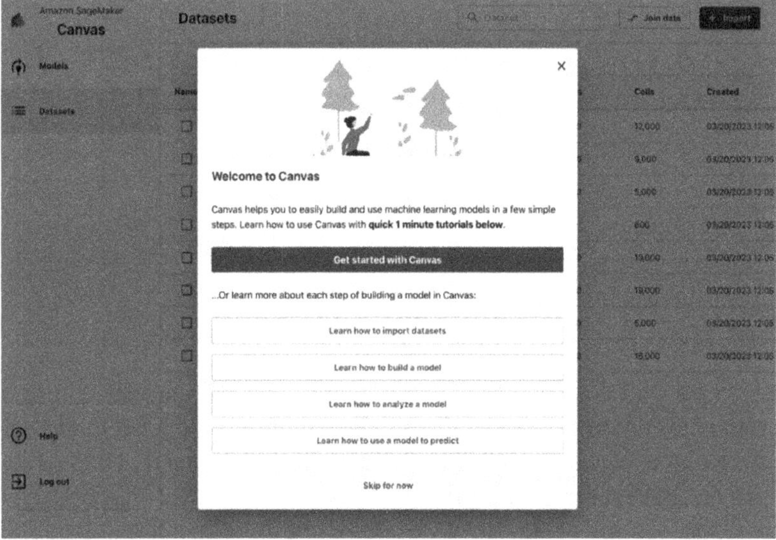

Figure 10.3 Canvas start screen.

information such as column names, types, the number of missing data, and average values can be identified.

In Canvas, users can click on each data column to visualize the corresponding information, and provide it to users, as shown in the image on the right.

Figure 10.4 SageMaker Canvas datasets.

Figure 10.5 Picking data from PC.

Let's create a model using the Titanic training data (Figure 10.12). To perform an accurate analysis on the model during data analysis, we need to exclude some data and create the model. In the Titanic data, the passenger ID column and name column do not have an impact on the likelihood of survival (Figure 10.13).

As you can see in Figure 10.14, to model the Titanic training data, you need to set the Target column. In this case, the Target column is set to "Survived" in order to predict the survival probability of the Titanic passengers.

If you need to set the model type, click on "Change type" in the "Model type" tab, as shown in Figure 10.15.

The Model type configuration window appears on Canvas as shown in Figure 10.16. Since the Titanic survivor prediction model aims to predict the survival of passengers, a 2 category model is selected.

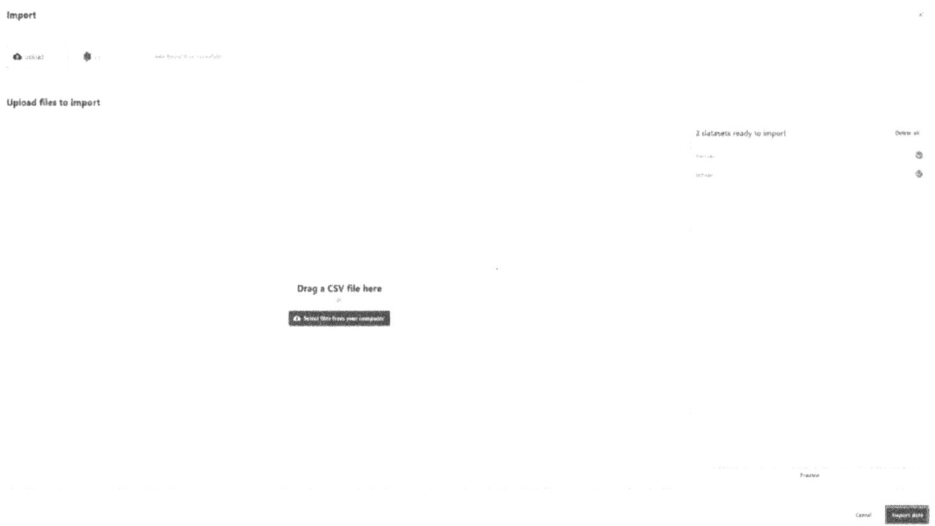

Figure 10.6 Import data in Canvas.

Amazon SageMaker Canvas

Models

Datasets

Datasets

Name	Source	Columns	Rows	Cells	Created	Status
test.csv	Local	11	418	4,598	02/01/2023 1:34 PM	Ready
train.csv	Local	12	891	10,692	02/01/2023 1:34 PM	Ready
canvas-sample-diabetic-readmission.csv	S3	16	1,000	16,000	01/13/2023 11:54 AM	Ready
canvas-sample-sales-forecasting.csv	S3	5	1,000	5,000	01/13/2023 11:54 AM	Ready
canvas-sample-loans-part-1.csv	S3	19	1,000	19,000	01/13/2023 11:54 AM	Ready
canvas-sample-product-descriptions.csv	S3	5	120	600	01/13/2023 11:54 AM	Ready
canvas-sample-shipping-logs.csv	S3	12	1,000	12,000	01/13/2023 11:54 AM	Ready
canvas-sample-maintenance.csv	S3	9	1,000	9,000	01/13/2023 11:54 AM	Ready
canvas-sample-housing.csv	S3	10	1,000	10,000	01/13/2023 11:54 AM	Ready
canvas-sample-loans-part-2.csv	S3	5	1,000	5,000	01/13/2023 11:54 AM	Ready

Figure 10.7 SageMaker Canvas datasets screen.

Once you have set the model type, click the "Quick build" button located in the top right corner to create the model.

When you click the "Quick build" button shown in Figure 10.17, the screen shown in Figure 10.18 will appear. It takes about 2–15 minutes to create a model.

The analysis screen of the Titanic training shows a brief result of the modeling. As shown in Figure 10.19, the accuracy for predicting survival on the Titanic was 88.83%. Each user may have a small margin of error regarding the

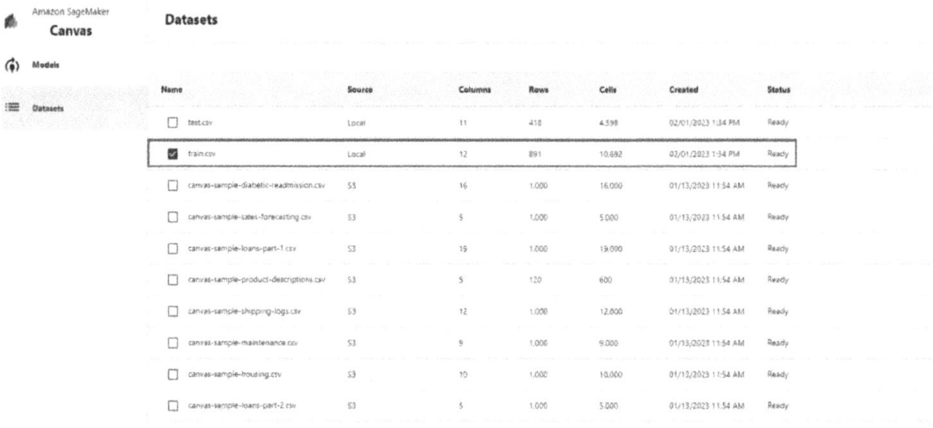

Amazon SageMaker
Canvas

Models
Datasets

Datasets

	Name	Source	Columns	Rows	Cells	Created	Status
☐	test.csv	Local	11	418	4,598	02/01/2023 1:34 PM	Ready
☑	train.csv	Local	12	891	10,692	02/01/2023 1:34 PM	Ready
☐	canvas-sample-diabetic-readmission.csv	S3	16	1,000	16,000	01/13/2023 11:54 AM	Ready
☐	canvas-sample-sales-forecasting.csv	S3	5	1,000	5,000	01/13/2023 11:54 AM	Ready
☐	canvas-sample-loans-part-1.csv	S3	19	1,000	19,000	01/13/2023 11:54 AM	Ready
☐	canvas-sample-product-descriptions.csv	S3	5	120	600	01/13/2023 11:54 AM	Ready
☐	canvas-sample-shipping-logs.csv	S3	12	1,000	12,000	01/13/2023 11:54 AM	Ready
☐	canvas-sample-maintenance.csv	S3	9	1,000	9,000	01/13/2023 11:54 AM	Ready
☐	canvas-sample-housing.csv	S3	10	1,000	10,000	01/13/2023 11:54 AM	Ready
☐	canvas-sample-loans-part-2.csv	S3	5	1,000	5,000	01/13/2023 11:54 AM	Ready

Figure 10.8 Datasets list screen.

Figure 10.9 Titanic train data.

probability. Additionally, the influence of each column on survival prediction can be seen listed on the left, as shown in Figure 10.19.

If you want to examine the results of the model built for predicting the survival chances of Titanic passengers in detail shown in Figure 10.20, you can click on “Scoring” to compare the actual values and the predicted values.

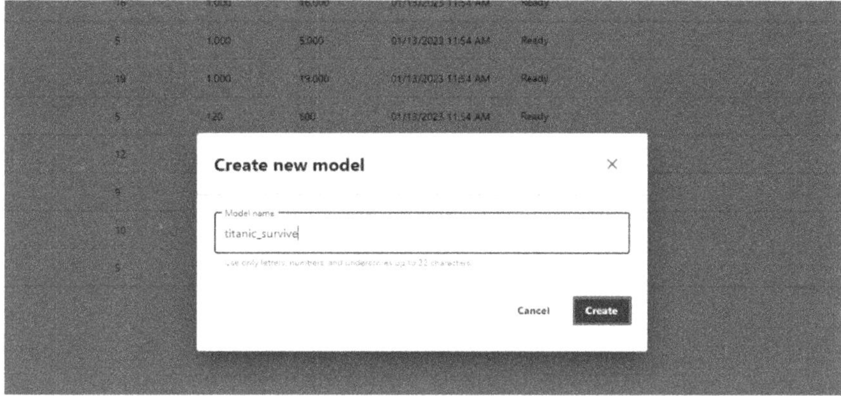

Figure 10.10 Model creation.

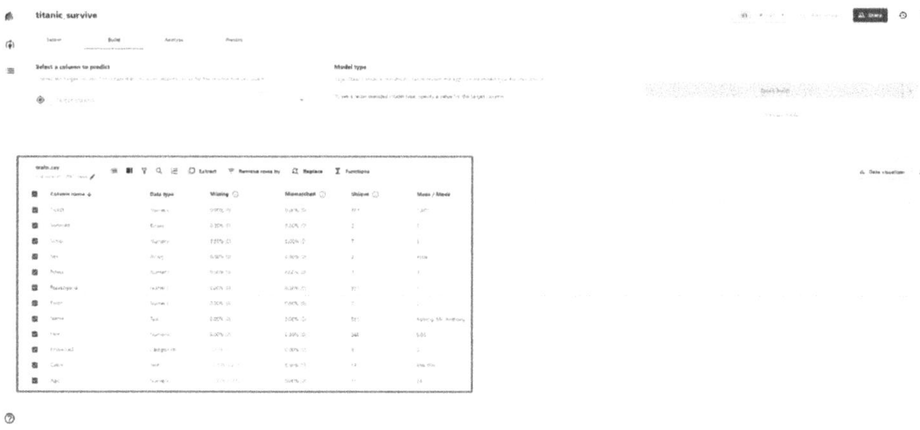

Figure 10.11 Titanic survive data.

Figure 10.12 Titanic train dataset 1.

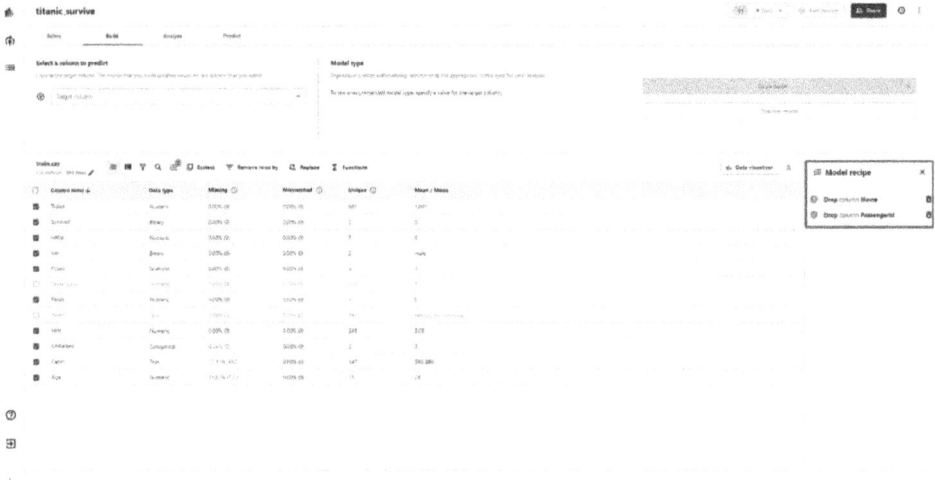

Figure 10.13 Titanic train dataset 2.

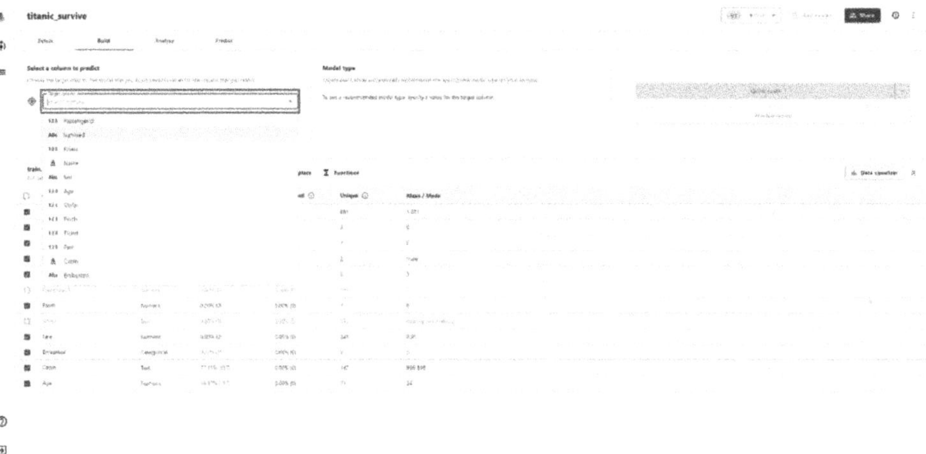

Figure 10.14 Titanic train modeling.

Let's take a look at the performance results of the Titanic survival prediction analysis model. By clicking on "Advanced metrics" shown in Figure 10.21, we can not only determine the values of TP, TN, FN, and FP but also understand indicators for data analysis such as F1 score, Accuracy, Precision, Recall, and AUC.

When you have built a survival prediction model on Canvas, let's evaluate new data using the built model. Click "Predict" in the top left corner of Figure 10.22 to proceed.

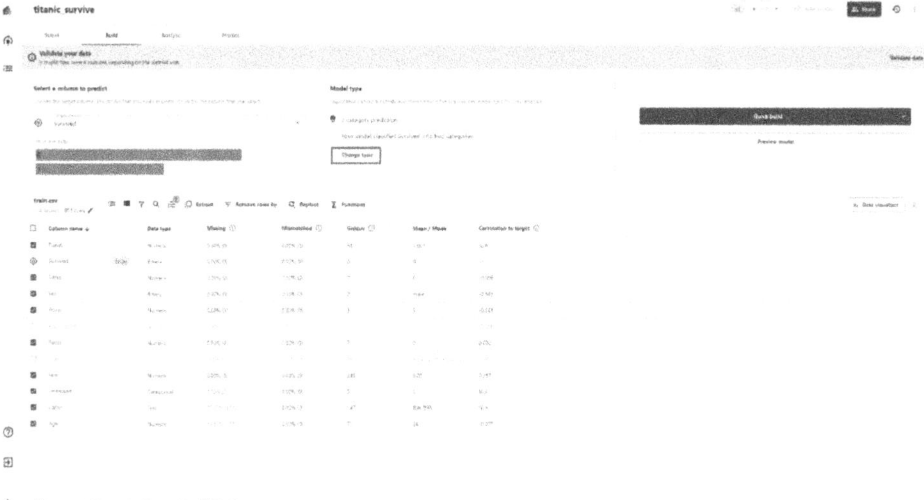

Figure 10.15 Canvas build screen.

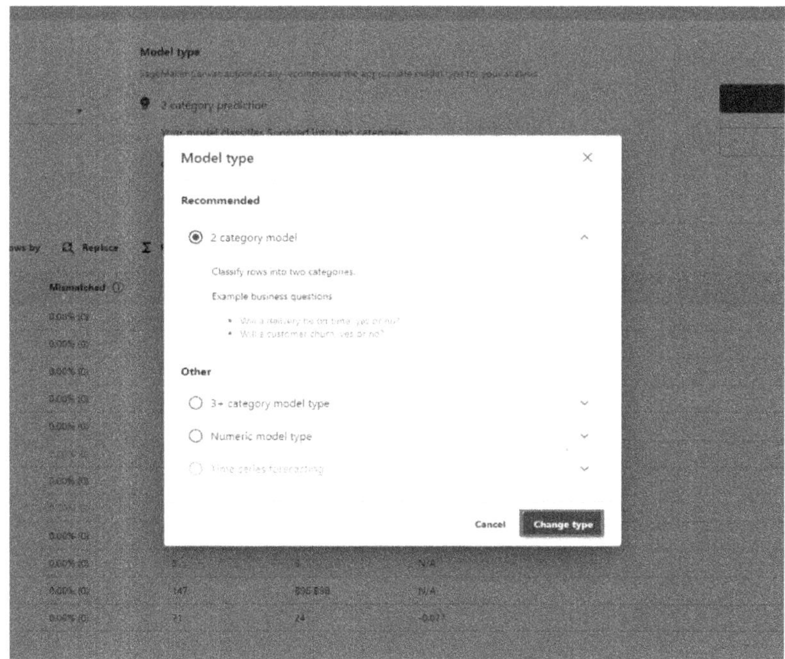

Figure 10.16 Setting the Canvas model type.

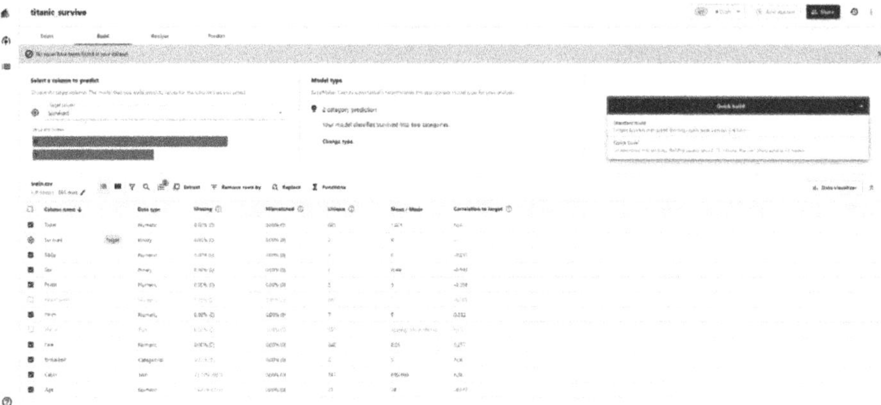

Figure 10.17 Quick build.

Figure 10.18 Model creation.

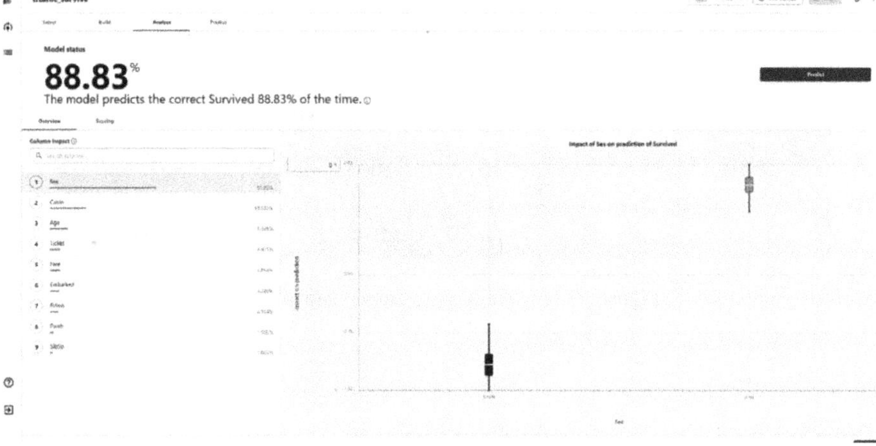

Figure 10.19 Titanic train analysis overview.

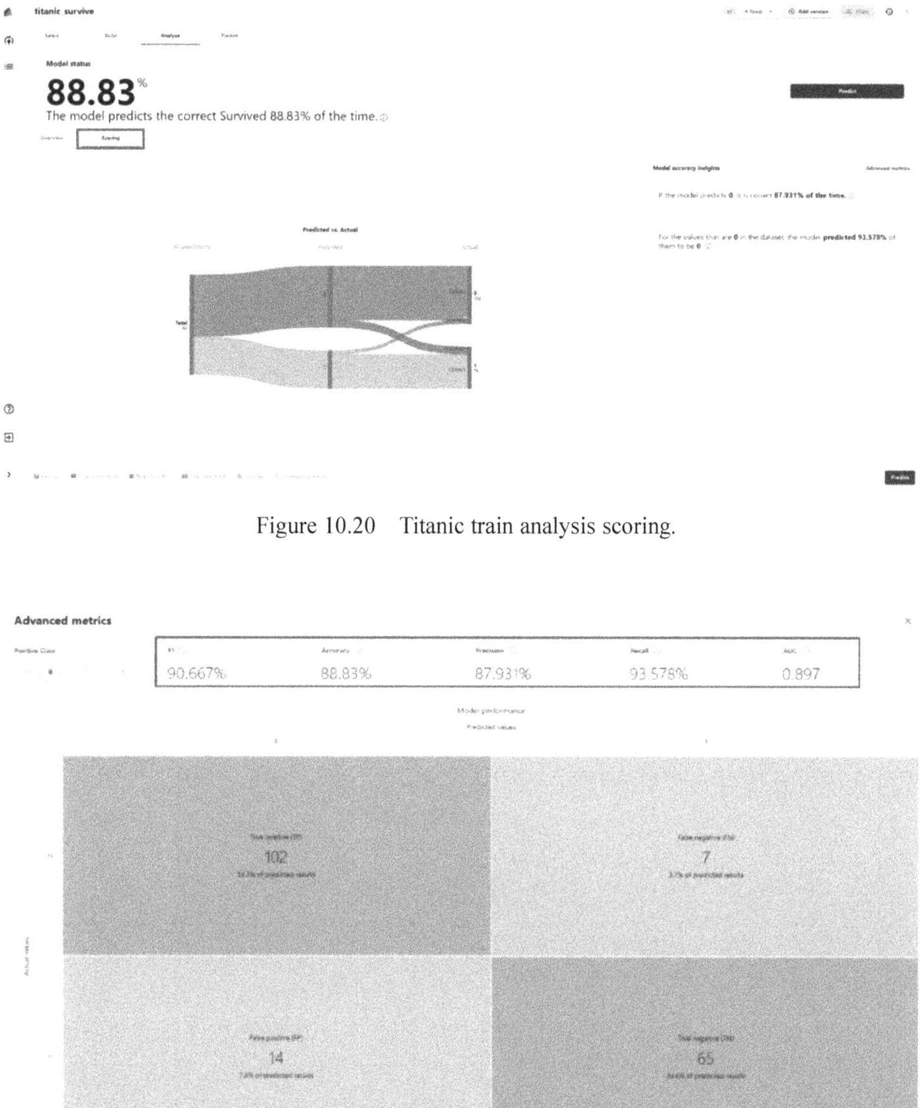

Figure 10.20 Titanic train analysis scoring.

Figure 10.21 Titanic train analysis advanced metrics.

Let's use a test dataset to make predictions using the model built on Canvas. Click on "Select dataset" and choose the test.csv file, as shown in Figure 10.23.

Once the prediction of the data to be tested through the Titanic prediction model is completed, a screen like Figure 10.24 will appear. If you want to make

Figure 10.22 Titanic survive predict.

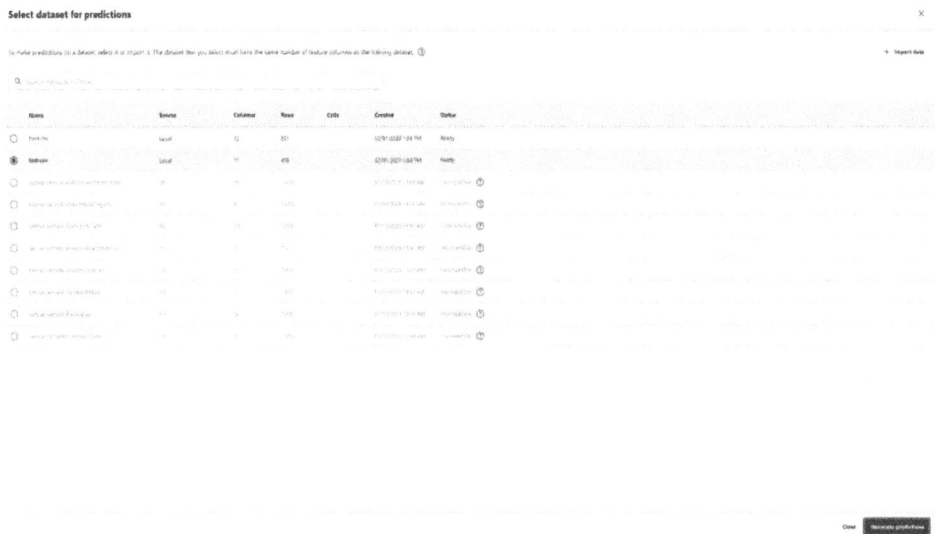

Figure 10.23 Dataset for predict.

predictions with the model created here, you can check them through Batch prediction and Single prediction in Predict Target values. First, Batch prediction is a method of predicting multiple data, so press the Batch prediction button to see the predicted results using the test data with the Batch Prediction method, and click the Preview button in the bottom dataset item.

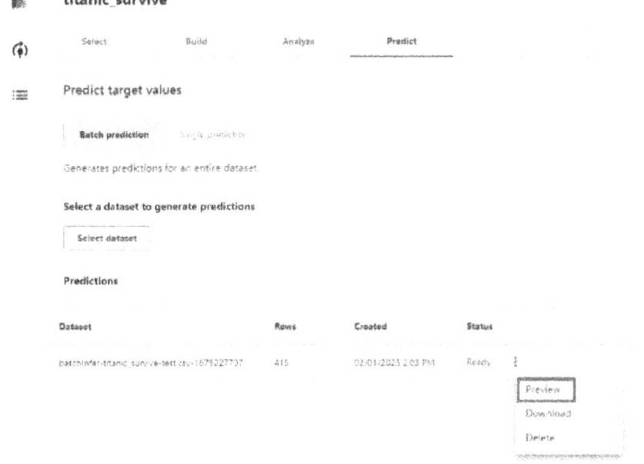

Figure 10.24 Batch prediction.

batchInfer-titanic_survive-test.csv-1675227797

Prediction (Survived)	Probability	PassengerId	Pclass	Name	Sex	Age	SibSp
0	77.2%	892	3	Kelly, Mr. James	male	34.5	0
0	52.3%	893	3	Wilkes, Mrs. Jame...	female	47	1
0	80.0%	894	2	Myles, Mr. Thoma...	male	62	0
0	75.8%	895	3	Wirz, Mr. Albert	male	27	0
1	56.8%	896	3	Hirvonen, Mrs. Ale...	female	22	1
0	73.2%	897	3	Svensson, Mr. Joh...	male	14	0
1	70.2%	898	3	Connolly, Miss. Kate	female	30	0
0	61.5%	899	2	Caldwell, Mr. Albe...	male	26	1
1	69.4%	900	3	Abrahim, Mrs. Jos...	female	18	0
0	80.1%	901	3	Davies, Mr. John S...	male	21	2
0	80.7%	902	3	Ilieff, Mr. Ylio	male		0
0	63.5%	903	1	Jones, Mr. Charles...	male	46	0
1	87.7%	904	1	Snyder, Mrs. John ...	female	23	1
0	78.9%	905	2	Howard, Mr. Benja...	male	63	1
1	86.6%	906	1	Chaffee, Mrs. Her...	female	47	1
1	79.2%	907	2	del Carlo, Mrs. Se...	female	24	1
0	78.6%	908	2	Keane, Mr. Daniel	male	35	0

Download CSV

Figure 10.25 Test data preview.

In Figure 10.25, you can see that each test data is a result created based on the trained model. It can be observed that the test data generally has a probability value of over 60%.

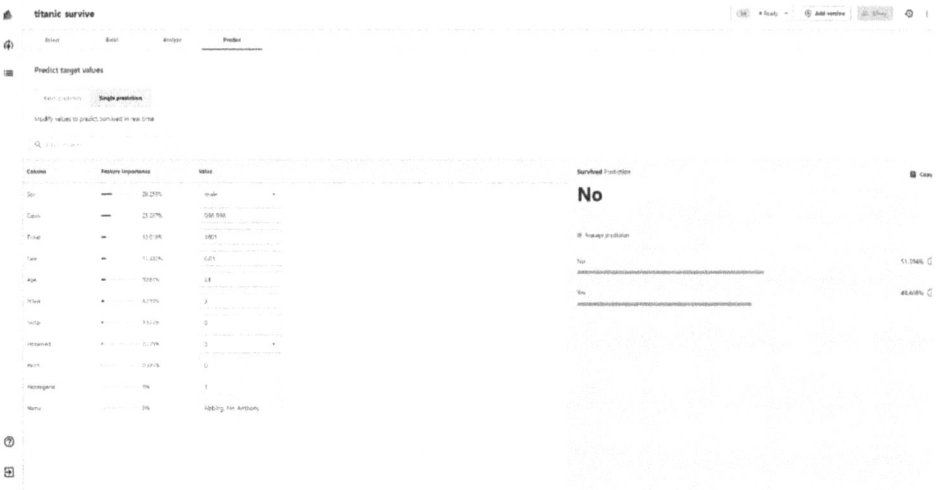

Figure 10.26 Single prediction.

Single prediction is when a user inputs a single piece of data based on the created model and tries to predict the outcome. The user can set values for each column as desired and then make a prediction (Figure 10.26).

10.2 Creating a Diabetes Prediction Model for the Pima Indian Region

This time let's create a predictive model using the diabetes outcome data of Native Americans in the Pima region provided by Kaggle, and try to predict the presence of diabetes based on it. You can check the data at the following URL (https://www.kaggle.com/datasets/uciml/pima-indians-diabetes-database) (Figure 10.27).

On the previous page, the data can be downloaded through the "Download" button located in the top right corner.

The data utilizes the Diabetes dataset, consisting of 768 rows and 9 attributes (Figure 10.28). The attribute names and detailed explanations of the dataset can be confirmed through Table 10.2.

Figure 10.29 is a scene that imports Diabetes data received from Kaggle. You can do this by using the Import button located in the top right corner.

In Figure 10.30, you can see that you can import the data from Diabetes.csv file by dragging it onto the screen.

After confirming that the data has been uploaded correctly, you can import it to Canvas by clicking the Import data button at the bottom right (Figure 10.31).

You can check the imported data in the datasets (Figure 10.32).

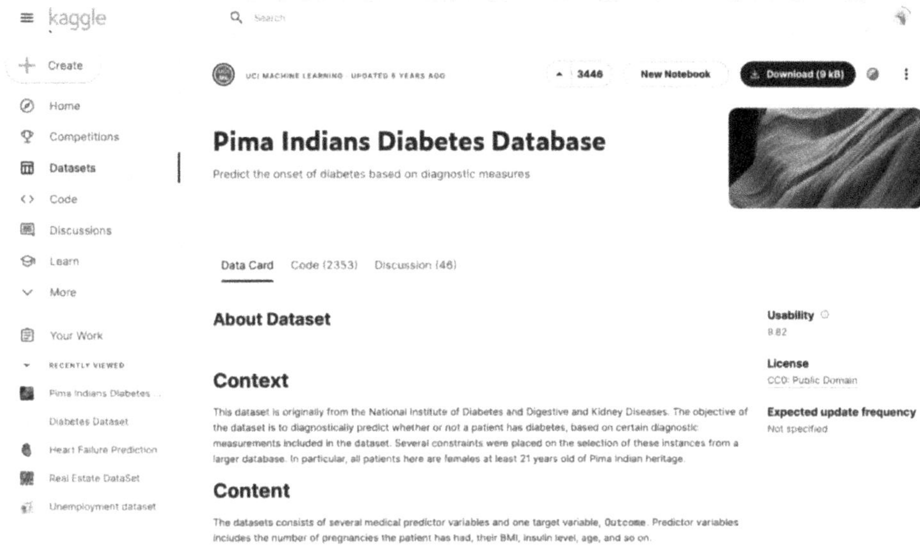

Figure 10.27 Kaggle Pima Indians diabetes data screen.

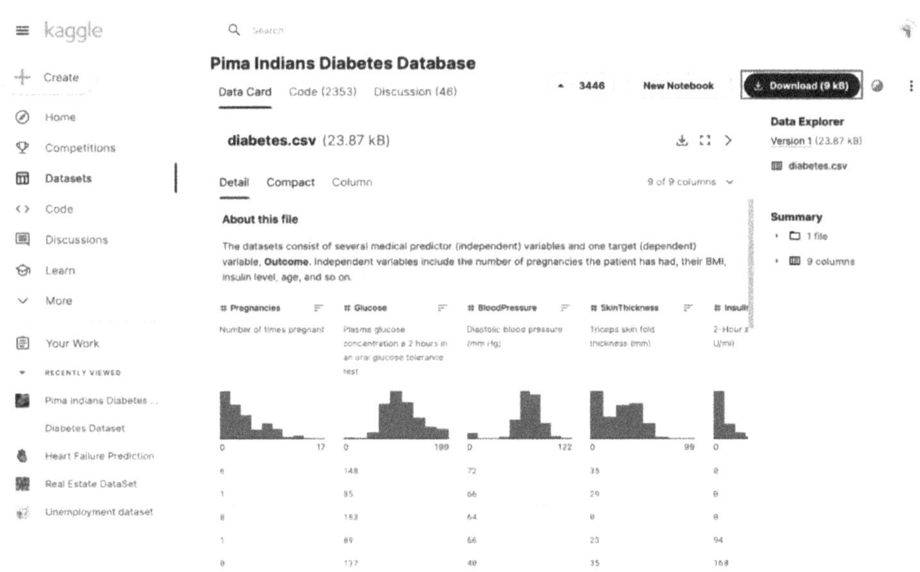

Figure 10.28 Kaggle diabetes dataset.

In order to make predictions using the Pima diabetes data on Canvas, it is necessary to create a model as shown in Figure 10.33. Therefore, we check the diabetes.csv file, which is the data to be trained.

Table 10.2 Diabetes dataset description.

Variable name	Definition	Detailed explanation
Pregnancies	Pregnancies	Number of pregnancies
Glucose	Glucose	Glycemic load test results
Blood pressure	Blood pressure	Blood pressure measurement value (mm Hg)
Skin thickness	Measurement of body fat percentage	Measurement value of subcutaneous fat in the back of the triceps (mm)
Insulin	Insulin prices	Serum insulin (mu U/mL)
BMI	Body mass index	Body weight (kg)/(Height (m))^2
Diabetes Pedigree function	Diabetes history weighting value	The genetic factors of diabetes are weighted to determine their values
Age	Age	Age
Outcome	The value of determination	Determining the presence of diabetes (0: no diabetes or 1: diabetes present)

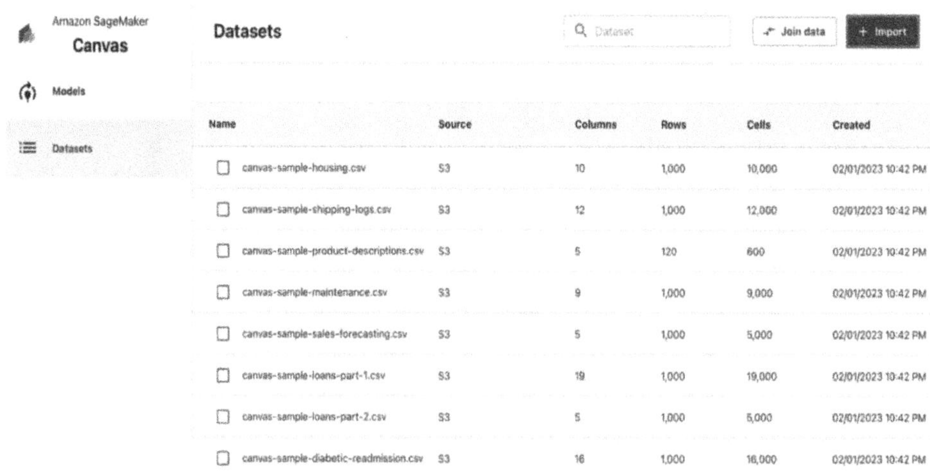

Figure 10.29 SageMaker Canvas datasets.

Import

Upload S3 Add Redshift or Snowflake

Upload files to import

Drag a CSV file here

or

Select files from your computer

Cancel Import data

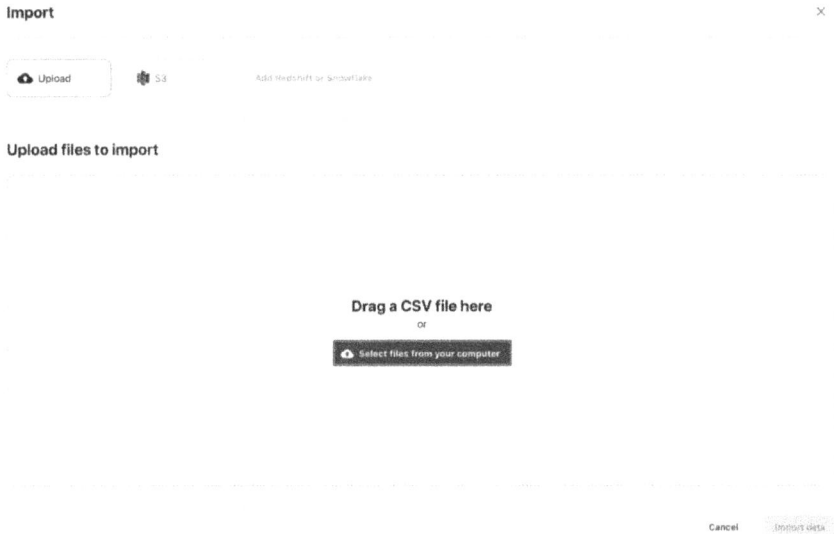

Figure 10.30 Loading data in SageMaker Canvas.

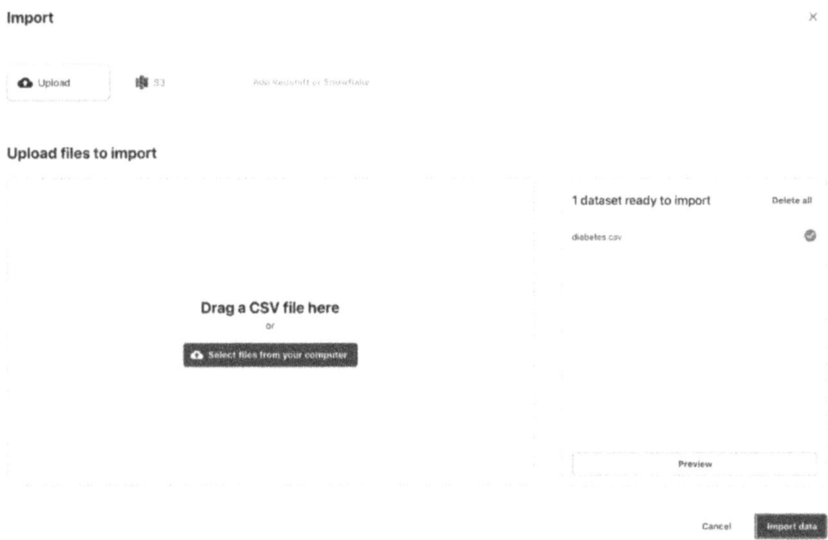

Figure 10.31 Loading diabetes data.

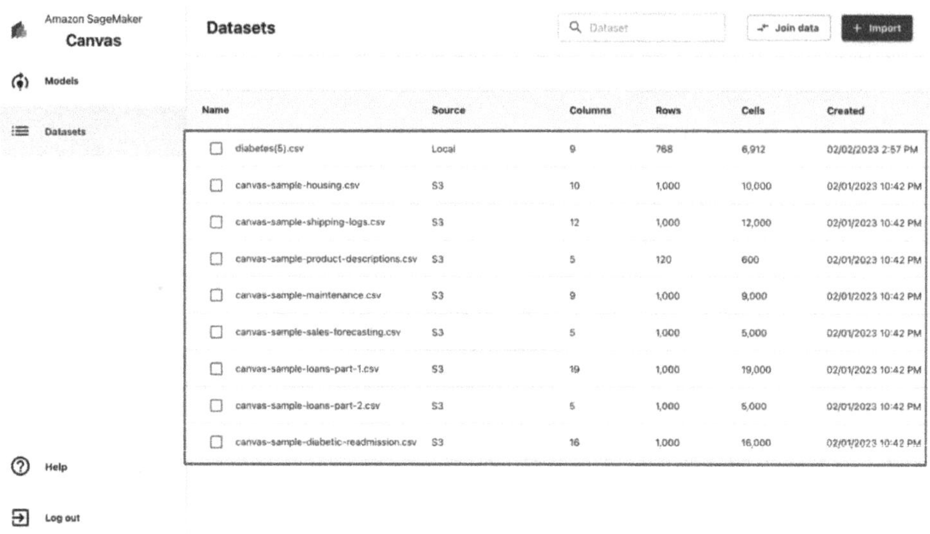

Amazon SageMaker **Canvas**

Models

Datasets

Datasets

Dataset | Join data | + Import

Name	Source	Columns	Rows	Cells	Created
diabetes(5).csv	Local	9	768	6,912	02/02/2023 2:57 PM
canvas-sample-housing.csv	S3	10	1,000	10,000	02/01/2023 10:42 PM
canvas-sample-shipping-logs.csv	S3	12	1,000	12,000	02/01/2023 10:42 PM
canvas-sample-product-descriptions.csv	S3	5	120	600	02/01/2023 10:42 PM
canvas-sample-maintenance.csv	S3	9	1,000	9,000	02/01/2023 10:42 PM
canvas-sample-sales-forecasting.csv	S3	5	1,000	5,000	02/01/2023 10:42 PM
canvas-sample-loans-part-1.csv	S3	19	1,000	19,000	02/01/2023 10:42 PM
canvas-sample-loans-part-2.csv	S3	5	1,000	5,000	02/01/2023 10:42 PM
canvas-sample-diabetic-readmission.csv	S3	16	1,000	16,000	02/01/2023 10:42 PM

Help

Log out

Figure 10.32 SageMaker Canvas datasets screen.

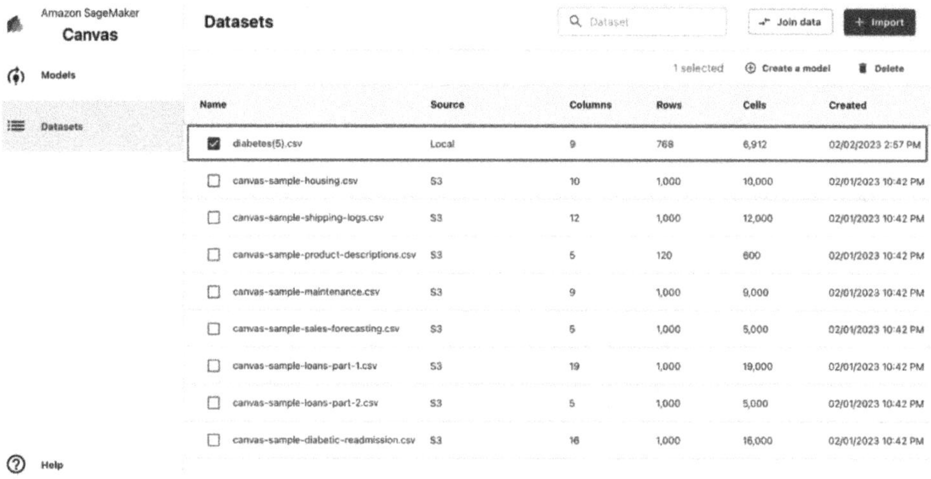

Amazon SageMaker **Canvas**

Models

Datasets

Datasets

Dataset | Join data | + Import

1 selected | Create a model | Delete

Name	Source	Columns	Rows	Cells	Created
diabetes(5).csv	Local	9	768	6,912	02/02/2023 2:57 PM
canvas-sample-housing.csv	S3	10	1,000	10,000	02/01/2023 10:42 PM
canvas-sample-shipping-logs.csv	S3	12	1,000	12,000	02/01/2023 10:42 PM
canvas-sample-product-descriptions.csv	S3	5	120	600	02/01/2023 10:42 PM
canvas-sample-maintenance.csv	S3	9	1,000	9,000	02/01/2023 10:42 PM
canvas-sample-sales-forecasting.csv	S3	5	1,000	5,000	02/01/2023 10:42 PM
canvas-sample-loans-part-1.csv	S3	19	1,000	19,000	02/01/2023 10:42 PM
canvas-sample-loans-part-2.csv	S3	5	1,000	5,000	02/01/2023 10:42 PM
canvas-sample-diabetic-readmission.csv	S3	16	1,000	16,000	02/01/2023 10:42 PM

Help

Figure 10.33 Datasets list screen.

Before building a model in Canvas, take a look at the diabetes data in advance. By referring to Figure 10.34, you can understand the attributes and values of the uploaded data.

As shown in Figure 10.35, when the "Create new model" window appears in the top right corner, set the model's name and then press the "Create" button to create the model.

diabetes(5).csv Previewing first 100 rows + Create a model ×

Pregnancies	Glucose	BloodPressure	SkinThickness	Insulin	BMI	DiabetesPedigr...	Age	Outcome
6	148	72	35	0	33.6	0.627	50	1
1	85	66	29	0	26.6	0.351	31	0
8	183	64	0	0	23.3	0.672	32	1
1	89	66	23	94	28.1	0.167	21	0
0	137	40	35	168	43.1	2.288	33	1
5	116	74	0	0	25.6	0.201	30	0
3	78	50	32	88	31	0.248	26	1
10	115	0	0	0	35.3	0.134	29	0
2	197	70	45	543	30.5	0.158	53	1
8	125	96	0	0	0	0.232	54	1
4	110	92	0	0	37.6	0.191	30	0
10	168	74	0	0	38	0.537	34	1
10	139	80	0	0	27.1	1.441	57	0
1	189	60	23	846	30.1	0.398	59	1
5	166	72	19	175	25.8	0.587	51	1
7	100	0	0	0	30	0.484	32	1

Figure 10.34 Diabetes data.

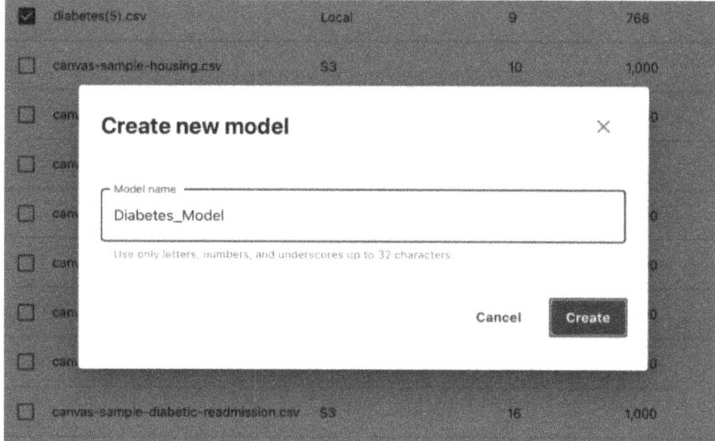

Figure 10.35 Diabetes model creation.

After generating the model shown in Figure 10.35, the screen shown in Figure 10.36 will appear. When examining the diabetes data, a variety of information such as column names, data types, the number of missing values, and mean values can be identified.

If you look at Figure 10.37, in Canvas, each data column has been checked to visualize the information for the user shown on the right side of the screen.

If you look at Figure 10.38, in order to create a model for diabetes training data, you need to set the Target Column.

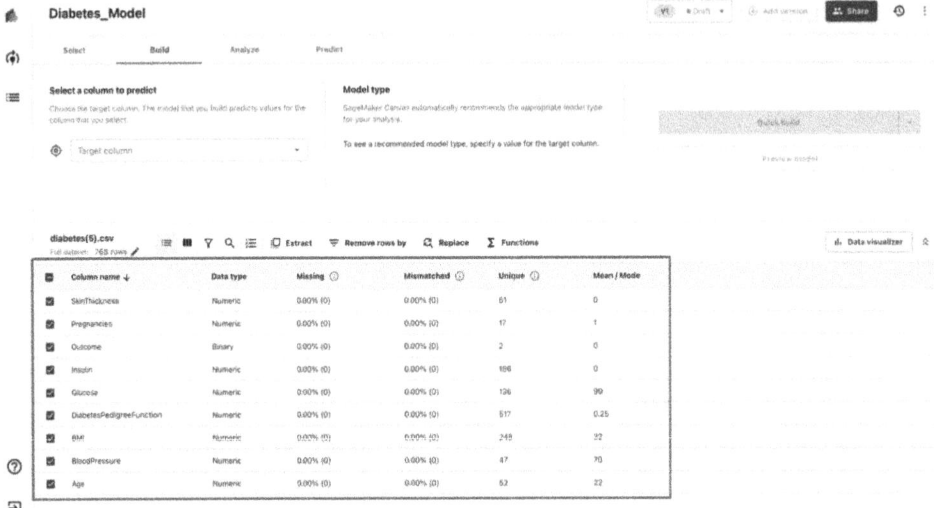

Figure 10.36 Diabetes data.

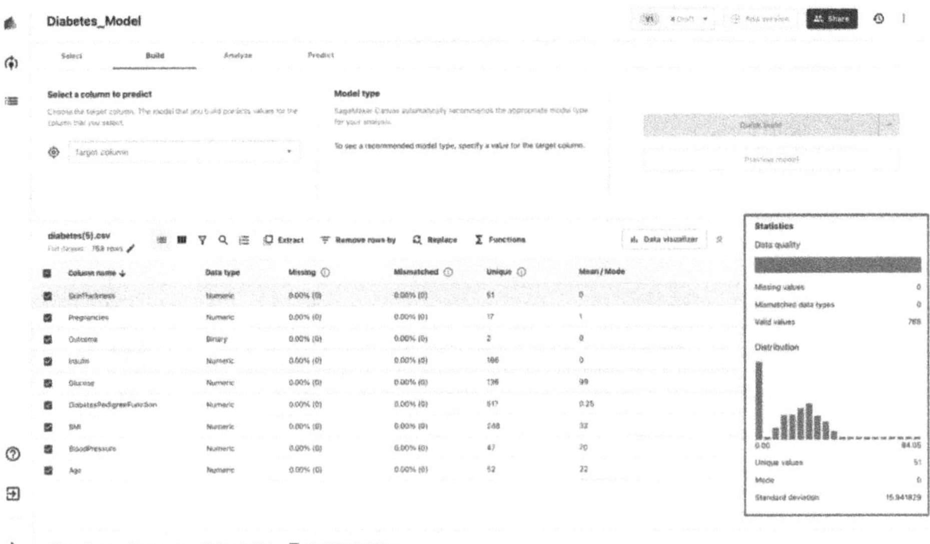

Figure 10.37 Diabetes data statistics.

To create a model on Canvas, you need to set the Target Column. In this case, the Target Column is set as "Outcome" to predict diabetes in the Pima Indian population. If the user needs to set the model type, they can click on "Change type" as shown in Figure 10.39.

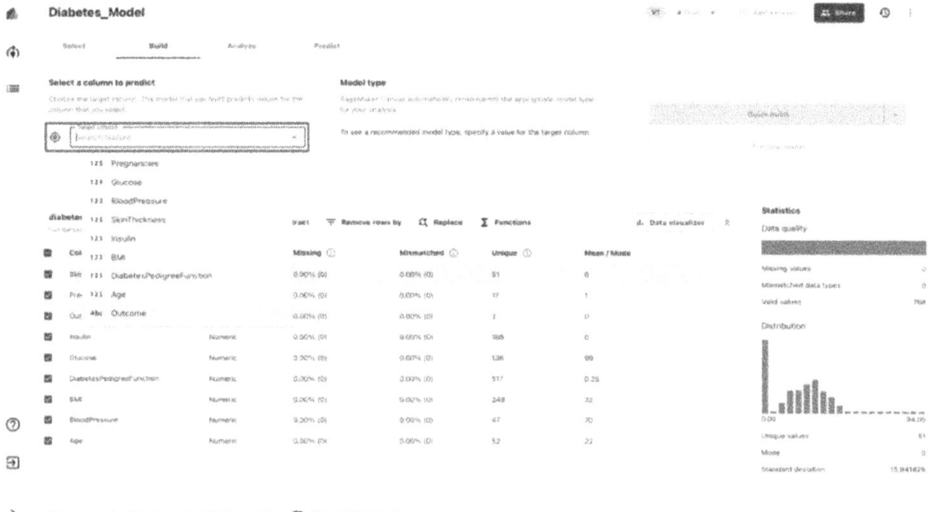

Figure 10.38 Diabetes model creation.

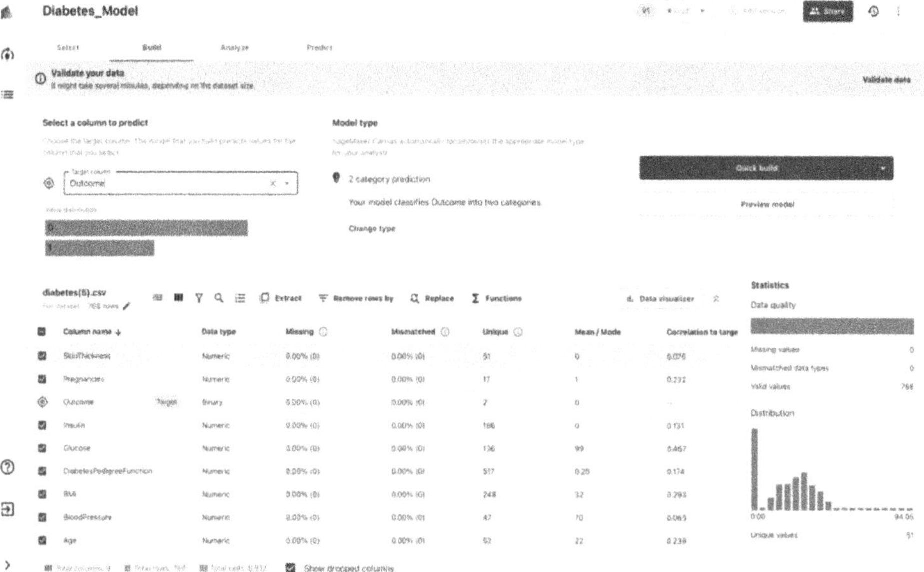

Figure 10.39 Model type screen.

The models type settings when creating a model in Canvas are shown in Figure 10.40. Since the diabetes prediction model is about predicting the presence of diabetes in the Pima Indian population, we will select the 2-category model.

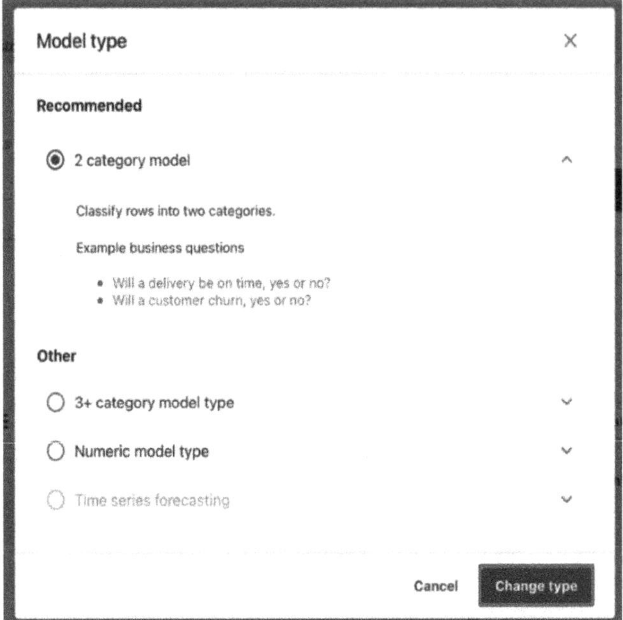

Figure 10.40 Canvas model type setting.

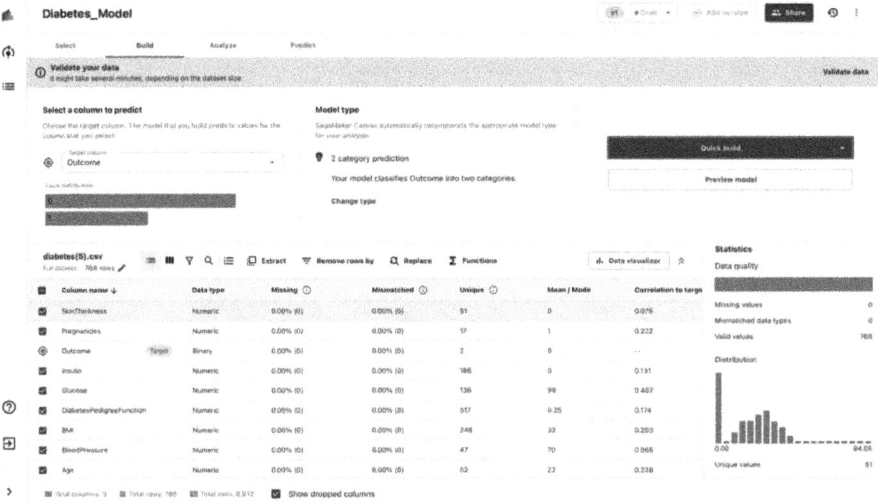

Figure 10.41 Quick build.

If you have set the model creation type, click the "Quick build" button in the top right corner to create the model.

When you click on the "Quick build" button shown in Figure 10.41, the screen shown in Figure 10.42 will appear. It takes approximately 2–15 minutes to create the model.

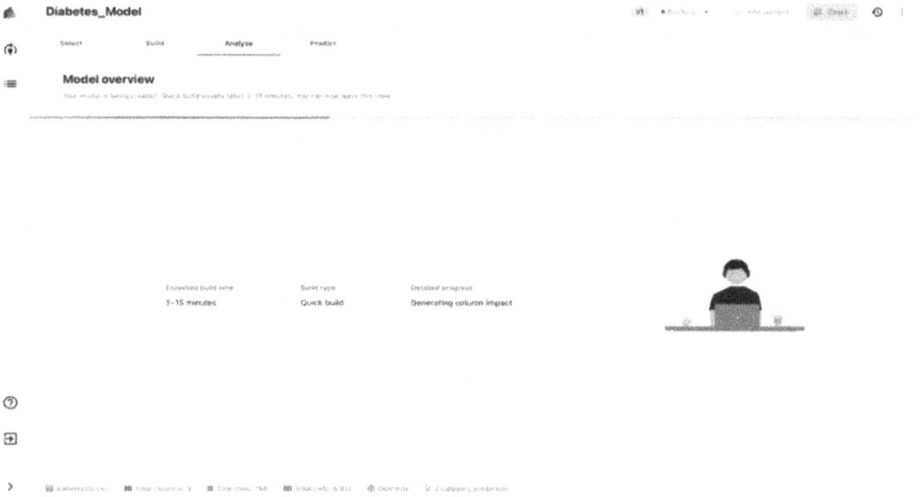

Figure 10.42 Quick model generation.

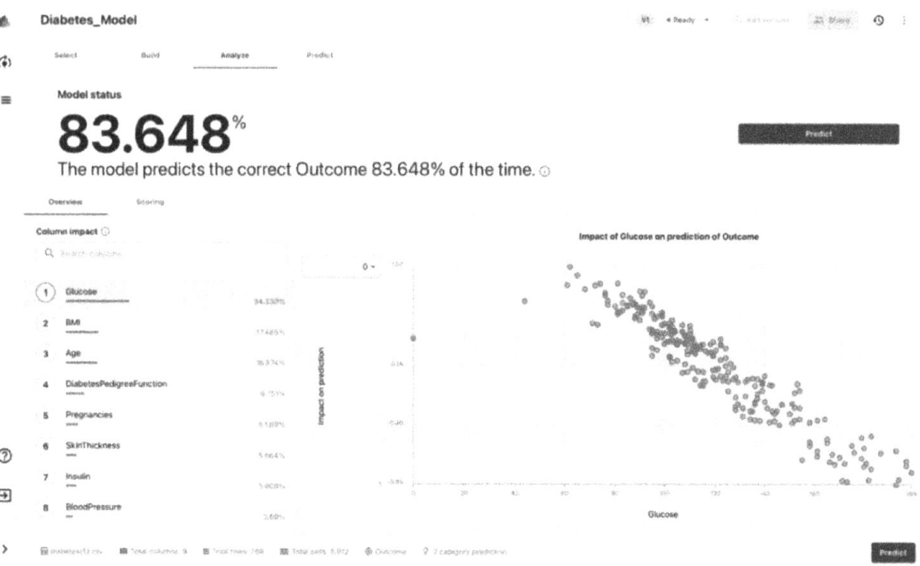

Figure 10.43 Diabetes model overview.

The overall results for the model are shown in the diabetes training analysis screen. As seen in Figure 10.43, the accuracy for predicting diabetes was 83.64%. It is possible to assess the impact of each column on predicting diabetes.

If you want to assess the detailed analysis results of the generated model as shown in Figure 10.44, you can compare the actual values and the predicted values by clicking on “Scoring.”

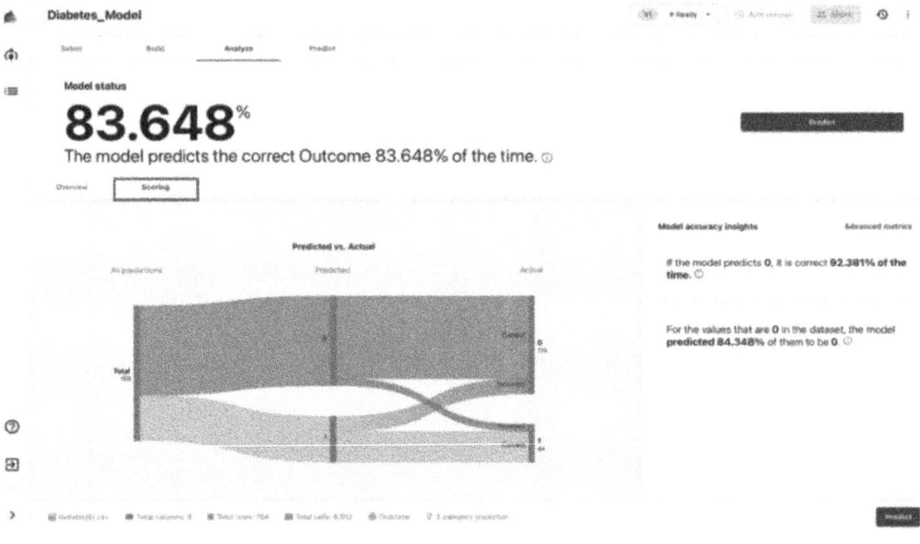

Figure 10.44 Diabetes analysis scoring.

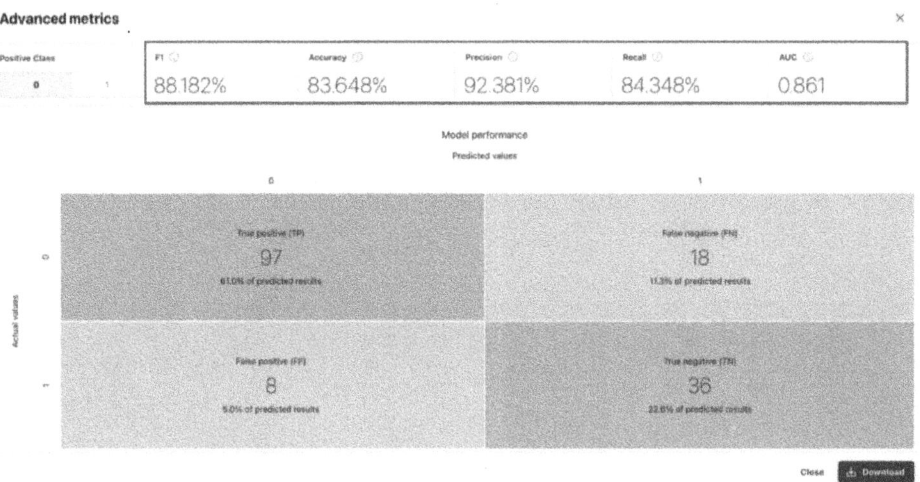

Figure 10.45 Diabetes analysis advanced metrics.

To view the results of the performance of the Diabetes prediction analysis model, click on "Advanced metrics" as shown in Figure 10.44. As shown in Figure 10.45, you can identify TP, TN, FN, and FP values, as well as metrics related to data analysis such as F1 score, Accuracy, Precision, Recall, and AUC.

If you have built a diabetes prediction model in Canvas, clicking on the "Predict" item in Figure 10.46 will display the screen shown in Figure 10.46. Let's use the generated prediction model to make predictions on the next screen.

Figure 10.46 Diabetes prediction.

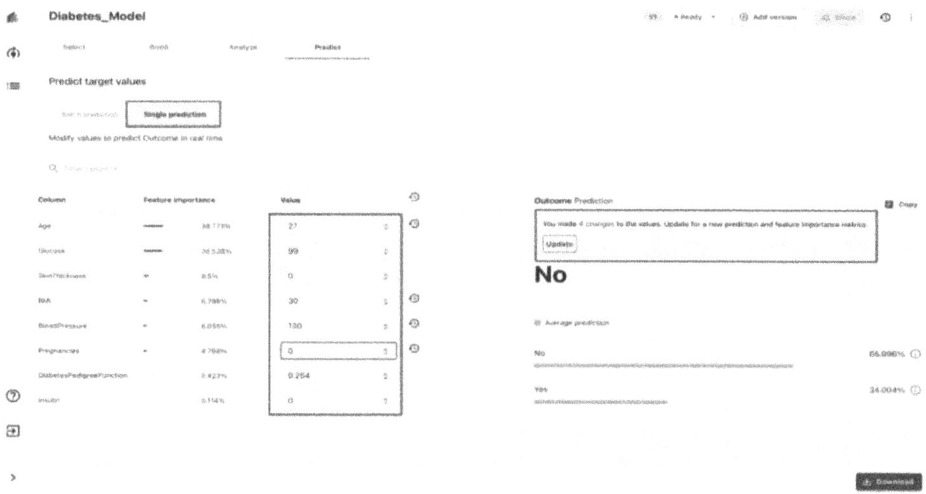

Figure 10.47 Single prediction.

Next, let's try predicting using the data values to be tested through the diabetes prediction model that the user has set. In Figure 10.47, you can see the results predicted using the Single Prediction method. After pressing the "Single prediction" button, you can enter the desired values for each column's value. Once you input a new value, you can click the "Update" button in the Outcome Prediction section to make a prediction.

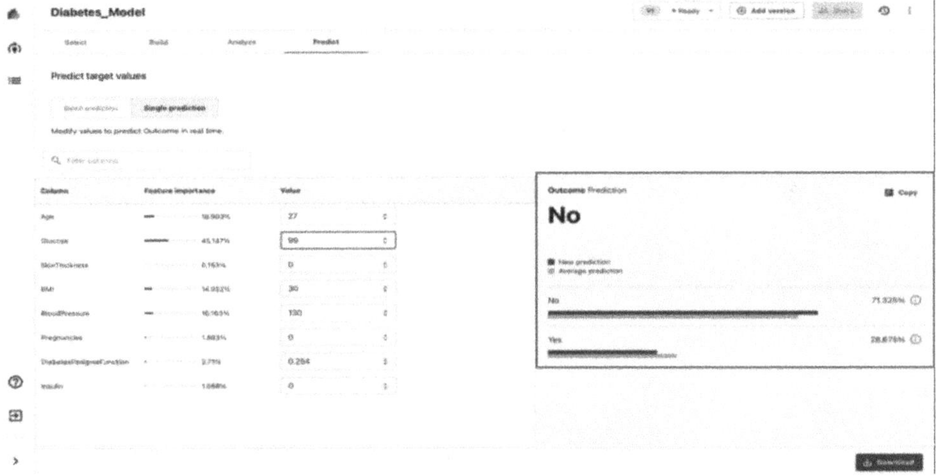

Figure 10.48 Test data preview.

Figure 10.48 shows the predicted results based on the values entered by the user. Looking at the new predicted values, the probability of having diabetes is predicted to be 28.675%, while the probability of not having diabetes is predicted to be 71.325%.

10.3 Practice Predicting the State of the Patient's Cardiovascular Disease

In this example, we will utilize the heart disease health indicators (blood pressure, cholesterol, BMI, smoking, chronic conditions) provided by Kaggle to build a heart disease prediction model using SageMaker Canvas. You can check the Heart Disease Health data at the following URL: https://www.kaggle.com/datasets/alexteboul/heart-disease-health-indicators-dataset.

As shown in Figure 11.49, on the Kaggle site, there is a button on the top right to download the file. Click the "Download" button on the top right of the screen to download the file.

If you have downloaded the file, upload it to Canvas. Once the upload is complete, the screen shown in Figure 10.50 will appear. Then, click the "Import data" button at the bottom right to upload the data.

When the cardiovascular disease prediction data upload is complete, select the file to generate the model on the Datasets screen. In Figure 10.51, select the Heart Disease Health Indicators CSV file and click "Create a model" in the top right corner to generate the model.

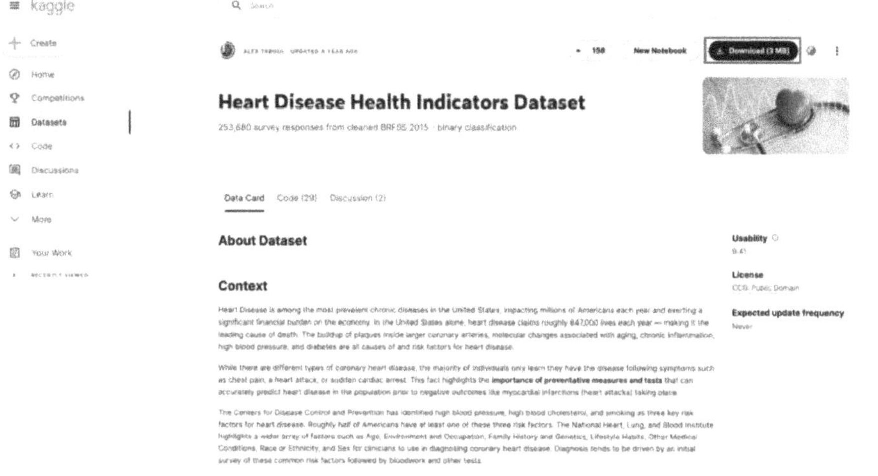

Figure 10.49 Kaggle heart disease.

Table 10.3 Heart disease health dataset.

Variable name	
Blood pressure (high)	Alcohol consumption
Cholesterol	BMI
Smoking	Household income
Diabetes	Sleep
Obesity	Education
Age	Marital status
Sex	Time since last checkup
Race	Health care coverage
Diet	Mental health
Exercise	

Once the data file upload is complete as shown in Figure 10.51, the screen shown in Figure 10.52 will appear.

After completing the exploration of the data, you need to set the target column and then build the model. To predict the presence of heart disease, set the Target Column as Heart Disease or Attack. After that, select “Standard build” instead of “Quick build” to generate the model. It takes approximately 30 minutes for the model to be created.

When the Standard Model build is completed, the screen shown in Figure 10.54 will appear. The result of building the Standard Model showed an accuracy of 85%.

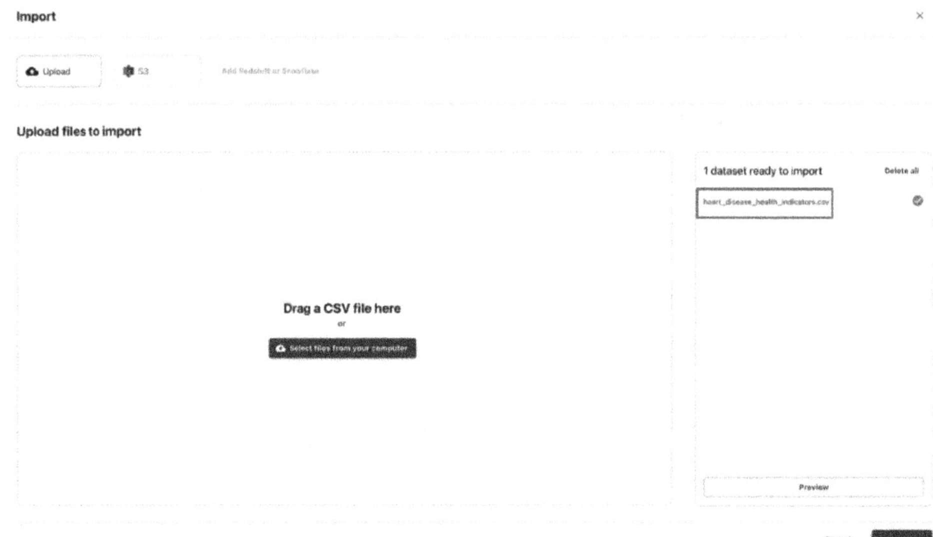

Figure 10.50 Kaggle heart disease dataset.

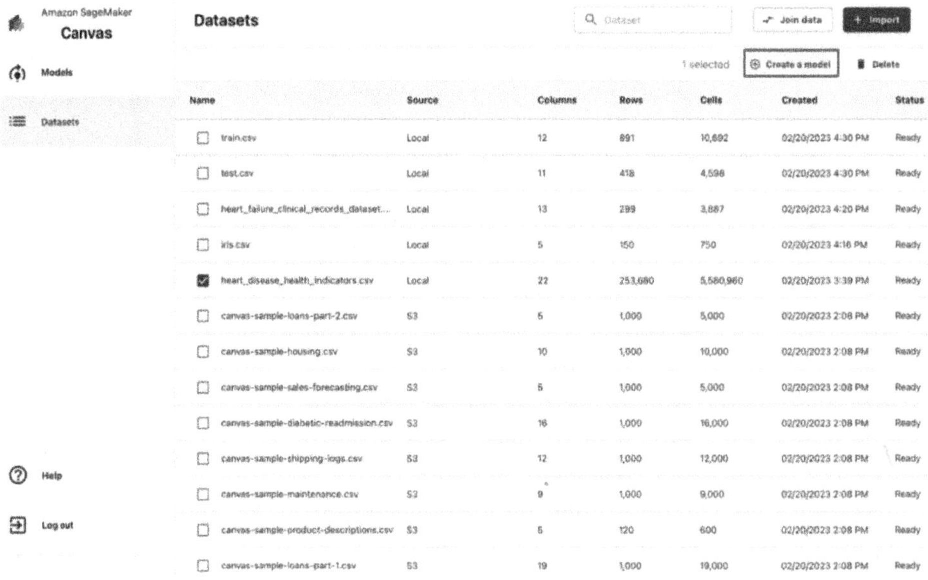

Amazon SageMaker
Canvas
Models
Datasets
Help
Log out
Datasets
Join data
+ Import
1 selected
Create a model
Delete
Name
Source
Columns
Rows
Cells
Created
Status
train.csv Local 12 891 10,692 02/20/2023 4:30 PM Ready
test.csv Local 11 418 4,598 02/20/2023 4:30 PM Ready
heart_failure_clinical_records_dataset... Local 13 299 3,887 02/20/2023 4:20 PM Ready
iris.csv Local 5 150 750 02/20/2023 4:16 PM Ready
heart_disease_health_indicators.csv Local 22 253,680 5,580,960 02/20/2023 3:39 PM Ready
canvas-sample-loans-part-2.csv S3 5 1,000 5,000 02/20/2023 2:08 PM Ready
canvas-sample-housing.csv S3 10 1,000 10,000 02/20/2023 2:08 PM Ready
canvas-sample-sales-forecasting.csv S3 5 1,000 5,000 02/20/2023 2:08 PM Ready
canvas-sample-diabetic-readmission.csv S3 16 1,000 16,000 02/20/2023 2:08 PM Ready
canvas-sample-shipping-logs.csv S3 12 1,000 12,000 02/20/2023 2:08 PM Ready
canvas-sample-maintenance.csv S3 9 1,000 9,000 02/20/2023 2:08 PM Ready
canvas-sample-product-descriptions.csv S3 5 120 600 02/20/2023 2:08 PM Ready
canvas-sample-loans-part-1.csv S3 19 1,000 19,000 02/20/2023 2:08 PM Ready

Figure 10.51 Creating a heart disease model.

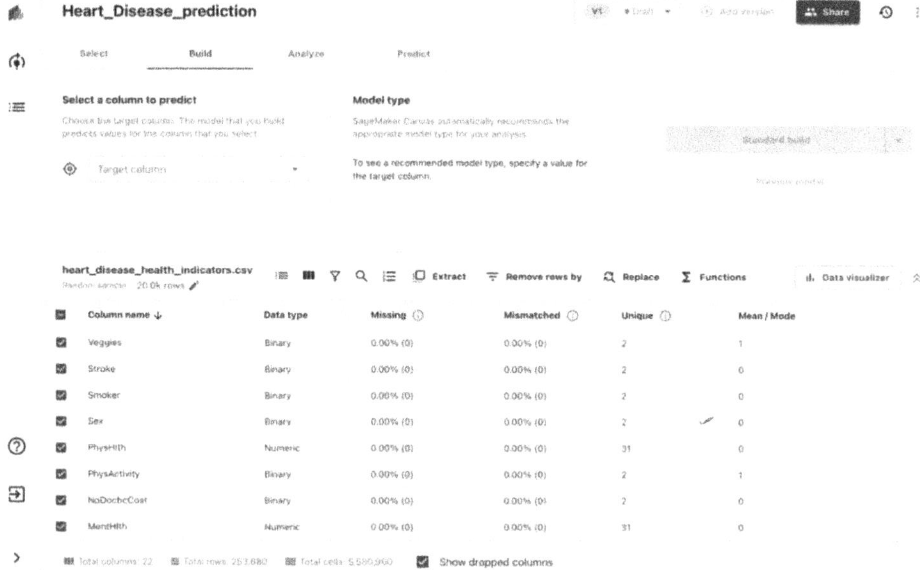

Figure 10.52 Heart disease dataset preview.

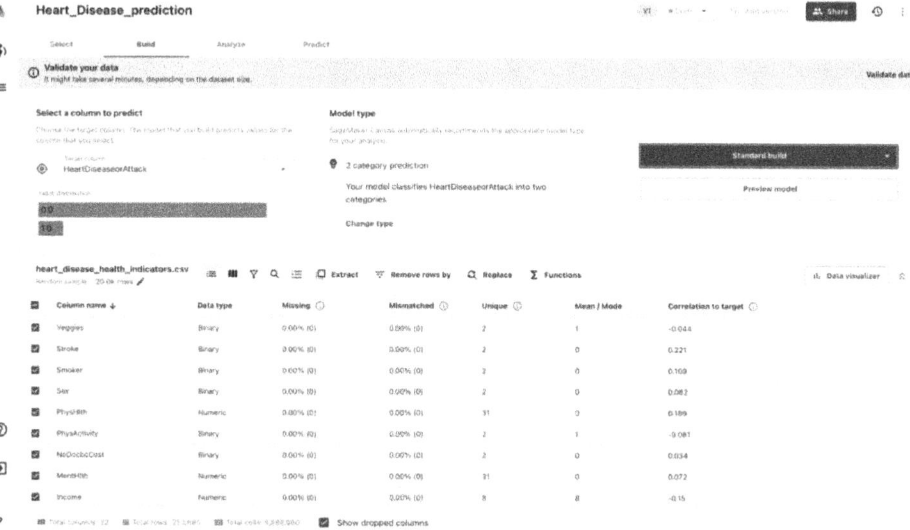

Figure 10.53 Standard structure of heart disease.

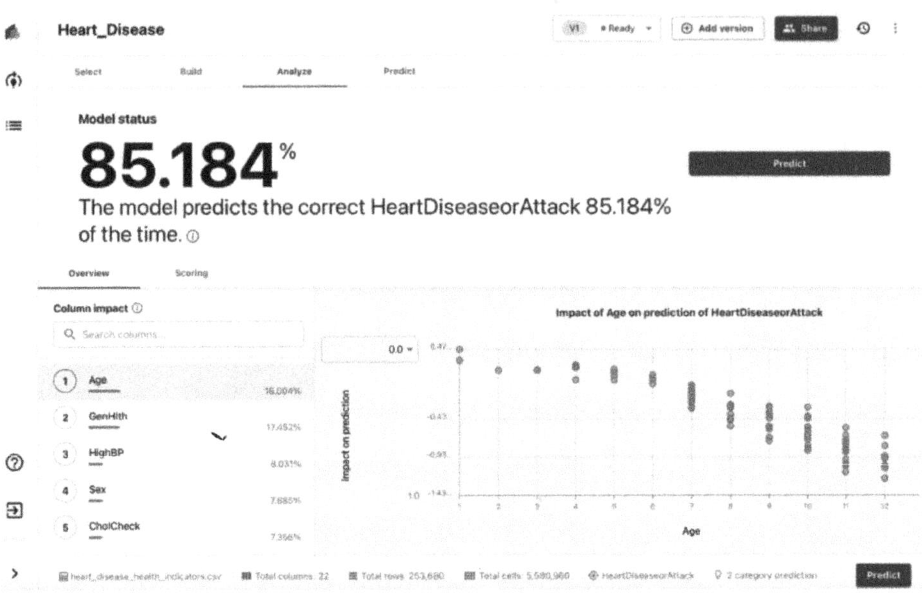

Figure 10.54 Heart disease analysis overview.

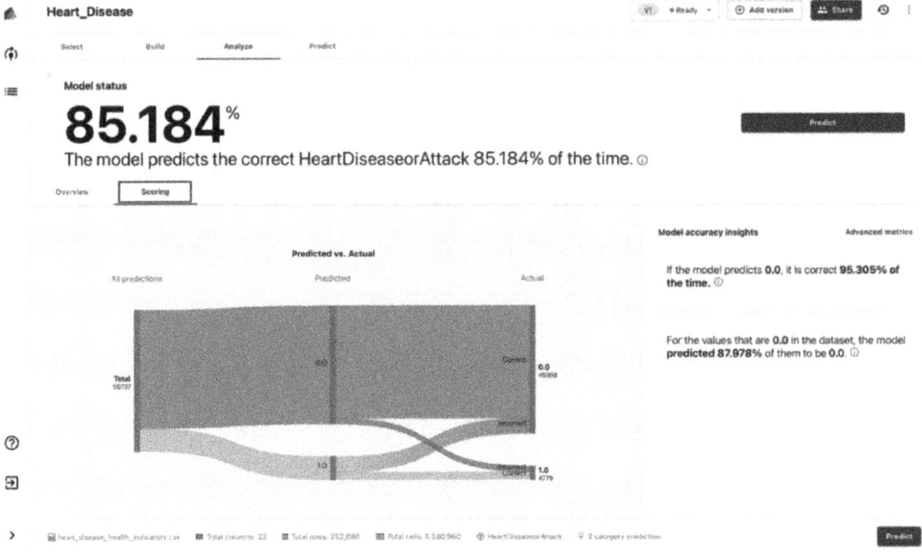

Figure 10.55 Heart disease analysis scoring.

If you want to examine the model built for heart disease in detail as shown in Figure 10.55, you can click on “Scoring” to compare the actual values with the predicted values.

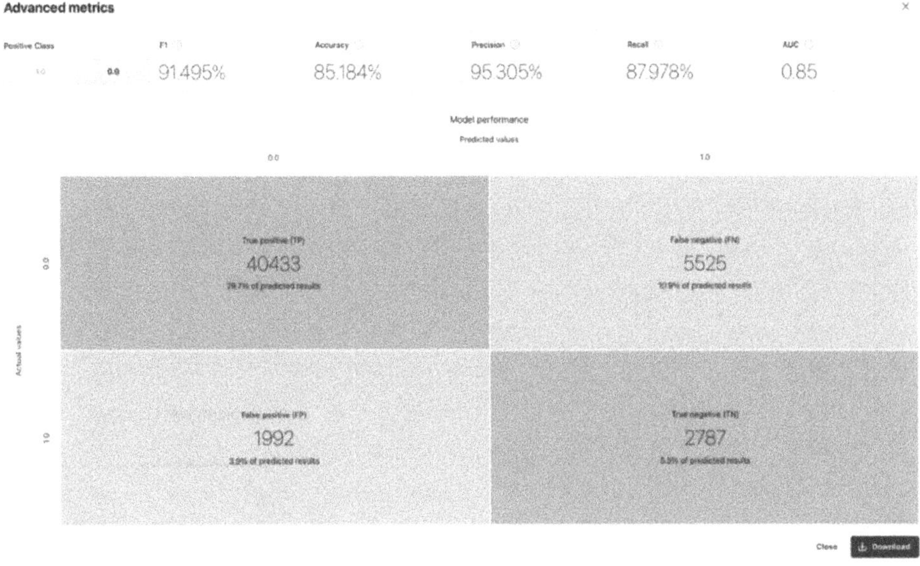

Figure 10.56 Heart disease analysis advanced metrics.

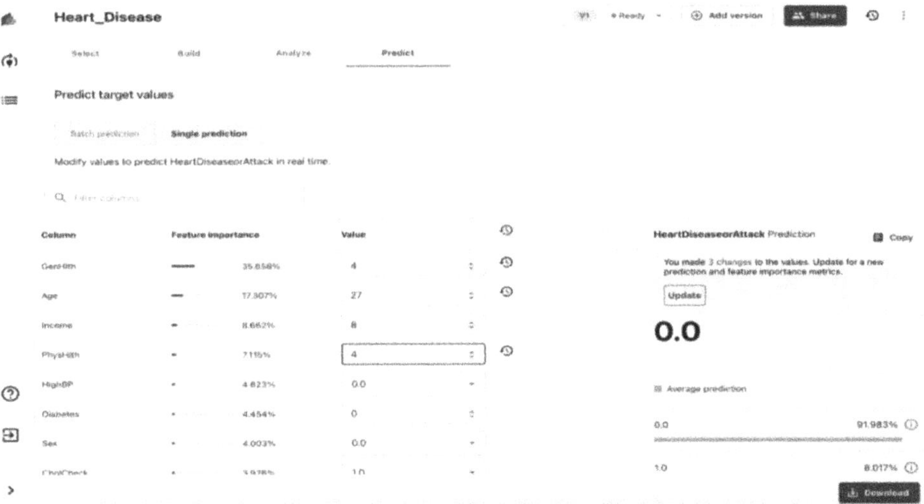

Figure 10.57 Heart disease single prediction.

If you want to check F1 Score, Accuracy, Precision, Recall, and AUC values, they can click on "Advanced metrics" in Figure 10.56 to view the values. Additionally, you can also determine the values of TP, FN, FP, and TN individually.

After finishing the analysis on the results, try to make predictions once the model is created. As shown in Figure 10.57, select "Single prediction" and make

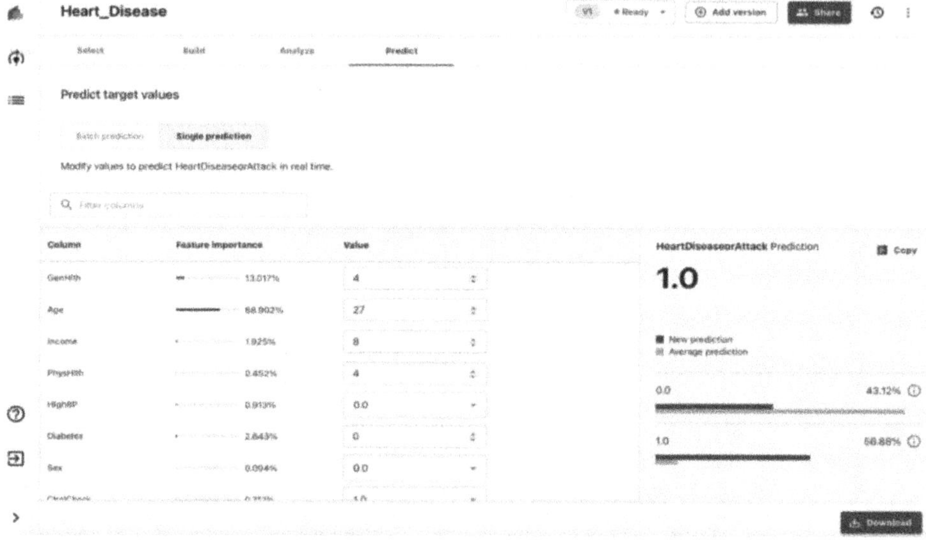

Figure 10.58 Heart disease single prediction result.

predictions. Enter the value for each column, then click on the "Update" button on the right to make the prediction.

If you clicked the "Update" button shown in Figure 10.57, the screen shown in Figure 10.58 will appear. The user's Single Prediction result is displayed in the bottom right section. The current prediction is made in a binary classification form for heart disease, using 0 and 1.

10.4 Practice Questions

Q1. How does Amazon SageMaker Canvas enable users to predict the survival probability of Titanic passengers without writing code?

Q2. What steps are involved in creating a diabetes prediction model for the Pima Indian region using SageMaker Canvas?

Q3. Describe the process of building a cardiovascular disease prediction model in SageMaker Canvas.

Chapter 11

SageMaker Studio

Amazon SageMaker Studio is a web-based integrated development environment that allows you to manage the entire lifecycle of machine learning models, from data collection to model building, training, and deployment, all in one place. Specifically, through the JupyterLab-based interface, you can easily access the necessary tools and libraries, allowing you to efficiently perform tasks such as uploading data, training, and tuning models, and adjusting experiments. Additionally, through the code sharing feature, you can collaborate smoothly with team members. SageMaker Studio also supports integration with various machine learning services provided by AWS, allowing users to effectively proceed with machine learning projects by utilizing AWS services in an integrated manner.

11.1 SageMaker Studio Notebook

To start using the Notebook in SageMaker Studio, go to the main home of SageMaker, as shown in Figure 11.1. SageMaker Studio can be launched by clicking on the “Studio” tab on the left side of the AWS SageMaker Console main screen.

If you have clicked on the “Studio” tab in Figure 11.1, the screen shown in Figure 11.2 will appear. Log in and click the “Open Studio” button on the right side of the screen to launch Studio, provided that a domain has been created.

If you click on “Studio” in Figure 11.2, the screen shown in Figure 11.3 will appear. It takes about 1–2 minutes to start Studio. Create a Notebook on the SageMaker Studio screen and run Studio.

Click on “New” in the “File” tab, as shown in Figure 11.4. Then, click on “Notebook” to create a notebook file.

When creating a Notebook in SageMaker Studio, you need to configure the notebook environment as shown in Figure 11.5. In this lab, set Image as Data Science, Kernel as Python 3, and Instance type as ml.t3.medium. Please note that

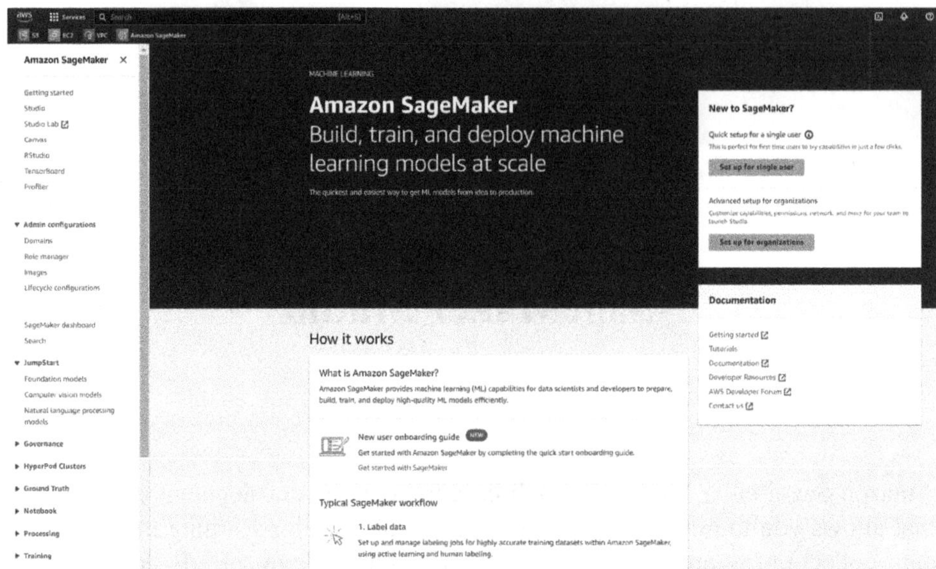

Figure 11.1 SageMaker main screen.

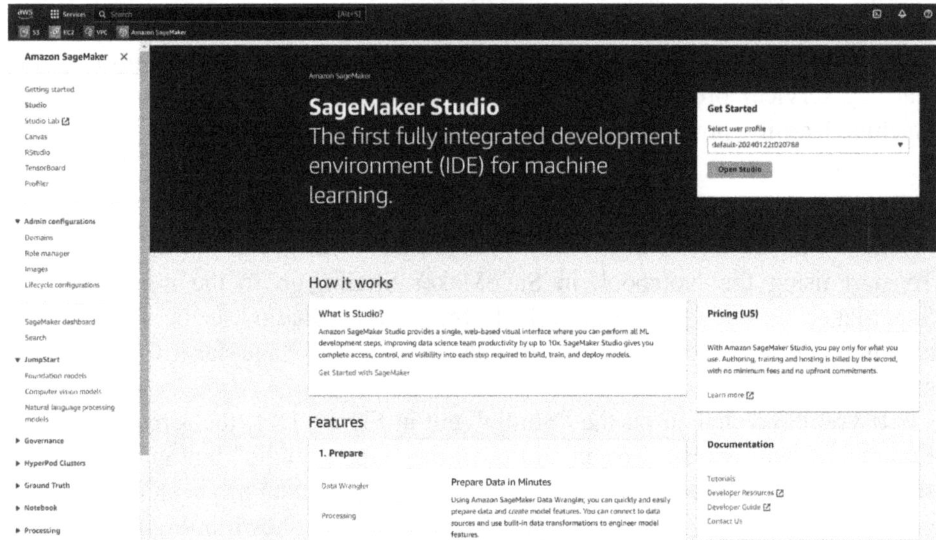

Figure 11.2 SageMaker Studio screen 1.

when configuring the Instance type, if the user does not select an appropriate instance type, excessive costs may be incurred.

SageMaker Studio Notebook provides an environment that allows collaboration without the need to pre-configure instance types and file storage capacities.

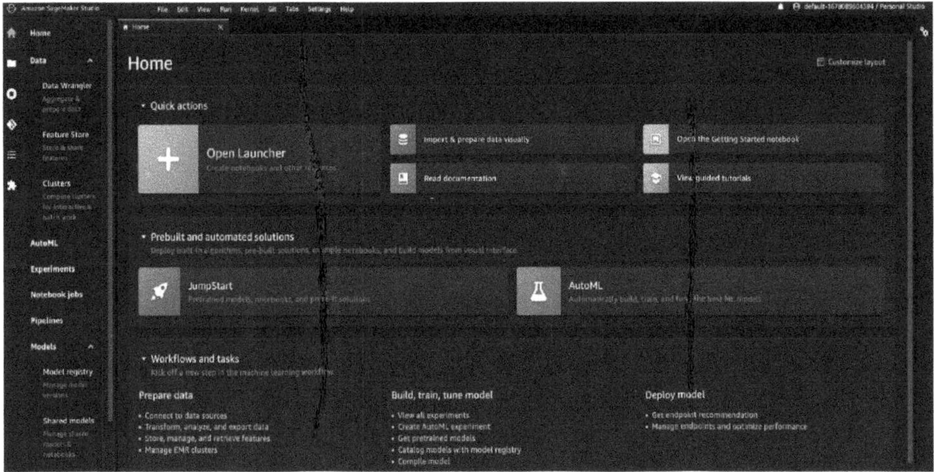

Figure 11.3 SageMaker Studio screen 2.

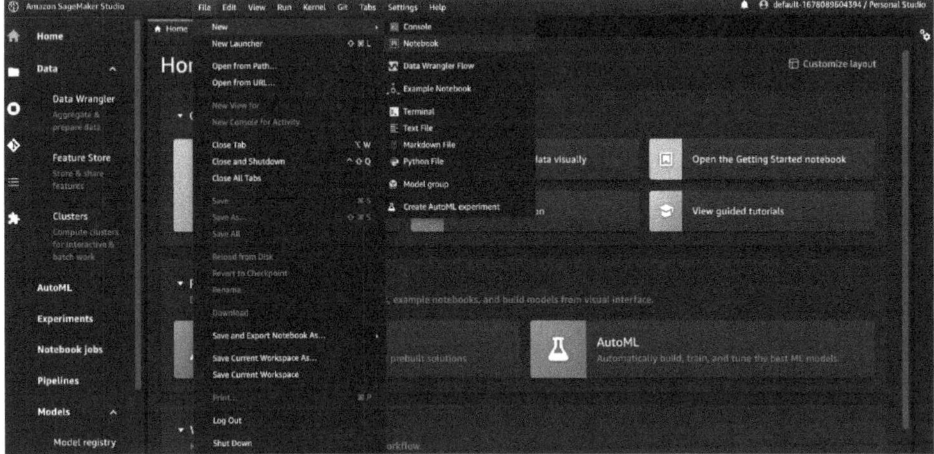

Figure 11.4 Creating a SageMaker Studio Notebook 1.

By default, it provides persistent storage space, so even if the instance running the notebook is terminated, the notebook can still be viewed and shared. Additionally, SageMaker Studio Notebook allows for easy replication of results while generating models and exploring data, enabling collaborative work with others. Access to read-only copies of the notebook is provided through a secure URL. Dependent items for the notebook are included in the notebook's metadata.

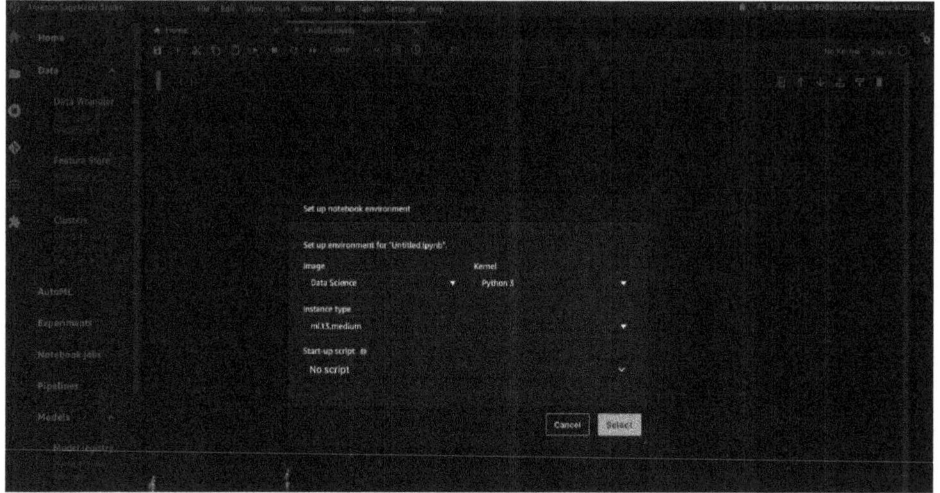

Figure 11.5 Creating a SageMaker Studio Notebook 2.

11.2 Image Classification Practical with Convolutional Neural Networks

In this example, let's try image classification using a convolutional neural network, one of the deep learning techniques, using the Fashion MNIST dataset provided by Kaggle. The dataset consists of a collection of small images such as sneakers, shirts, and sandals, as shown in Figure 11.6. It is composed of 70,000 images with a size of 28 × 28 pixels, and can be classified into 10 categories, just like the basic MNIST dataset.

As you can see in Figure 11.6, it is a code that loads the Fashion MNIST dataset built into TensorFlow. First, import the TensorFlow library, and then import the Keras library within TensorFlow. Once you have imported the Fashion MNIST dataset, proceed to divide the Fashion MNIST data into training data and test data. When you call the `load_data()` function, four NumPy arrays will be returned (Figure 11.7).

- The train_input and train_target arrays are used for model training.
- The test_input and test_target arrays are used for model testing.

If you print out the shape of each array, the training data and the test data each have an array of 28 × 28 with 60,000 and 10,000 images, respectively. Therefore, we can understand that each data consists of an array of 60,000 or 10,000 integers ranging from 0 to 9.

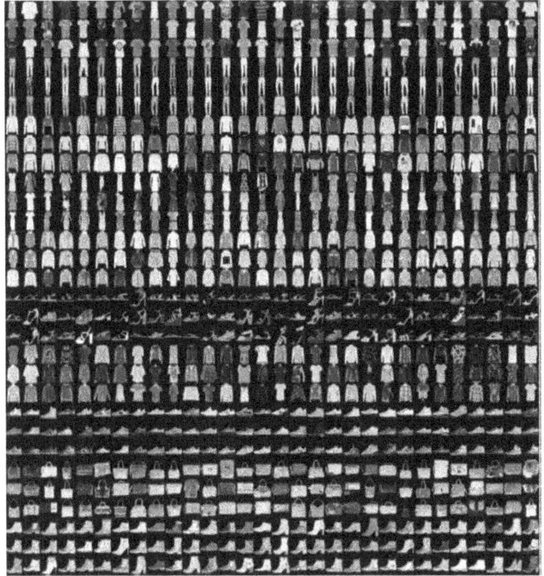

Figure 11.6 Fashion MNIST image dataset.

Figure 11.7 Upload fashion MNIST dataset.

```
from tensorflow import keras
import tensorflow as tf
# 1. Loading the Fashion MNIST dataset.
fashion_mnist = tf.keras.datasets.fashion_mnist
from sklearn.model_selection import train_test_split
```

```
(train_input, train_target), (test_input,
test_target) = fashion_mnist.load_data()
# Training dataset
train_input.shape
# Test Data Set
test_input.shape
len(test_target)
```

As shown in Figure 11.8, after importing the train_test_split function from sklearn's model_selection, the data is separated into train set and target set. The ratio of training data to test data is set to 8:2 using the test_size option with 0.2. Additionally, random_state=42, and seed is set to 42. The Fashion MNIST data is then converted to have values ranging from 0 to 1 for both the training and test data. The following libraries are used next.

```
# Visualization of training dataset
plt.figure()
plt.imshow(train_input[0])
plt.colorbar()
plt.grid(False)
plt.show()
```

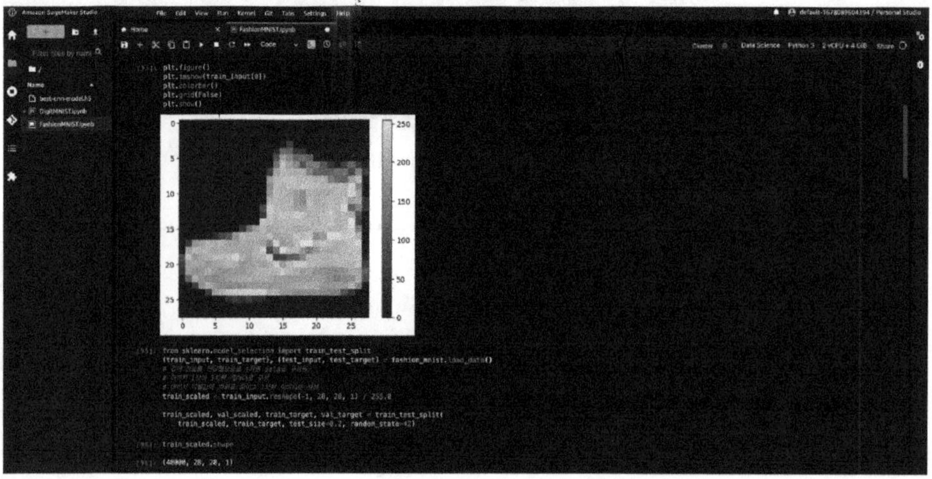

Figure 11.8 Fashion MNIST data transformation.

```
# Separation of Fashion MNIST dataset
from sklearn.model_selection import train_test_split
(train_input, train_target), (test_input, test_
target) = fashion_mnist.load_data()
```

It is composed of 4-dimensional data because it conveys in-depth information.

Reduce the range of pixel values in the image and generate a 3D image.

```
train_scaled = train_input.reshape(-1, 28, 28, 1) /
255.0
train_scaled, val_scaled, train_target, val_target =
train_test_split(train_scaled, train_target, test_
size=0.2, random_state=42)
#Check scaling dataset
train_scaled.shape
val_scaled.shape
```

As shown in Figure 11.9, the function for generating the model is as follows. First, the model is declared using the function `model.Sequential()`. Then, using the function `model.add()`, layers are created one by one. Looking at the parameters inside `model.add()` for `Conv2D(32, kernel_size=3, activation='relu', padding='same', input_shape=(28,28,1))`, it can be seen that 32 masks are being used. The kernel size is set to 3, and the activation function is set to "relu."

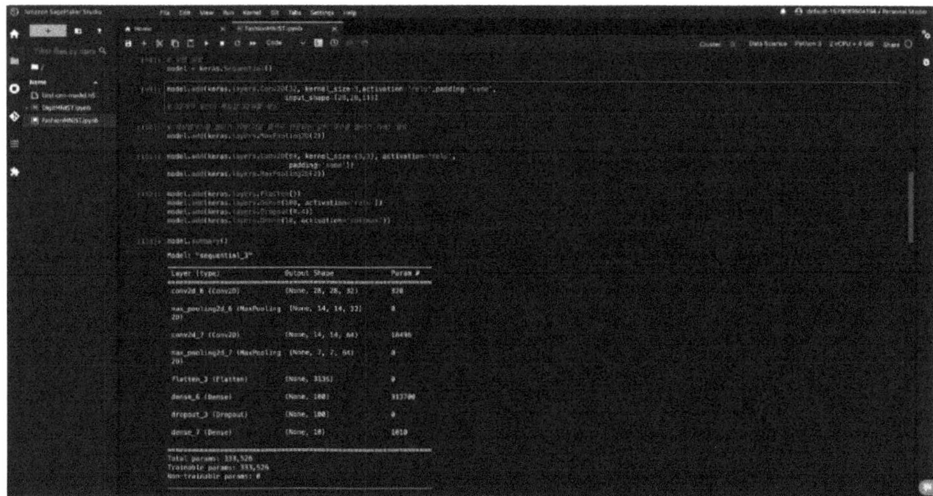

Figure 11.9 Creating the fashion MNIST model.

```
# Model creation

model = keras.Sequential()

model.add(keras.layers.Conv2D(32,
kernel_size=3,activation='relu',padding='same',
input_shape=(28,28,1)))

# 32 filters generate 32 feature maps.

# Pooling in order to reduce the size of the feature
map (to decrease the number of values passed to the
next layer)

model.add(keras.layers.MaxPooling2D(2))

model.add(keras.layers.Conv2D(64, kernel_size=(3,3),
activation='relu', padding='same'))

model.add(keras.layers.MaxPooling2D(2))
model.add(keras.layers.Flatten())
model.add(keras.layers.Dense(100, activation='relu'))

model.add(keras.layers.Dropout(0.4))

model.add(keras.layers.Dense(10,
  activation='softmax'))

# Model Summary

model.summary()
```

As you can see in Figure 11.10, the `compile()` method is used to set the optimizer, loss function, and metrics to be used for training the model.

- The loss function is used to measure the error of the model during the training process.

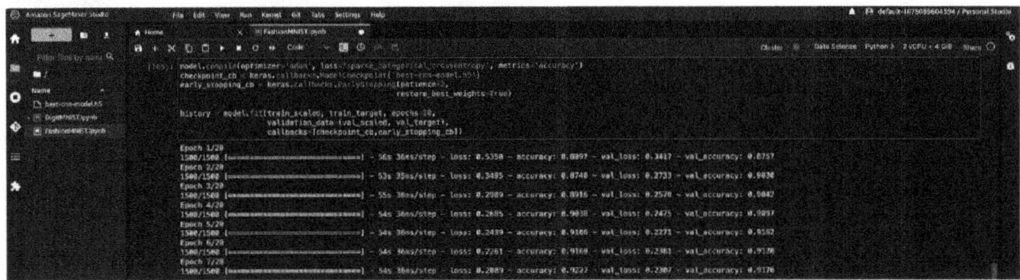

Figure 11.10 Fashion MNIST model compilation.

- Optimizer refers to the method of updating a model based on data and a loss function.
- The metric is used to evaluate the training and testing phases. Setting it to "accuracy" evaluates the model based on the proportion of images correctly classified.

```
# Model Compilation

model.compile(optimizer='adam',
loss='sparse_categorical_crossentropy',
  metrics='accuracy')

checkpoint_cb = keras.callbacks.
  ModelCheckpoint('best-cnn-model.h5')

early_stopping_cb = keras.callbacks.
  EarlyStopping(patience=2, restore_best_weights=True)

history = model.fit(train_scaled, train_target,
epochs=10,

validation_data=(val_scaled, val_target),

callbacks=[checkpoint_cb,early_stopping_cb])
```

Figure 11.11 is a visualization of the result of model compilation from Figure 11.10. To assess the performance of the Fashion MNIST model, first examine the training dataset and observe that the loss function value decreases as the model is repeated. In the case of the test dataset, it can be seen that the loss

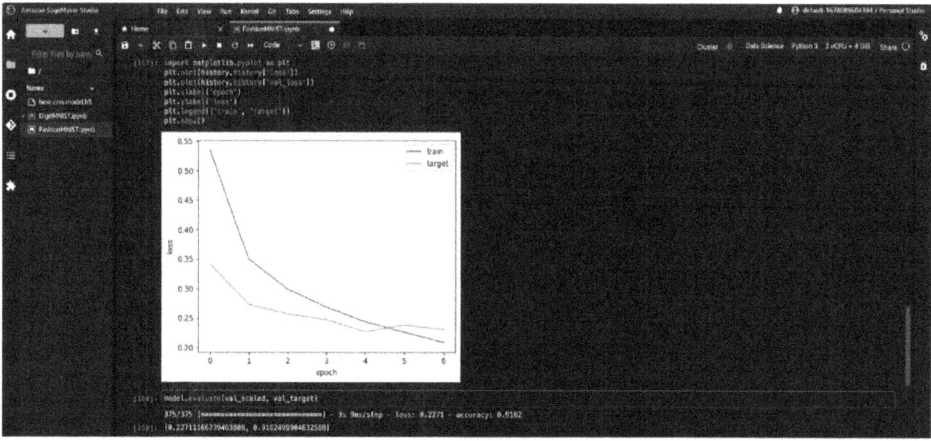

Figure 11.11 Fashion MNIST model performance.

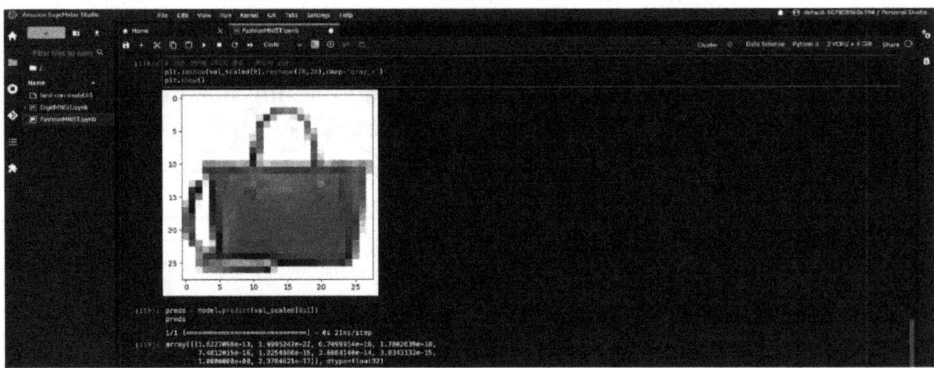

Figure 11.12 Output result of the first image in fashion MNIST.

function value slightly increases after four repetitions. Using the `evaluate()` method below, you can obtain the values of loss and accuracy.

```
# Loss Function Visualization
import matplotlib.pyplot as plt
plt.plot(history.history['loss'])
plt.plot(history.history['val_loss'])
plt.xlabel('epoch')
plt.ylabel('loss')
plt.legend(['train', 'target'])
plt.show()
```

Next, let's examine how the model made predictions by outputting the first verification image. As you can see in Figure 11.12, if we use the `predict()` method in the code `preds = model.predict(val_scaled[0:1])`, we can confirm the results of the model predicting the classes of each image. The results of `preds` are arrays representing the output values of the neural network for each `val_scaled`. If we print the first prediction, `val_scaled[0]`, it is an array with 10 values.

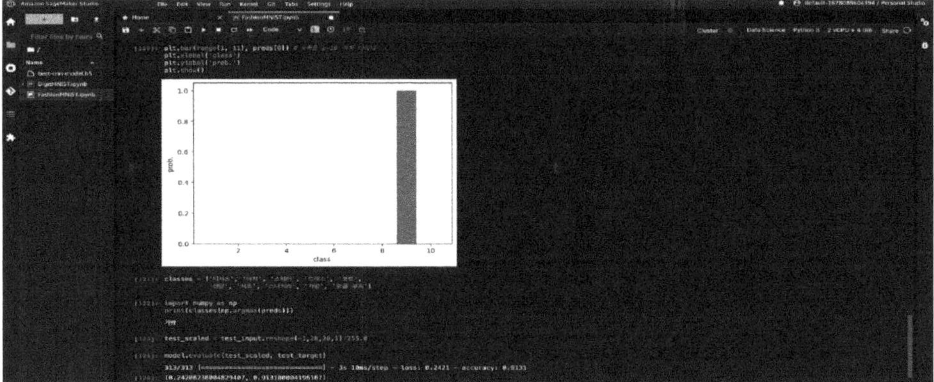

Figure 11.13 Fashion MNIST prediction results.

```
# Validation First Image Display - Similar to Handbag
plt.imshow(val_scaled[0].reshape(28,28),cmap='gray_r')
plt.show()
# Predictive Model Generation
preds = model.predict(val_scaled[0:1])
preds
```

As you can see in Figure 11.13, if you use the `np.argmax()` function to find the index with the highest value, it will output 9. In other words, the trained neural network predicted "bag."

```
plt.bar(range(1, 11), preds[0]) # The x-axis
  represents numbers from 1 to 10.
plt.xlabel('class')
plt.ylabel('prob.')
plt.show()
# Fashion Items
classes = ['t-shirt', 'pants', 'sweater', 'dress',
  'coat', 'sandal', 'shirt', 'sneakers', 'bag',
  'ankle boots']
```

```
import numpy as np

print(classes[np.argmax(preds)])
test_scaled = test_input.reshape(-1,28,28,1)/255.0

model.evaluate(test_scaled, test_target)
```

11.3 KNN-based Iris Species Prediction Algorithm

Let's try predicting the variety using the boto3 integration with S3 bucket, based on the KNN algorithm described in Chapter 2, using the functions provided in AWS SageMaker Studio (see Figures 11.14–11.20).

11.3.1 *Predicting iris varieties*

```
import urllib

import pandas as pd

import numpy as np

import seaborn as sns

import matplotlib.pyplot as plt

download_url = "https://archive.ics.uci.edu/ml/
  machine-learning-databases/iris/iris.data"

file_name = "iris.data"

urllib.request.urlretrieve (download_url, file_name)

# Reading data based on Pandas DataFrame

# 'sepal_length', 'sepal_width', 'petal_length',
  'petal_width', and 'species' are variables.

df=pd.read_csv('./{}'.format(file_name),
  names=['sepal_length','sepal_width',
  'petal_length','petal_width', 'species'])

df.head()
```

```
# Number of variables

df.count()
```

	sepal_length	sepal_width	petal_length	petal_width	species
0	5.1	3.5	1.4	0.2	Iris-setosa
1	4.9	3.0	1.4	0.2	Iris-setosa
2	4.7	3.2	1.3	0.2	Iris-setosa
3	4.6	3.1	1.5	0.2	Iris-setosa
4	5.0	3.6	1.4	0.2	Iris-setosa

Figure 11.14 Iris dataset.

```
sepal_length    150
sepal_width     150
petal_length    150
petal_width     150
species         150
dtype: int64
```

Figure 11.15 Number of iris data.

```
# Value graphs for each variable
plt.figure(1 , figsize = (15 , 6))
n = 0
for x in ['sepal_length' , 'sepal_width' , 'petal_
length', 'petal_width']:
 n += 1
 plt.subplot(1 , 4 , n)
plt.subplots_adjust(hspace =0.5 , wspace = 0.5)
 sns.distplot(df[x] , bins = 20)
plt.title('Distplot of {}'.format(x))
plt.show()
```

```
# Correlation relationship (heatmap): The length and
width of the petals have the highest level of
correlation.

plt.figure(figsize=(7,4))
sns.heatmap(df.corr(),annot=True)
plt.show()
```

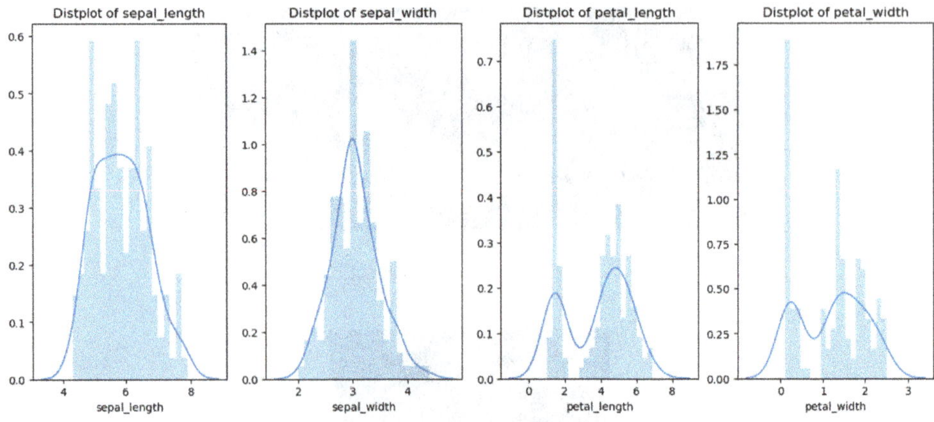

Figure 11.16 Visualization of iris data distribution.

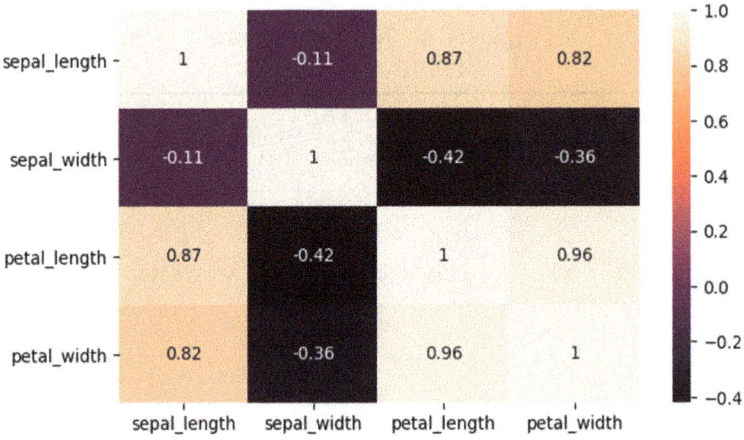

Figure 11.17 Iris heatmap.

```
# Alternative correlation view: Shows a strong
correlation between the length and width of petals.
g = sns.pairplot(df, hue='species', markers='+')
plt.show()
```

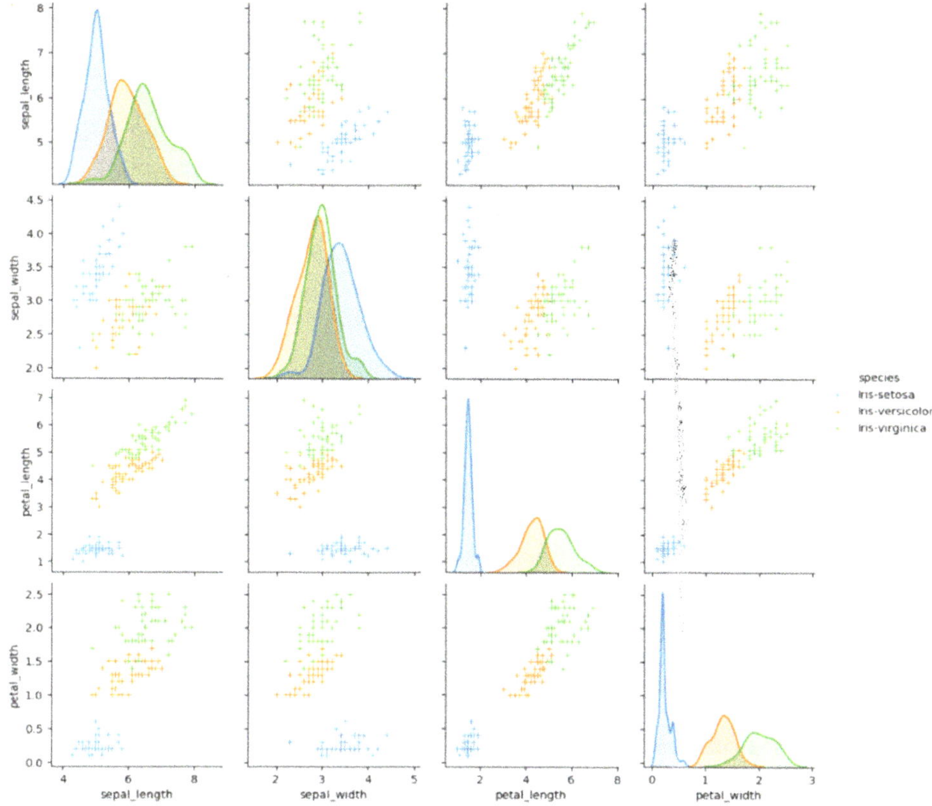

Figure 11.18 Visualization of substitute correlation in iris data.

```
# Create Train and Test data randomly in a ratio of
  7:3.
train_data, test_data = np.split(df.sample(frac=1,
  random_state= np.random.RandomState()), [int(0.7 *
  len(df))])
# Creating a KNN algorithm model
from sklearn.neighbors import KNeighborsClassifier
from sklearn.neighbors import KNeighborsClassifier
# knn classifier declaration: classification based
  on the most common classifications of the 3
  closest neighbors.
```

	sepal_length	sepal_width	petal_length	petal_width	species	prediction
0	5.4	3.0	4.5	1.5	Iris-versicolor	Iris-versicolor
1	6.9	3.1	5.1	2.3	Iris-virginica	Iris-virginica
2	4.4	2.9	1.4	0.2	Iris-setosa	Iris-setosa
3	7.7	2.6	6.9	2.3	Iris-virginica	Iris-virginica
4	4.6	3.1	1.5	0.2	Iris-setosa	Iris-setosa
5	5.2	2.7	3.9	1.4	Iris-versicolor	Iris-versicolor
6	5.0	3.5	1.6	0.6	Iris-setosa	Iris-setosa
7	6.7	3.1	5.6	2.4	Iris-virginica	Iris-virginica
8	7.9	3.8	6.4	2.0	Iris-virginica	Iris-virginica
9	4.8	3.0	1.4	0.3	Iris-setosa	Iris-setosa
10	5.4	3.4	1.7	0.2	Iris-setosa	Iris-setosa
11	5.4	3.4	1.5	0.4	Iris-setosa	Iris-setosa
12	5.2	4.1	1.5	0.1	Iris-setosa	Iris-setosa
13	6.0	2.7	5.1	1.6	Iris-versicolor	Iris-virginica
14	6.7	3.1	4.7	1.5	Iris-versicolor	Iris-versicolor
15	4.9	2.4	3.3	1.0	Iris-versicolor	Iris-versicolor
16	6.8	3.0	5.5	2.1	Iris-virginica	Iris-virginica
17	7.2	3.6	6.1	2.5	Iris-virginica	Iris-virginica
18	6.4	2.8	5.6	2.1	Iris-virginica	Iris-virginica
19	5.1	3.7	1.5	0.4	Iris-setosa	Iris-setosa

Figure 11.19 Iris variety prediction results.

```
knn = KNeighborsClassifier(n_neighbors=3)

# Creating KNN model

knn.fit(train_data[['sepal_length', 'sepal_width',
'petal_length', 'petal_width']],
train_data["species"])

# Use the Model and Test data to make predictions.

preds_array = knn.predict(test_data[['sepal_length',
'sepal_width', 'petal_length', 'petal_width']])

# Convert an array to a data frame with a single
column called pred.

preds_df = pd.DataFrame(preds_array,
columns=['prediction'])
```

Predicted species	Iris-setosa	Iris-versicolor	Iris-virginica
Actual species			
Iris-setosa	15	0	0
Iris-versicolor	0	14	1
Iris-virginica	0	0	15

Figure 11.20 Summary of iris cultivar data prediction results.

```
#Consecutively arrange and add the preceding data
frame to the Test data frame.
combined_df = test_data.reset_index(drop=True).
join(preds_df)
# Seems to have better predictions compared to known
classes.
combined_df.head(20)
```

```
# The prediction of the variety based on observed
species is generally accurate.

pd.crosstab(combined_df['species'], combined_
df['prediction'], rownames=['Actual species'],
colnames=['Predicted species'])
```

Upload the iris varietal data back into a CSV file inside a Pandas dataframe.

```
# Reloading a CSV file into a Pandas dataframe
df = pd.read_csv('./{}'.format(file_name),
names=['sepal_length', 'sepal_width', 'petal_length',
'petal_width', 'species'])
# Remap species values to integers
df['species'] = df['species'].replace({'Iris-setosa':
0, 'Iris-versicolor': 1, 'Iris-virginica': 2})
# Move species column to first position
```

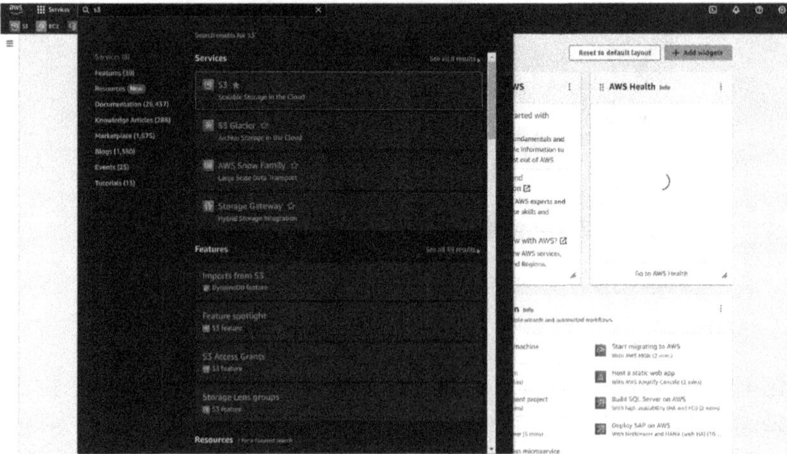

Figure 11.21 Amazon S3 bucket.

```
df = pd.concat([df['species'], df.drop(['species'],
axis=1)], axis=1)

# Create training and test DataFrames based on a
random 70/30 split

train_data, test_data = np.split(df.sample(frac=1,
random_state= np.random.RandomState()), [int(0.7 *
len(df))])
```

To import a dataset within AWS, you need to configure an S3 bucket. In order to use the iris-test.csv file and iris-train.csv data in the existing S3 bucket that was set up in Chapter 9, or a newly created bucket, you need to configure the settings. Return to the main AWS homepage and search for S3 in the search bar, then select it as shown in Figure 11.21.

Copy the existing bucket name and paste it into the bucket part in the code below, as shown in Figure 11.22. When the data is uploaded to the S3 bucket, proceed with the practice again in SageMaker Studio (see Figure 11.23).

```
import boto3

from datetime import datetime

import sagemaker

from sagemaker import get_execution_role
```

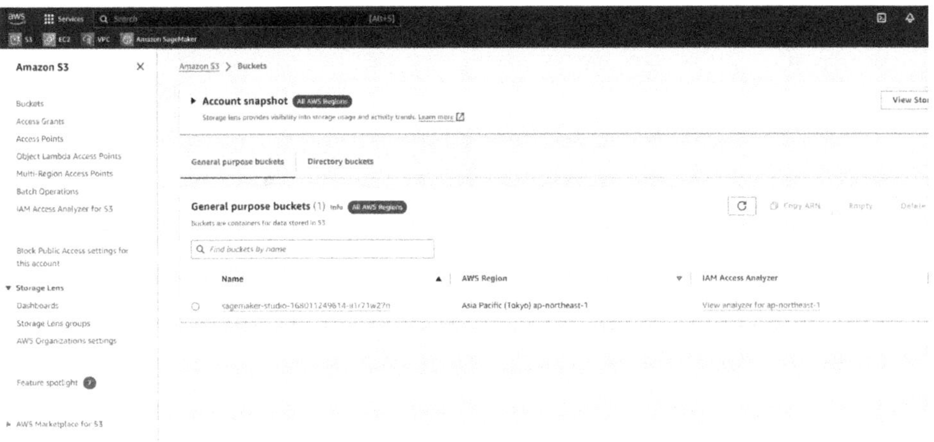

Figure 11.22 Amazon S3 bucket Information.

```
from sagemaker.predictor import csv_serializer,
json_deserializer

from sagemaker.amazon.amazon_estimator import
get_image_uri

# S3 Configuration

bucket = 'sagemaker-studio-23011249614-ii1r71w27n' #
Change to the bucket name you set

bucket = 'sagemaker-studio-23011249614-ii1r71w27n' #
change to the designated bucket name.

train_fname = 'iris-train.csv'

test_fname = 'iris-test.csv'

output_path = 's3://{}/output'.format(bucket)

# Save the training dataset to the notebook
instance.

train_data.to_csv(train_fname, index=False,
header=False)

# Store the test dataset in the notebook instance.

test_data.to_csv(test_fname, index=False,
header=False)
```

```
# Upload training dataset and test dataset to the s3
bucket.

boto3.Session().resource('s3').Bucket(bucket).
Object("{}/{}".format('train', train_fname)).
upload_file(train_fname)

boto3.Session().resource('s3').Bucket(bucket).
Object("{}/{}".format('test', test_fname)).
upload_file(test_fname)

# Setting Up the Training Environment

job_name = 'iris-job-{}'.format(datetime.now().
strftime("%Y%m%d%H%M%S"))

# Setting the KNN Estimator

knn = sagemaker.estimator.Estimator(get_image_
uri(boto3.Session().region_name, "knn"),

get_execution_role()

train_instance_count=1,

train_instance_type='ml.m4.xlarge',

output_path=output_path,
sagemaker_session=sagemaker.Session())

# Hyperparameter setting: Classification based on
the closest 3 neighbors typically.

knn.set_hyperparameters(predictor_type='classifier',

 feature_dim=4,

 k=3,

 sample_size=len(train_data))

# Setting data types and paths for training and
testing data

content_type = "text/csv"

train_input=sagemaker.session.s3_input(s3_data="s3://
  {}/{}/".format(bucket,'train'),content_type=
  content_type)
```

```
The method get_image_uri has been renamed in sagemaker>=2.
See: https://sagemaker.readthedocs.io/en/stable/v2.html for details.
train_instance_count has been renamed in sagemaker>=2.
See: https://sagemaker.readthedocs.io/en/stable/v2.html for details.
train_instance_type has been renamed in sagemaker>=2.
See: https://sagemaker.readthedocs.io/en/stable/v2.html for details.
The class sagemaker.session.s3_input has been renamed in sagemaker>=2.
See: https://sagemaker.readthedocs.io/en/stable/v2.html for details.
The class sagemaker.session.s3_input has been renamed in sagemaker>=2.
See: https://sagemaker.readthedocs.io/en/stable/v2.html for details.
INFO:sagemaker:Creating training-job with name: iris-job-20230307065926
2023-03-07 06:59:27 Starting - Starting the training job...
2023-03-07 06:59:55 Starting - Preparing the instances for training......
2023-03-07 07:00:47 Downloading - Downloading input data...
2023-03-07 07:01:12 Training - Downloading the training image..............
2023-03-07 07:03:48 Training - Training image download completed. Training in progress....
2023-03-07 07:04:24 Uploading - Uploading generated training modelDocker entrypoint called with argument(s): train
```

Figure 11.23 Setting up the training and testing dataset environment.

```
INFO:sagemaker:Creating model with name: knn-2023-03-07-07-05-10-478
INFO:sagemaker:Creating endpoint-config with name knn-2023-03-07-07-05-10-478
INFO:sagemaker:Creating endpoint with name knn-2023-03-07-07-05-10-478
----------!
```

Figure 11.24 Endpoint Model deployment completed.

```
test_input = sagemaker.session.s3_input(s3_
data="s3://{}/{}/".format(bucket, 'test'),
content_type=content_type)

#Train a model using only training data

knn.fit({'train': train_input}, job_name=job_name)

# Train the KNN model using training and validation
data.

knn.fit({'train': train_input, 'test': test_input},
job_name=job_name)
```

To deploy the model, the model is deployed to the endpoint within SageMaker (see Figure 11.24).

```
# Deploying the model to SageMaker Endpoint

knn_predictor = knn.deploy(initial_instance_
count=1,instance_type='ml.m4.xlarge')

knn_predictor.serializer = csv_serializer

knn_predictor.deserializer = json_deserializer
```

```
# Delete the column "품종" from the test data.

test_data_array = test_data.drop(['species'],
axis=1).values

# Predicting Test Data

preds = knn_predictor.predict(test_data_array)

# Predicting the value of a single iris species
dataset.

preds = knn_predictor.predict([4.8, 3.0, 1.4, 0.1])

# Convert JSON predictions to an array

preds_array=np.array([preds['predictions'][i]
['predicted_label']for i in range
(len(preds['predictions']))])

# Convert the data format of the 'Prediction' column
from an array to a data frame.

preds_df = pd.DataFrame(preds_array,
columns=['prediction']).

# Add the predicted dataframe to the test dataframe.

combined_df = test_data.reset_index(drop=True).
join(preds_df)

# Mapping back the values to their original values
by breed.

combined_df['species'] = combined_df['species'].
replace({0: 'Iris-setosa', 1: 'Iris-versicolor', 2:
'Iris-virginica'})

combined_df['prediction'] = combined_
df['prediction'].replace({0: 'Iris-setosa', 1: 'Iris-
versicolor', 2: 'Iris-virginica'})

# Check the predicted variety values properly
compared to the existing iris varieties.

combined_df.head(20)
```

To prevent possible charges that may occur when maintaining a session, we finalize by deleting the endpoint (see Figure 11.25).

	species	sepal_length	sepal_width	petal_length	petal_width	prediction
0	Iris-versicolor	6.9	3.1	4.9	1.5	Iris-versicolor
1	Iris-versicolor	6.2	2.2	4.5	1.5	Iris-versicolor
2	Iris-setosa	5.4	3.4	1.7	0.2	Iris-setosa
3	Iris-versicolor	6.0	2.7	5.1	1.6	Iris-virginica
4	Iris-setosa	4.7	3.2	1.6	0.2	Iris-setosa
5	Iris-versicolor	6.1	2.8	4.7	1.2	Iris-versicolor
6	Iris-versicolor	5.1	2.5	3.0	1.1	Iris-versicolor
7	Iris-versicolor	5.4	3.0	4.5	1.5	Iris-versicolor
8	Iris-virginica	7.7	3.8	6.7	2.2	Iris-virginica
9	Iris-setosa	5.7	4.4	1.5	0.4	Iris-setosa
10	Iris-setosa	5.0	3.3	1.4	0.2	Iris-setosa
11	Iris-versicolor	5.5	2.4	3.8	1.1	Iris-versicolor
12	Iris-setosa	5.5	3.5	1.3	0.2	Iris-setosa
13	Iris-versicolor	6.0	2.2	4.0	1.0	Iris-versicolor
14	Iris-versicolor	5.9	3.2	4.8	1.8	Iris-versicolor
15	Iris-setosa	5.1	3.8	1.9	0.4	Iris-setosa
16	Iris-virginica	6.0	3.0	4.8	1.8	Iris-virginica
17	Iris-versicolor	5.8	2.7	4.1	1.0	Iris-versicolor
18	Iris-virginica	6.3	2.5	5.0	1.9	Iris-virginica
19	Iris-virginica	6.3	3.4	5.6	2.4	Iris-virginica

Figure 11.25 Predicts iris cultivar prediction result.

```
# Delete Endpoint
sagemaker.Session().delete_endpoint(knn_predictor.
endpoint)
```

11.4 Practice Questions

Q1. Explain how Amazon SageMaker Studio enhances collaboration among machine learning project team members.

Q2. Describe the practical application of convolutional neural networks for image classification using the Fashion MNIST dataset in SageMaker Studio.

Q3. How does the KNN-based Iris Species Prediction Algorithm demonstrate the integration of AWS services within SageMaker Studio for machine learning projects?

Chapter 12

SageMaker Autopilot

Autopilot, introduced to SageMaker at the end of 2019, is a service that automates the development of machine learning models. It automatically processes the entire process from data preprocessing to model selection, training, and hyperparameter tuning. This service identifies the optimal model and suggests it to the user, allowing easy access through the SageMaker Studio interface. Autopilot creates and evaluates models using open-source frameworks such as TensorFlow and Scikit-learn. The operation of Autopilot is simple enough that users only need to upload the dataset to Amazon S3 and specify the columns to be trained. Additionally, Autopilot tasks can be started using either the GUI of SageMaker Studio or simple code using the SageMaker SDK. The strength of Autopilot lies in its convenience for users and the transparency it provides in the model building process. This allows users to understand how the model works and make improvements if necessary. In this chapter, we will explore the features of SageMaker Autopilot and its use in data analysis, feature engineering, and model tuning.

1. **Data analysis:** Amazon SageMaker Autopilot has become an essential tool in modern machine learning applications. This service supports key tasks such as linear regression, binary classification, and multi-class classification, and selects appropriate models based on the characteristics of the data. For example, if the target attribute only has two values, a binary classification model would be most suitable. In addition, SageMaker Autopilot has the capability to calculate statistics for various columns within a dataset. This feature includes key statistics such as the number of unique values, mean, and median, allowing machine learning experts to gain deep insights into the data. This automated data analysis is extremely helpful during the initial data understanding phase. After the data analysis, SageMaker Autopilot generates 10 pipeline candidates for effective model training. These candidates attempt to integrate various data preprocessing and feature engineering strategies to achieve optimal performance.

2. **Feature engineering:** Feature engineering is one of the key stages in the machine learning process, and it plays a role in transforming the input dataset into a suitable format for model training. SageMaker Autopilot offers 10 pipelines, which not only serve as a simple reference, but also include specific instructions for users to manually execute each stage, replicating the process performed by Autopilot. This approach provides high transparency and control levels, enabling a deep understanding of how the model is constructed. With this understanding, users can evaluate the performance and accuracy of the model, and perform additional optimizations or adjustments if necessary. This transparency is particularly useful when explaining or validating the model to other stakeholders.

Ultimately, SageMaker Autopilot provides users with a high level of flexibility and control, serving as an important tool for the success of machine learning projects.

3. **Model tuning:** This stage of Autopilot focuses primarily on training and tuning the model based on the defined pipeline in data analysis. SageMaker Autopilot performs automatic model tuning for each configured pipeline, using hyperparameter optimization techniques to build the most suitable model for the given dataset. This process takes place in a fully managed infrastructure. Once model tuning is completed, you can review detailed information and performance metrics of the trained model using Amazon SageMaker Studio. Additionally, you can evaluate the model's performance intuitively by utilizing various visualization tools. If a programming approach is required, you can automate these tasks using Amazon SageMaker Experiments SDK.

12.1 Utilizing SageMaker Autopilot

In this chapter, let's use Autopilot in SageMaker Studio to build a model without coding directly.

12.1.1 *Starting Autopilot*

Amazon SageMaker Autopilot is a powerful tool that automates the creation and evaluation of machine learning models. It explores the optimal model through various algorithms, feature engineering strategies, and hyperparameter combinations. Among the various models generated in this way, it selects the optimal model that meets the given performance metrics and constraints. In addition, through the console, users can monitor the progress of the job in real-time and access detailed information about the models being generated or evaluated.

Furthermore, users can customize the details of the generated models or explore the data characteristics of those models.

The ultimately selected optimal model can be deployed to SageMaker endpoints, enabling it to be used for inference tasks in real-world environments. The key advantage of Amazon SageMaker Autopilot is that it provides an efficient and intuitive process while reducing the complexity of the machine learning workflow. Through it, users can easily automate complex tasks related to building machine learning models. Next, we will explore how to use Autopilot by using a marketing dataset. This dataset represents whether customers will accept marketing proposals as a binary classification problem. The data includes approximately 41,000 labeled customer samples.

1. Open SageMaker Studio and create a new Python3 notebook using the Data Science kernel (Figure 12.1).
2. Download the dataset and decompress it as follows.

```
%%sh

apt-get -y install unzip

wget -N https://sagemaker-sample-data-us-west-2s3.-
us-west2.amazonaws.com/autopilot/direct_marketing/
bank-additional.zip

unzip -o bank-additional.zip
```

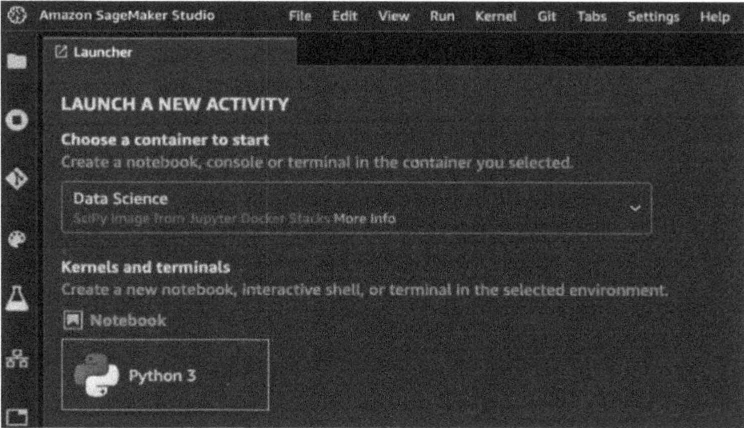

Figure 12.1 Creating a notebook.

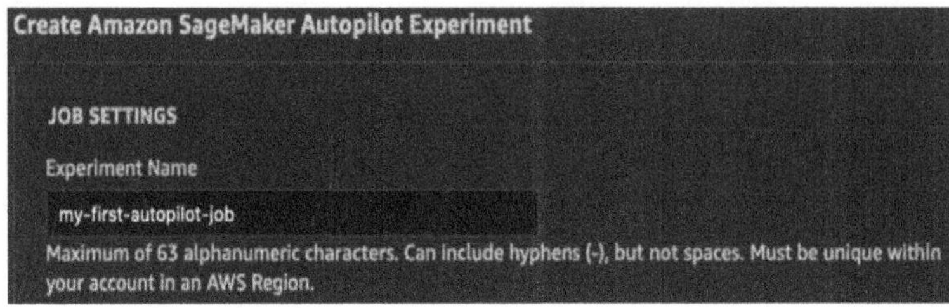

Figure 12.2 Creating an Autopilot experiment.

3. Executed the feature engineering script using Amazon SageMaker Processing. You can simply upload the dataset to S3. It will be stored in the default bucket created by SageMaker.

```
import sagemaker
prefix = 'sagemaker/DEMO-autopilot/input'
sess = sagemaker.Session()
uri = sess.upload_data(path="./bank-additional/bank-addition-full.csv", key_prefix=prefix
print(uri)
```

The dataset will be located on S3 at the following location.

```
s3://sagemaker-us-east-2-123456789012/sagemaker/DEMO-autopilot/input/bank-additional-full.csv
```

4. Now, click on the experiment icon in the vertical icon bar on the left. By doing so, the Experiments tab will open, and then click on the "Create experiment" button to create a new Autopilot task.
5. Configure the operation on the next screen. Specify the name as "my-first-autopilot-job" as shown in Figure 12.2.

   Set the location of the input dataset using the location returned by the path. You can explore S3 buckets or directly enter S3 locations. In the next step, set the S3 location of the dataset. As shown in Figure 12.3, you can search for S3 buckets or directly enter S3 locations.
6. The next step is to define the names of the target attribute. The name of the column that stores the yes or no labels, as shown in Figure 12.4, is called "y".

   Figure 12.5 Set the location where the result of the operation will be copied to.

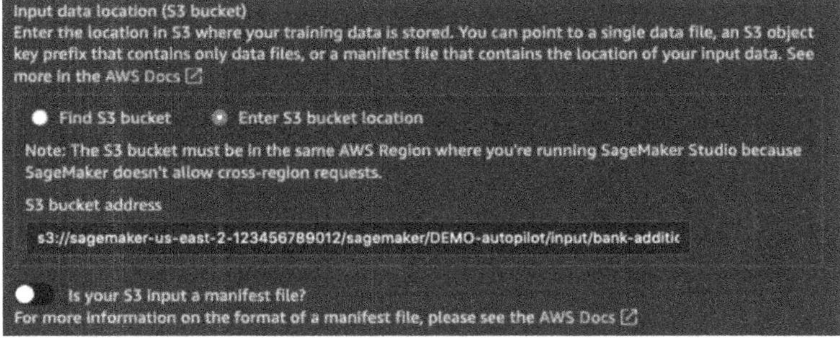

Figure 12.3 Defining the input location.

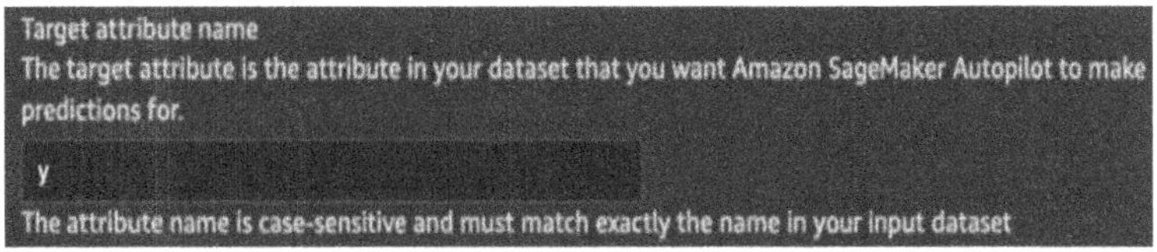

Figure 12.4 Defining the target attribute.

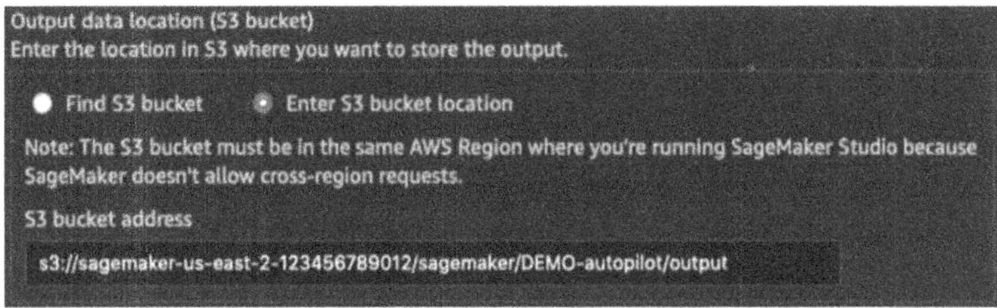

Figure 12.5 Defining the output location.

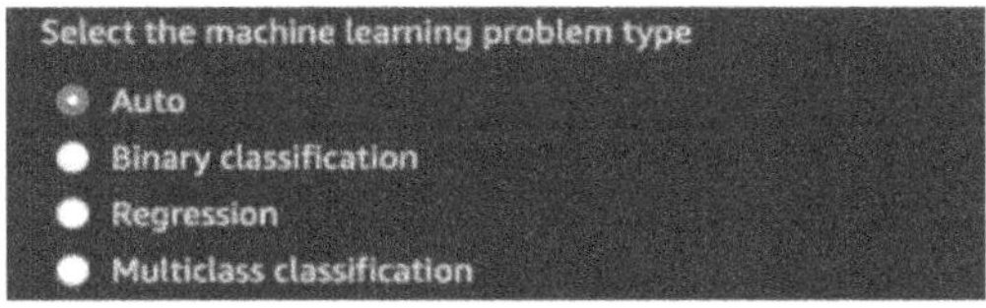

Figure 12.6 Setting the problem type.

7. In the next step, set the S3 location of the downloaded and extracted dataset. You can search for the S3 bucket as shown in Figure 12.6 or enter the S3 location directly.
8. Finally, decide whether to execute the entire operation or simply create a notebook. Choose to execute the entire operation as shown in Figure 12.7.

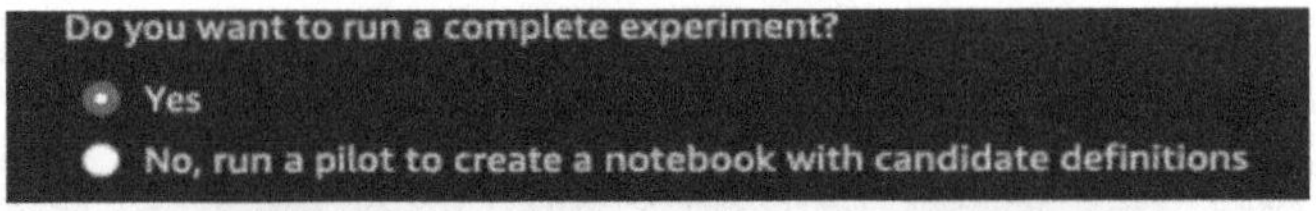

Figure 12.7 Running a complete experiment.

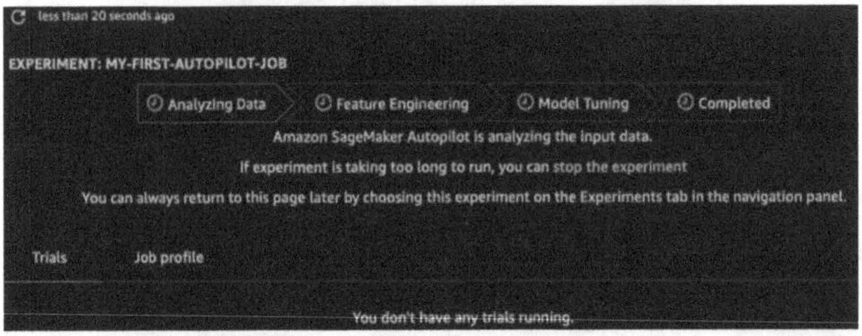

Figure 12.8 Viewing job progress and analyzing data.

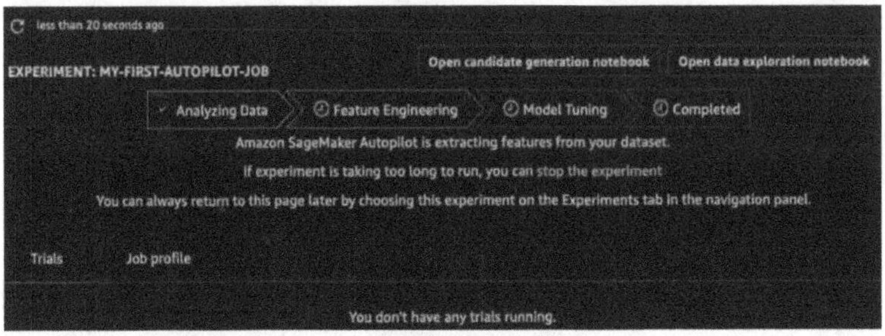

Figure 12.9 Viewing job progress feature engineering.

9. In the advanced settings section, you can change the IAM role, set encryption keys for operational assets, and define the VPC to launch the working instances. In this case, it is recommended to keep the default values.
10. Work setup is complete. If you check "Yes," the work will start.

12.1.2 *Monitoring task*

Monitoring the tasks on Amazon SageMaker Autopilot is an important step to ensure that the tasks are being executed smoothly and to identify any issues. When exploring the functionality, new experiments are listed under the Experiments tab, and you can click on the task description with the right mouse button to check its current status. As expected, the work starts with data analysis and proceeds as highlighted in Figure 12.8.

About 10 minutes later, the data analysis is completed, and the work moves to Feature Engineering (Figure 12.9).

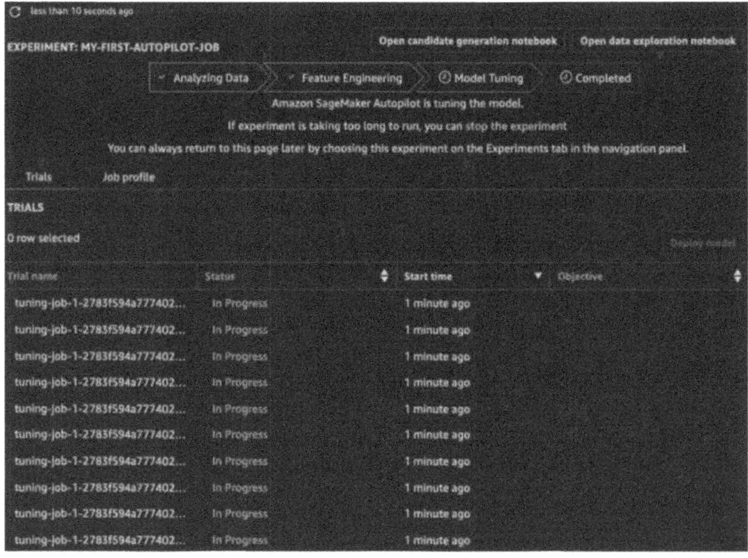

Figure 12.10 Viewing job progress model tuning.

Trial name	Status	Start time	Objective
★ *Best:* tuning-job-1-2783f594a...	Completed	7 minutes ago	0.9178199768066406
tuning-job-1-2783f594a777402...	Completed	3 minutes ago	0.9164850115776062
tuning-job-1-2783f594a777402...	Completed	7 minutes ago	0.9151489734649658
tuning-job-1-2783f594a777402...	Completed	3 minutes ago	0.9134498834609985

Figure 12.11 Viewing result.

Once Feature Engineering is complete, the process moves on to model tuning. A trial is a name given to a collection of related tasks (such as preprocessing, batch transformation, and training tasks) in Amazon SageMaker Experiments (Figure 12.10).

First, you can see ten tuning operations. These correspond to ten candidate pipelines. Once the operations are completed, you can see the objective, which is the metric the operations attempted to optimize. You can sort the operations based on the metric, and the best performing tuning operation will be highlighted with an asterisk (Figure 12.11).

Select the task and click the right mouse button to open the test details, where you can view additional information about other tasks belonging to the same test. Next, in Figure 12.12, you can see the actual algorithms and hyperparameter values used in XGBoost model training.

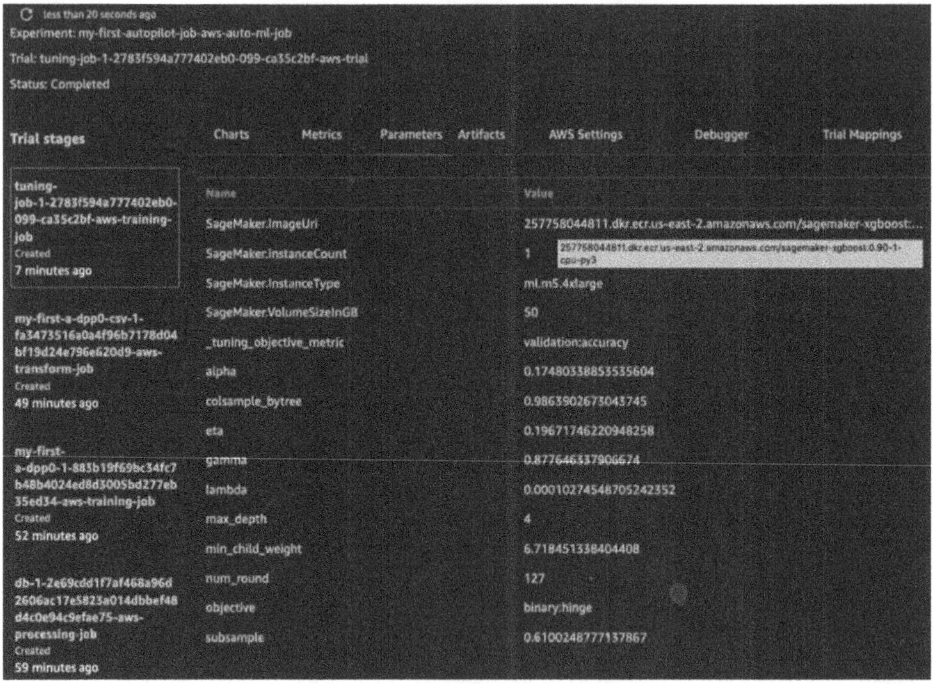

Figure 12.12 Examining a job.

The work is completed, and 500 tasks have been trained. This is the default value that can be changed in the SageMaker API, and at this point, the tasks can be deployed. Let's compare the top 10 tasks using the built-in visualization tools in SageMaker Studio.

12.1.3 *Comparing monitoring tasks*

Comparing monitoring tasks in Amazon SageMaker Autopilot can be helpful in evaluating the performance of different machine learning model configurations, and the method is as follows.

- **Metrics and logs:** By analyzing the metrics and logs generated during each operation, you can view detailed information about real-time indicators such as accuracy, precision, and recall, as well as the training process. By comparing the metrics and logs of various tasks, you can identify patterns and trends that help you understand the performance of the model.
- **Experiment:** SageMaker provides experiment functionality that allows users to organize, track, and compare multiple training jobs. Therefore, using SageMaker Experiments, you can group relevant training jobs into

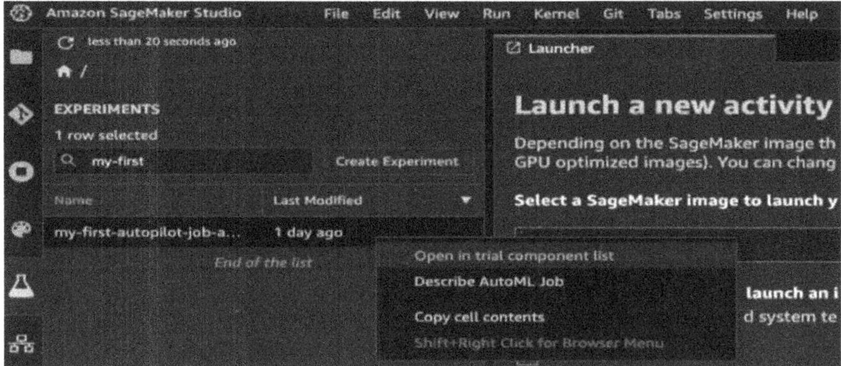

Figure 12.13 Comparing jobs.

experiments, track parameters and metrics, and compare the performance of different experiments. You can also tag experiments with metadata for later organization and search purposes.

- **Debugger:** By using SageMaker Debugger, you can capture real-time data in training tasks such as gradients, weights, and activations, and visualize the data to identify differences between different tasks. In addition, you can gain insights into the performance by comparing the internal operations of the model and identify areas that need improvement.
- **Hyperparameter tuning:** Users can specify the range of hyperparameters to explore using SageMaker's built-in hyperparameter tuning feature. Additionally, they can automatically run multiple training jobs using various combinations of hyperparameters. Then, users can compare the performance of the different jobs to identify the optimal hyperparameters for the model.

By comparing work using these tools, users can gain insights into the performance of machine learning models, optimize them, and improve accuracy and performance. As time goes by, thousands of tasks may arise, and you may want to compare the attributes of these tasks. The following is an example of a monitoring task.

1. Move to the Experiments tab on the left, find the task, and then right-click with the mouse to select "Open" from the list of experiment components (Figure 12.13).
2. Doing this will open the test component list, as shown in Figure 12.14.

   To open the Table Attributes panel on the right, click on the gear-shaped icon, then select the Objective Metric box in the Metric section. Finally, in the Type Filter section, select only Training jobs. Click on the arrow in the main panel to sort the tasks in descending order for the objective metrics.

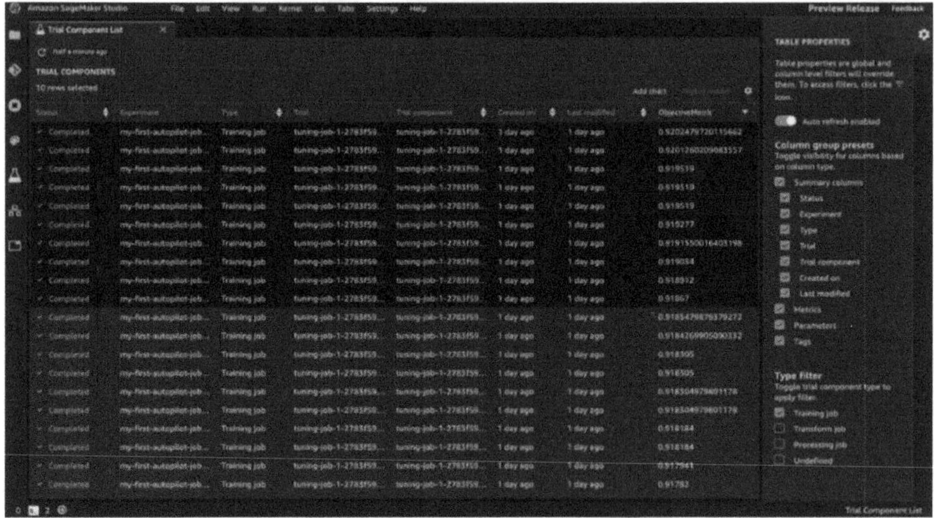

Figure 12.14 Comparing jobs 2.

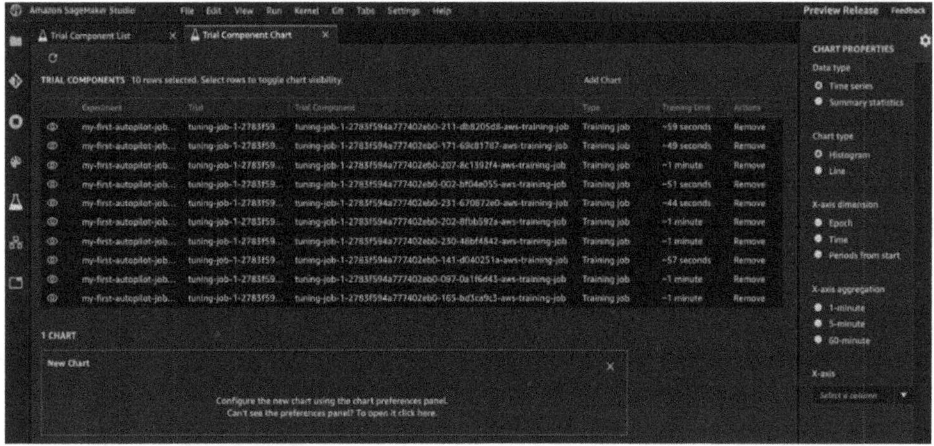

Figure 12.15 Comparing jobs 3.

After that, press Shift to select the top 10 tasks, and click the Add Chart button.

3. Doing this will display test component chart tab, as shown in Figure 12.15. Click on the chart box below to open the chart properties panel on the right side.

   Because the training process is very short, there is not enough data for time series charts. Therefore, we will choose summary statistics. We will create a scatter plot comparing training accuracy and validation accuracy, as shown in Figure 12.16.

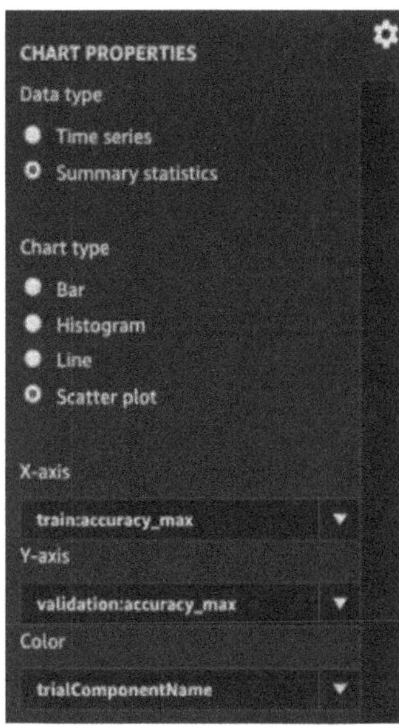

Figure 12.16 Creating a chart.

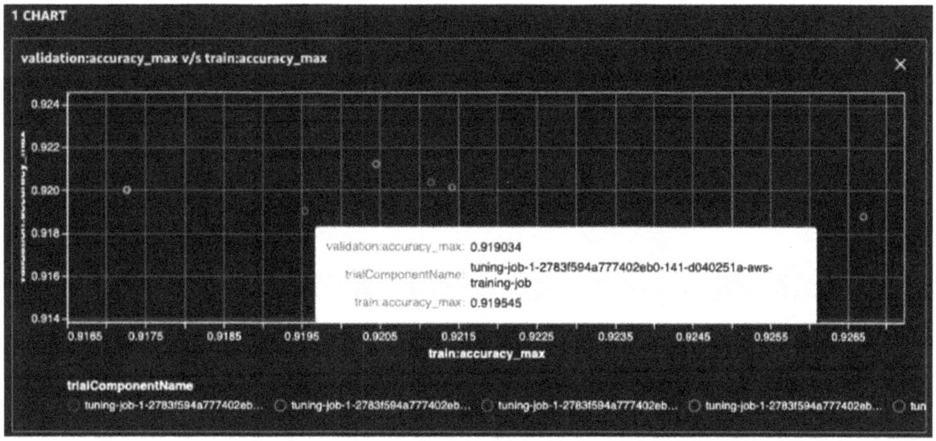

Figure 12.17 Plotting accuracies.

4. If you enlarge the following chart, you can visualize the tasks and their corresponding metrics.

   You can also create a chart showing the impact of some hyperparameters on accuracy to select the model and conduct additional tests. You may consider some models as candidates for ensemble predictions (Figure 12.17).

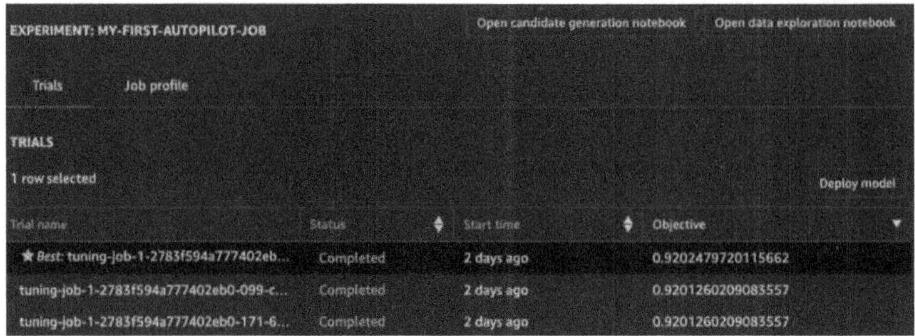

Figure 12.18 Deploying a model 1.

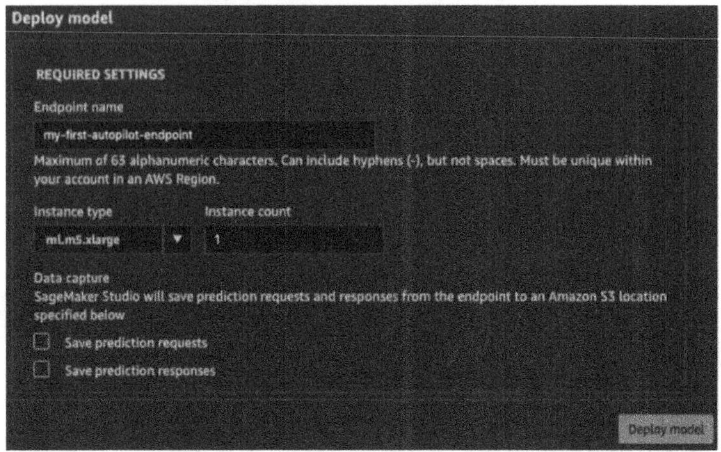

Figure 12.19 Deploying a model 2.

12.1.4 *Model deployment and invocation*

With its architecture and various built-in tools and features, SageMaker Autopilot allows for the construction, training, deployment, and management of high-quality machine learning models in the cloud through a simple process.

1. Go back to the Experiments tab and right-click on the experiment name, then select Describe AutoML run.

   This will open the training run list. Check if the list is sorted in descending order according to your goals, then choose the top-performing run. Finally, click on the Deploy model button (Figure 12.18).
2. In the screen displayed in Figure 12.19, only the endpoint name is specified and all other settings are kept as they are. The model is deployed to a real-time HTTPS endpoint that backs up ml.m5.xlarge instances.

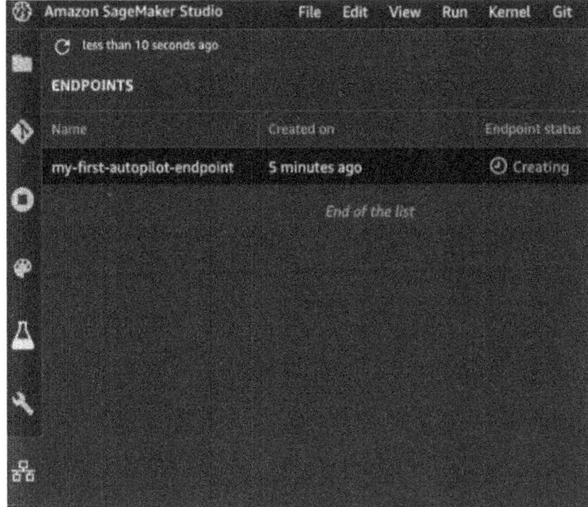

Figure 12.20 Creating an endpoint.

3. If you move to the endpoint section of the left vertical panel, you can see that endpoints are created. As shown in Figure 12.20, initially, it will be in the Creating state. After a few minutes, it will change to the InService state.
4. Go to the Jupyter notebook. Define the name of the endpoint and the sample to predict, and use the first line of the dataset.
5. Create a boto3 client for SageMaker runtime. This runtime includes an API called invoke_endpoint, which allows efficient embedding in client applications that only need to make model invocations. The code below is used to create the client.

```
ep_name = 'my-first-autopilot-endpoint'

sample='56,housemaid,married,basic.4y,no,no,no,teleph
one,may,mon,261,1,999,0,nonexistent,1.1,93.994,
  -36.4,3.857,5191.0'
```

6. Send the sample to the endpoint and deliver the input and output content types.

```
response=sm_rt.invoke_endpoint(EndpointName=ep_name,
  ContentType='text/csv', Accept='text/csv',
  Body=sample)
```

7. Decode and output the prediction results. It is anticipated that the customer will respond with "No," indicating that they will not accept the proposal.

```
response = response['Body'].read().decode("UTF-8")
print(response)
```

8. After the prediction is finished, the endpoint should be deleted to avoid unnecessary charges. This operation can be performed using the delete_endpoint API in boto3.

```
sm = boto3.Session().client('sagemaker')
sm.delete_endpoint(EndpointName=ep_name)
```

Utilizing Autopilot, we successfully built, trained, and deployed our first machine learning model on Amazon SageMaker. The code we just wrote downloads the dataset and makes predictions with the model. By using SageMaker Studio GUI, we can quickly experiment with new datasets and even less technically skilled users can create models directly.

12.2 SageMaker Autopilot SDK

SageMaker Autopilot SDK is a set of tools and libraries that allow you to interact with the SageMaker Autopilot service through code. By using the SDK, you can programmatically create and manage SageMaker Autopilot jobs, automating the process of building and training machine learning models. For developers or practitioners in the field of machine learning, using the SDK simplifies workflows and makes it easy to build machine learning models.

For more detailed information about the Amazon SageMaker SDK, you can visit https://sagemaker.readthedocs.io/en/stable/automl.html.

12.2.1 *Starting SageMaker SDK*

With the use of the SageMaker SDK, it is very easy to start Autopilot tasks. Simply upload the data to S3 and call a single API.

1. First, import the SageMaker SDK.

```
import sagemaker
sess = sagemaker.Session()
```

2. Download the dataset.

```
%%sh

apt-get -y install unzip

wget -N https://sagemaker-sample-data-us-west-2.s3-us-west2.amazonaws.com/autopilot/direct_marketing/bank-additional.zip

unzip -o bank-additional.zip
```

3. Upload the dataset to S3

```
bucket = sess.default_bucket()

prefix = 'sagemaker/DEMO-automl-dm'

s3_input_data=upload_data(path=“./bank-additional/bank-

additional-full.csv”, key_prefix=prefix+'input')
```

4. Configure the AutoML task. This task can be performed with a single line of code. Define the target attribute and the location to store the training results. Adjust the maximum execution time of the task, maximum execution time per task, or the number of candidate models to be adjusted. Please note that limiting the duration of the task too much may affect accuracy.

```
from sagemaker.automl.automl import AutoML

Auto_ml_job = AutoML(

role = sagemaker.get_execution_role(),

sagemaker_session = sess,

target_attribute_name = 'y',

output_path= 's3://{}/{}/output'.format(bucket,
  prefix), max_runtime_per_training_job_in_
  seconds=600,

max_candidates = 250,

total_job_runtime_in_seconds = 3600

)
```

5. Next, after delivering the location of the training set, start the Autopilot task and turn off the logs. Then, set the call to asynchronous mode to check the job status in the next cell.

```
auto_ml_job.fit(inputs=s3_input_data, logs=False,
wait=False)
```

12.2.2 *Job monitoring*

While the task is running, you can monitor the progress using the describe_auto_ml_job() API.

1. For example, the following code checks the job status every 30 seconds until the data analysis stage is completed.

```
from time import sleep

job = auto_ml_job.describe_auto_ml_job()

job_status = job['AutoMLJobStatus']

job_sec_status = job['AutoMLJobSecondaryStatus']

if job_status not in ('stopped','Failed'):

    while job_status in ('InProgress') and job_sec_
      status in ('AnalyzingData'):

        sleep(30)

        job = auto_ml_job.describe_auto_ml_job()

        job_status = job['AutoMLJobStatus']

    job_sec_status = job['AutoMLJobSecondaryStatus']

print(job_status, job_sec_status)
```

2. Once the data analysis is completed, you can use the two automatically generated notebooks. You can find the location using the same API.

```
The above code prints the S3 paths of two laptops.

job = auto_ml_job.describe_auto_ml_job()

job_condidate_notebook = job['AutoMLJobArtifacts']

['CandidateDefinitionNotebookLocation']
```

```
job_data_notebook = job['AutoMLJobArtifacts']
['DataExplorationNotebookLocation']
print(job_candidate_notebook)
print(job_data_notebook)
```

The above code prints the S3 paths of two laptops.

```
s3://sagemaker-us-east-2-123456789012/sagemaker/DEMO-
automl-dm/output/automl-2020-04-24-14-21-112-938/
sagemaker-automl-candidates/pr-1-a99cb67acb5945d695c0
e74afe8ffe3ddaebafa95f394655ac973432d1/notebooks/
SageMakerAutopilotCandidateDefinitionNotebook.ipynb

s3://sagemaker-us-east-2-123456789012/sagemaker/DEMO-
automl-dm/output/automl-2020-04-24-14-21-112-938/
sagemaker-automl-candidates/pr-1-a99cb67acb5945d695c0
e74afe8ffe3ddaebafa95f394655ac973432d1/notebooks/
SageMakerAutopilotDataExplorationNotebook.ipynb
```

3. You can use the AWS CLI to copy two notebooks to the local environment.

```
%%sh -s $job_candidate_notebook $job_data_notebook
aws s3 cp $1 .
aws s3 cp $2 .
```

4. During the execution of feature engineering, you can wait for the operation using the same code mentioned earlier. If the “AutoMLJobSecondaryStatus” field in the Job_info dictionary matches “FeatureEngineering,” you wait until the operation is complete.
5. Once feature engineering is complete, model tuning begins.

   During this process, you can track your work using Amazon SageMaker Experiments SDK. We will cover SageMaker Experiments in detail in a later chapter, but let’s examine the code in this chapter. It is sufficient to pass the experiment name to the Experiment Analytics object. We can retrieve information about all the tuning jobs performed so far from the pandas DataFrame. We can also conveniently display the number of jobs executed so far and the top 5 jobs.

Number of jobs: 109

	ObjectiveMetric - Max	TrialComponentName	DisplayName
35	0.918594	tuning-job-1-57d7f377bfe54b40b1-050-b8c34b30-a...	tuning-job-1-57d7f377bfe54b40b1-050-b8c34b30-a...
43	0.917700	tuning-job-1-57d7f377bfe54b40b1-045-20ddd705-a...	tuning-job-1-57d7f377bfe54b40b1-045-20ddd705-a...
21	0.917316	tuning-job-1-57d7f377bfe54b40b1-065-2d7d46ad-a...	tuning-job-1-57d7f377bfe54b40b1-065-2d7d46ad-a...
17	0.916933	tuning-job-1-57d7f377bfe54b40b1-071-a0ce585e-a...	tuning-job-1-57d7f377bfe54b40b1-071-a0ce585e-a...
49	0.915911	tuning-job-1-57d7f377bfe54b40b1-039-024a3819-a...	tuning-job-1-57d7f377bfe54b40b1-039-024a3819-a...

Figure 12.21 Viewing a job.

```
import pandas as pd
from sagemaker.analytics import ExperimentAnalytics
exp = ExperimentAnalytics(
 Sagemaker_session=sess,
 Experiment_name=job['AutoMLJobName'] +
'-aws-auto-ml-job')
df = exp.dataframe()
print("Number of jobs: ", len(df))
df = pd.concat([df['ObjectiveMetric - Max'],
df.drop(['ObjectiveMetric - Max'], axis=1)], axis=1)
df.sort_values('ObjectiveMetric - Max', ascending=0)
[:5]
```

The above code execution results in a table being displayed, as shown in Figure 12.21.

6. Once the model tuning is complete, it is possible to find the best model candidate.

```
job_best_candidate = auto_ml_job.best_candidate()
print(job_best_candidate['CandidateName'])
print(job_best_candidate['FinalAutoMLJobObjectiveMet
ric'])
```

7. This outputs the name of the best tuning operation and the verification accuracy of that operation. You can use SageMaker SDK to deploy and test the model.

```
Tuning-job-1-57d7f377bfe54b40b1-030-c4f27053

{'MetricName' : 'validation: accuracy', 'Value':
0.9197599935531616}
```

12.2.3 *Summary*

SageMaker Autopilot automatically generates many sub-artifacts such as dataset splits, preprocessing scripts, preprocessed datasets, and models. To organize them, you can use code snippets, and the AWS CLI can also be used. Additionally, both SageMaker Studio GUI and SageMaker SDK can be used to facilitate model training.

12.3 Deep on SageMaker Autopilot

In this chapter, let's learn in detail about how SageMaker Autopilot processes data and trains models. First, let's take a look at the results produced by SageMaker Autopilot.

12.3.1 *Work result/outcome*

When you view S3 buckets, you can see multiple buckets that were created in the past.

```
$ aws s3 ls s3://sagemaker-us-east-2-123456789012/
sagemaker/DEMO-autopilot/output/

my-first-autopilot-job/
```

If you look at the code below, you can see many new prefixes. Let's try to identify what each of these prefixes is.

```
PRE data-processor-models/

PRE preprocessed-data/
```

```
PRE sagemaker-automl-candidates/
PRE transformed-data/
PRE tuning/
```

The prefix of preprocessed-data/tuning_data includes training and validation splits generated from the input dataset, divided into smaller CSV files.

- The prefix "preprocessed-data/tuning_data" contains the training and validation splits generated from the input dataset, with each split being divided into small CSV chunks.
- The sagemaker-automl-candidates prefix includes a total of 10 data preprocessing scripts (dpp [0-9] .py), one for each pipeline, along with the code for training on the input dataset using the respective script (trainer.py) and the code for handling each model (sagemaker_serve.py).
- The data-processor-models prefix includes 10 data processing models trained by the dpp script.
- The transformed-data prefix includes 10 processed versions of the training and validation splits.
- The prefix sagemaker-automl-candidates contains two automatically generated notebooks.
- Lastly, the tuning prefix includes the actual model trained during the model tuning phase. Figure 12.22 summarizes the relationships for the output.

12.3.2 *Data exploration and analysis*

Once the data analysis phase is complete, it can be accessed from Amazon S3. The first section, as seen in Figure 12.23, displays a sample of the dataset.

As shown in Figure 12.24, in the second section, the focus is on column analysis. This includes the percentage of missing values, the number of unique values, and descriptive statistics, among others. For example, the pdays field shows both the maximum and median values as 999. This may seem like an error, but as explained in the previous chapter, 999 is a placeholder value that signifies the customer has not been contacted previously.

Once the data analysis phase is complete, it can be accessed from Amazon S3. The first section simply displays a sample of the data set. The second section, which focuses on column analysis, covers the proportion of missing values, the number of unique values, and descriptive statistics. For example, in the "pdays" field, both the maximum and median values are shown as 999.

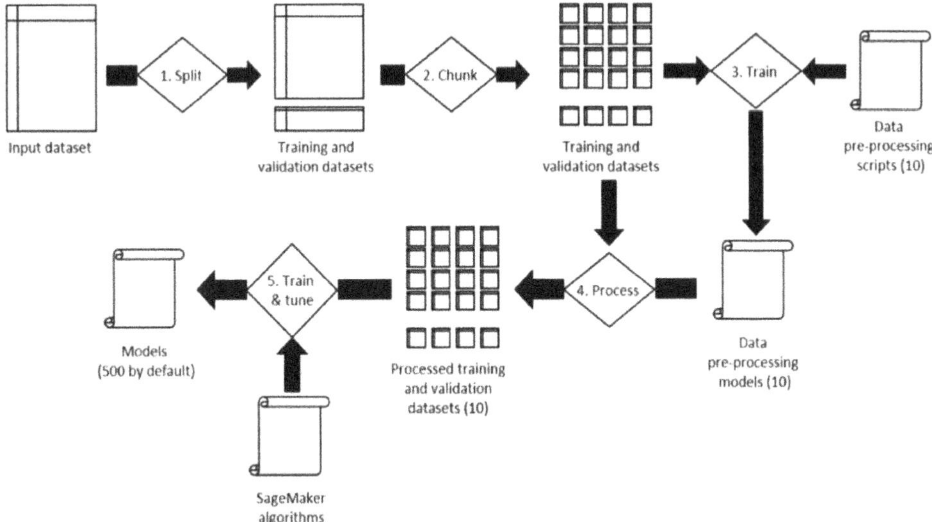

Figure 12.22 Summing up the Autopilot process.

Dataset Sample

The following table is a random sample of 10 rows from the training dataset. For ease of presentation, we are only showing **20 of the 21** columns of the dataset.

Suggested Action Items

- Verify the input headers correctly align with the columns of the dataset sample. If they are incorrect, update the header names of your input dataset in Amazon Simple Storage Service (Amazon S3).

	age	job	marital	education	default	housing	loan	contact	month	day_of_week	...	campaign	pdays	previous	poutcome
0	45	blue-collar	married	high.school	no	no	no	telephone	jun	wed	...	4	999	0	nonexistent
1	46	management	married	university.degree	no	yes	no	telephone	may	fri	...	5	999	0	nonexistent
2	39	admin.	married	university.degree	no	no	no	cellular	may	tue	...	7	999	0	nonexistent
3	25	student	single	high.school	no	yes	no	cellular	apr	wed	...	1	999	0	nonexistent
4	38	services	divorced	high.school	no	no	no	cellular	jul	thu	...	1	999	0	nonexistent
5	33	technician	married	university.degree	no	yes	no	telephone	may	thu	...	3	999	1	failure
6	47	management	married	high.school	no	no	yes	cellular	apr	fri	...	1	999	1	failure
7	48	technician	married	high.school	unknown	yes	no	telephone	may	fri	...	3	999	0	nonexistent
8	33	services	married	university.degree	no	unknown	unknown	telephone	may	thu	...	1	999	0	nonexistent
9	28	blue-collar	single	basic.9y	no	no	no	telephone	jun	tue	...	1	999	0	nonexistent

Figure 12.23 Viewing dataset statistics.

12.3.3 *Generate candidate pipeline*

This notebook includes definitions and training methods for 10 candidate pipelines. This notebook is immediately executable, and skilled experts can use it to rerun the process of machine learning and continuously improve experiments. Let's try manually running one of the pipelines as an example.

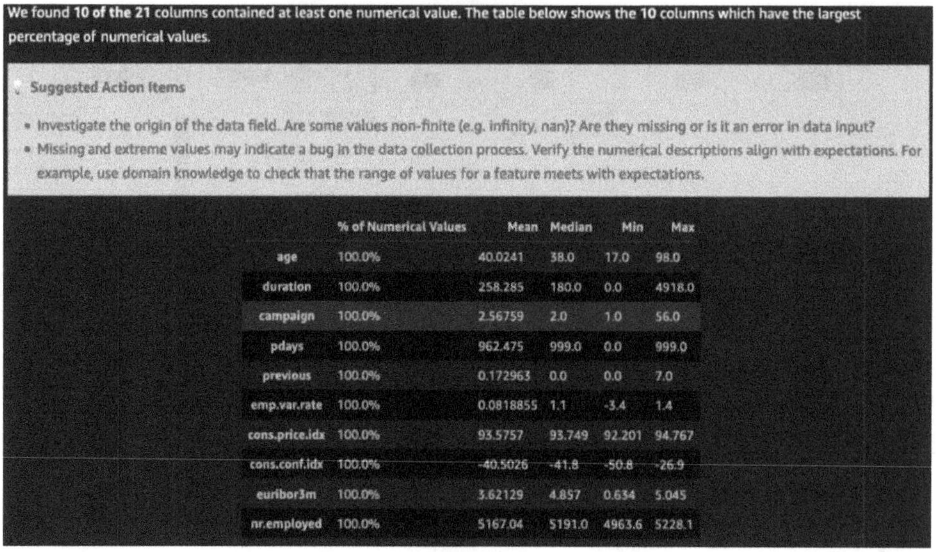

We found **10 of the 21** columns contained at least one numerical value. The table below shows the **10** columns which have the largest percentage of numerical values.

Suggested Action Items

- Investigate the origin of the data field. Are some values non-finite (e.g. infinity, nan)? Are they missing or is it an error in data input?
- Missing and extreme values may indicate a bug in the data collection process. Verify the numerical descriptions align with expectations. For example, use domain knowledge to check that the range of values for a feature meets with expectations.

	% of Numerical Values	Mean	Median	Min	Max
age	100.0%	40.0241	38.0	17.0	98.0
duration	100.0%	258.285	180.0	0.0	4918.0
campaign	100.0%	2.56759	2.0	1.0	56.0
pdays	100.0%	962.475	999.0	0.0	999.0
previous	100.0%	0.172963	0.0	0.0	7.0
emp.var.rate	100.0%	0.0818855	1.1	-3.4	1.4
cons.price.idx	100.0%	93.5757	93.749	92.201	94.767
cons.conf.idx	100.0%	-40.5026	-41.8	-50.8	-26.9
euribor3m	100.0%	3.62129	4.857	0.634	5.045
nr.employed	100.0%	5167.04	5191.0	4963.6	5228.1

Figure 12.24 Viewing dataset statistics 2.

1. Open the laptop and click on the “Import notebook” link in the top right corner to save a copy that allows reading and writing.
2. Execute the cell to retrieve all necessary outputs and parameters from the SageMaker configuration section.
3. Move to Candidate Pipelines, and create a runner object to execute the selected candidate pipeline.

```
from sagemaker_automl import AutoMLInteractiveRunner,
AutoMLLocalCandidate

automl_interactive_runner = AutoMLInteracriveRunner
  (AUTOML_LOCAL_RUN_CONFIG)
```

4. Next, add the first pipeline (dpp0).

Data transformation is done by first using RobustImputer to transform “numeric” features and then using Threshold One Hot Encoder to transform “categorical” features. All generated features are merged and then Robust StandardScaler is applied. The transformed data is used for tuning the XGBoost model. These objects are based on scikit-learn, so they should be very familiar. For example, RobustImputer can be seen as built on sklearn.impute.SimpleImputer with additional features.

```
automl_interactive_runner.select_candidate(
{"data_transformer":
 "name": "dpp0",
 }
)
```

5. If we take a look at other pipelines briefly, we can see various processing strategies and algorithms such as linear regression. This is one of the built-in algorithms in SageMaker, which will be covered in the next chapter.
6. Scroll down to the "Selected Candidates" section can confirm that only the first pipeline has been selected.

```
automl_interactive_runner.display_candidates()
```

Candidate Name	Algorithm	Feature Transformer
dpp0-xgboost	xgboost	dpp0.py

   This means that the data is processed with the dpp0.py script and that the model has been trained using the XGBoost algorithm.
7. When you click on the dpp0 hyperlink, you can see that the corresponding script opens and the scikit-learn transformer pipeline has been constructed.

   Missing values in numerical variables are supplemented using RobustImputer, and categorical variables are one-hot encoded using ThresholdOneHotEncoder. The generated features are scaled using RobustStandardScaler. Lastly, the labels are encoded using LabelEncoder. This preprocessed data is used to train the XGBoost model.

```
numeric_processors = Pipeline(
 steps = [('robustimputer', RobustImputer(strategy=
'constant', fill_values=nan))]
)
categorical_processors = Pipeline(
 steps=[('thresholdonehotencoder', ThresholdOneHotEnc
oder(threshold=301))]
)
```

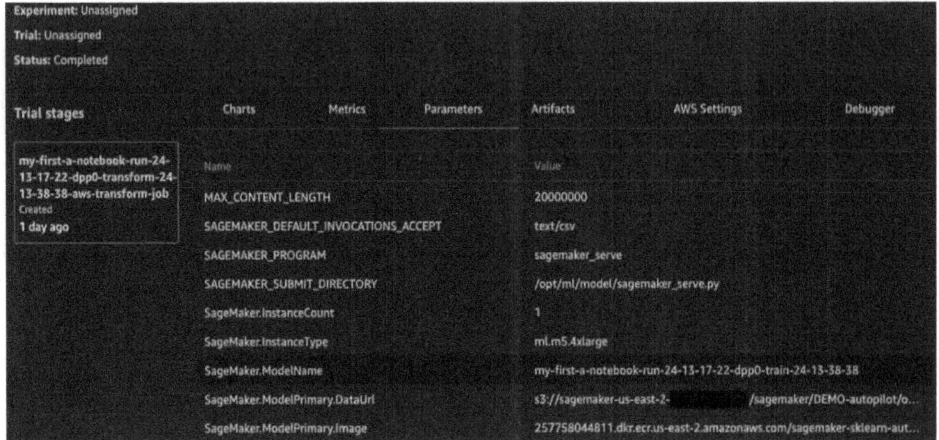

Figure 12.25 Describe a trial.

```
column_transformer = ColumnTransformer(

 transformers=[('numeric_processing', numeric_
processors, numeric), ('categorical_processing',
categorical_processors categorical)]

)
```

8. Go back and run the script in the Run Data Transformation Steps section.

```
automl_interactive_runner.fit_data_transformers
  (parallel_jobs=7)
```

9. Create sequential SageMaker jobs, and these results are stored in the newly generated prefix for notebook execution. The first job trains dpp0 transformers on the input dataset. The second job processes the input dataset using the resulting model.

```
$ aws s3 ls s3://sagemaker-us-east-2-123456789012/
sagemaker/DEMO-autopilot/output/

my-first-autopilot-job/my-first-a-notebook-run-
  24-13-17-22/
```

10. When you double-click the job name, a "Trial Component Description" window like Figure 12.25 will appear. In this window, you can find all the information you need to know about the job. Once data processing is complete, including parameters and the location of the output, the notebook will automatically proceed with model tuning and deployment.

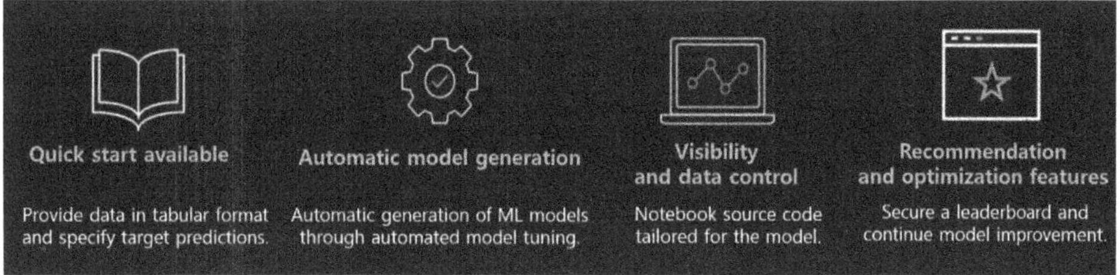

Figure 12.26 Features of SageMaker Autopilot.

12.3.4 *Summary*

Amazon SageMaker Autopilot is a tool that simplifies the process of building, training, and optimizing machine learning models. By using SageMaker Studio GUI and SageMaker SDK, you can effectively create classification models without complex coding. Users only need to visualize the data and specify the target column for prediction, and Autopilot automatically performs tasks such as data preprocessing, providing statistical insights, and extracting non-numerical information. Additionally, for further fine-tuning, Autopilot automatically provides Studio notebooks for all generated models, allowing for detailed analysis of the model's details and functionality. The various machine learning algorithms generated can be easily reviewed and compared within Autopilot. When evaluating the performance of a model, metrics such as accuracy, precision, ROC curve, and AUC are important, and using SageMaker Autopilot allows for comparing multiple models based on these metrics and understanding how each feature contributes to the prediction through feature importance. Thanks to the integration with Amazon SageMaker Clarify, Autopilot provides a report that clearly understands and explains the prediction method of the model. Through this, users can gain a deeper understanding of how the model operates. Figure 12.26 summarizes the key features of Autopilot.

12.4 Practice Questions

Q1. How does SageMaker Autopilot simplify the machine learning model development process for users?

Q2. What advantages does the automated data analysis feature of SageMaker Autopilot offer in the initial phases of machine learning projects?

Q3. Describe the role of feature engineering and model tuning in enhancing the performance of machine learning models within SageMaker Autopilot.

Chapter 13

Microsoft Azure

Azure is a cloud computing platform offered by Microsoft since 2010, and it provides over 600 services. Figure 13.1 shows the workflow of Azure Machine Learning (ML), which utilizes the Azure platform for data collection and management in the cloud. It allows users to easily create models using ML Studio and build web services, which can be utilized on various devices.

As shown in Figure 13.2, Azure ML Studio provides a user-friendly GUI environment that is different from traditional cloud platforms, ML libraries, and tools. It offers the convenience of accessing features that are difficult to utilize without coding by creating blocks and connecting them using a drag and drop system. Additionally, it allows users to insert scripts written in R or Python as blocks and visualize the results. Thanks to its easy-to-use structure, anyone with a simple working knowledge of the platform can easily create and deploy prediction models.

Through Figure 13.3 in the next chapter, you can see the overall functionality of Azure ML. In Azure ML, there is basic support for data input, output, and visualization, as well as popular ML algorithms used by data scientists.

- **Inserting experimental data:** You can bring the data to train the model and insert it into Azure Cloud for utilization.
- **Experiment data preprocessing:** When there are missing or typo errors in the data, it is necessary to preprocess the data in order to train the model.
- **Feature extraction:** Extracting features in accordance with the purpose and algorithm to train the model.
- **Learning and model evaluation:** Selecting algorithms, conducting training, evaluating the model, and checking for any errors.

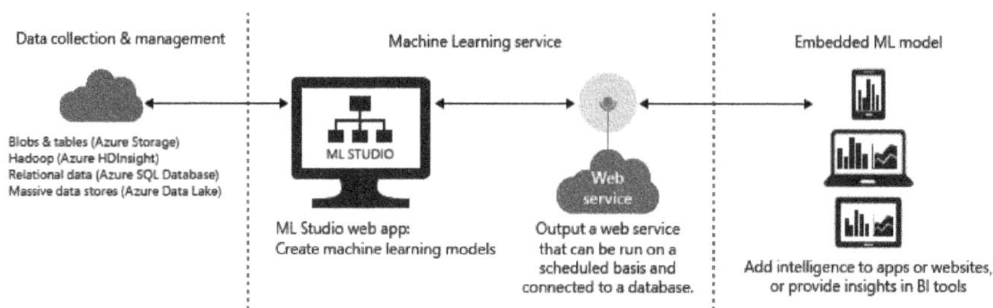

Figure 13.1 Azure ML basic workflow.

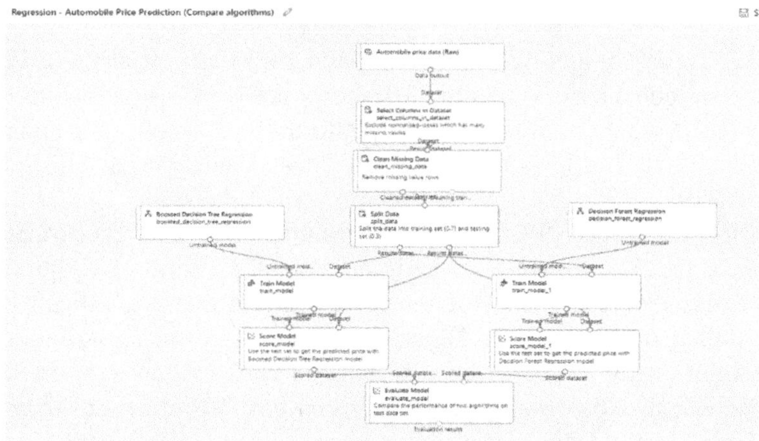

Figure 13.2 Azure ML Studio designer.

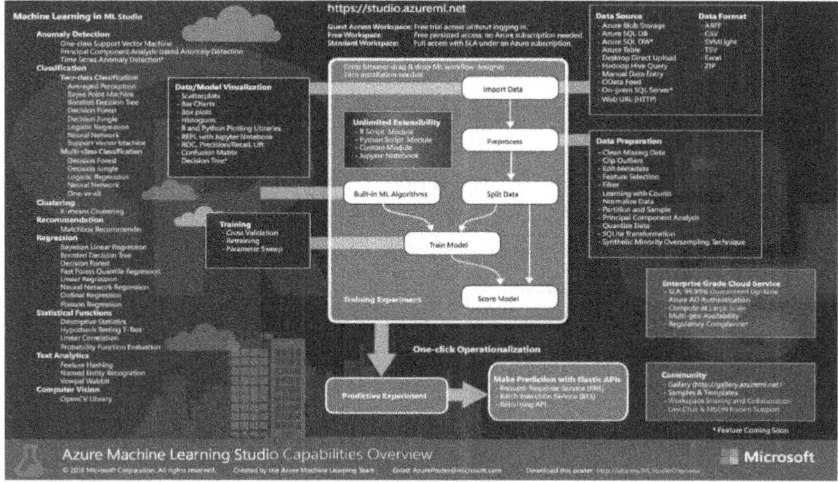

Figure 13.3 Azure ML Studio diagram.

- **Save trained model:** Save the completed trained model.
- **Applying and running the web service:** Apply and run the trained model to utilize it in a web service.

Using these components, you can develop and deploy models.

13.1 Registering Microsoft Azure Users

Go to https://azure.microsoft.com/ko-kr and click on the "Try Azure for free" link shown in Figure 13.4.

Figure 13.5 shows the home page for creating a trial account; click on the "Start free" link to create a trial account.

An Azure trial account provides a credit of 200 USD for 30 days, as well as free services for 12 months. Therefore, you can proceed with the later examples and exercises using the aforementioned account. Figure 13.6 shows the personal information consent section, where you should check the box and click the "Next" button to proceed.

Figure 13.7 is the card registration page that appears when checking out successfully. Enter a card, such as VISA, that supports international payments to proceed.

If the registration is completed normally as shown in Figure 13.8, a page will appear saying that preparation for starting Azure is complete.

If an Azure account has been created, you can go to URL to use Azure ML. (https://azure.microsoft.com/en-us/products/machine-learning)

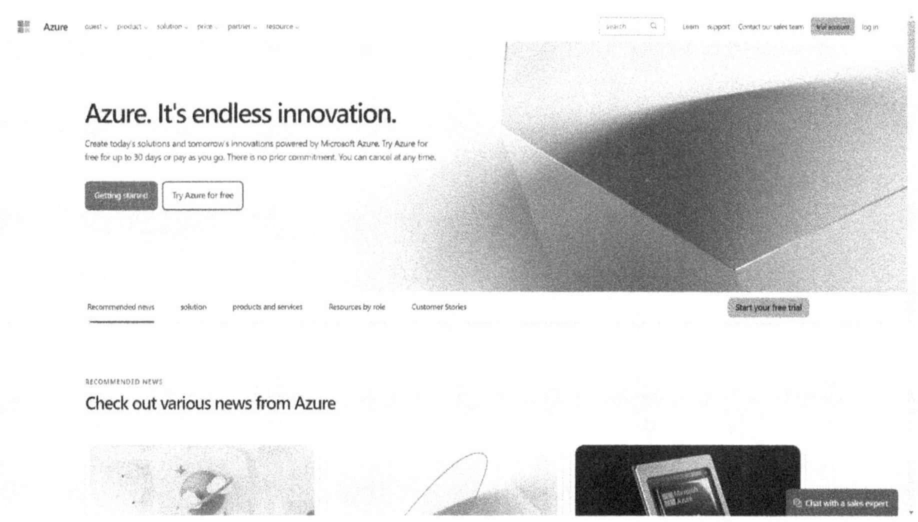

Figure 13.4 Azure homepage.

Build in the cloud with an Azure free account

Create, deploy, and manage applications across multiple clouds, on-premises, and at the edge.

Start your free trial

pay as you go

Popular services provided at no extra cost for 12 months

Over 55 other services always available for free

Starting from as little as USD200 Azure credits

Chat with a sales expert

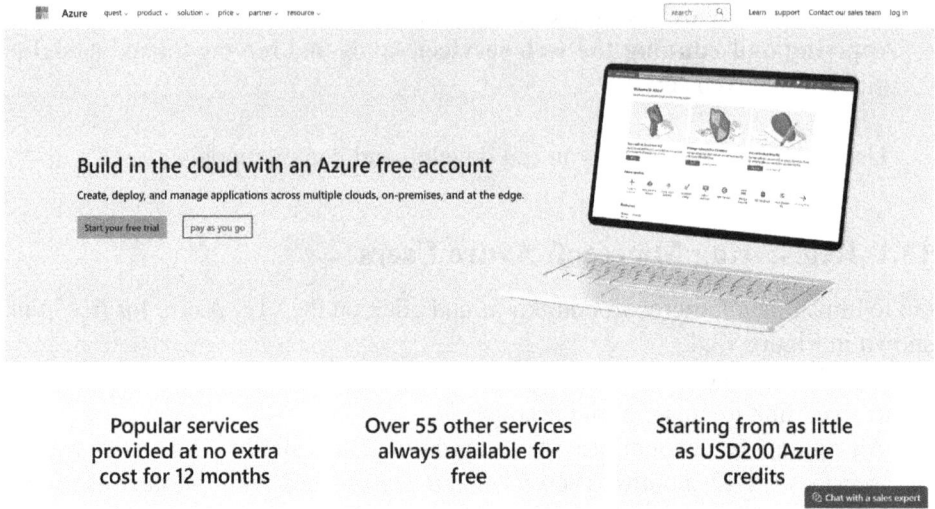

Figure 13.5 Creating an Azure trial account.

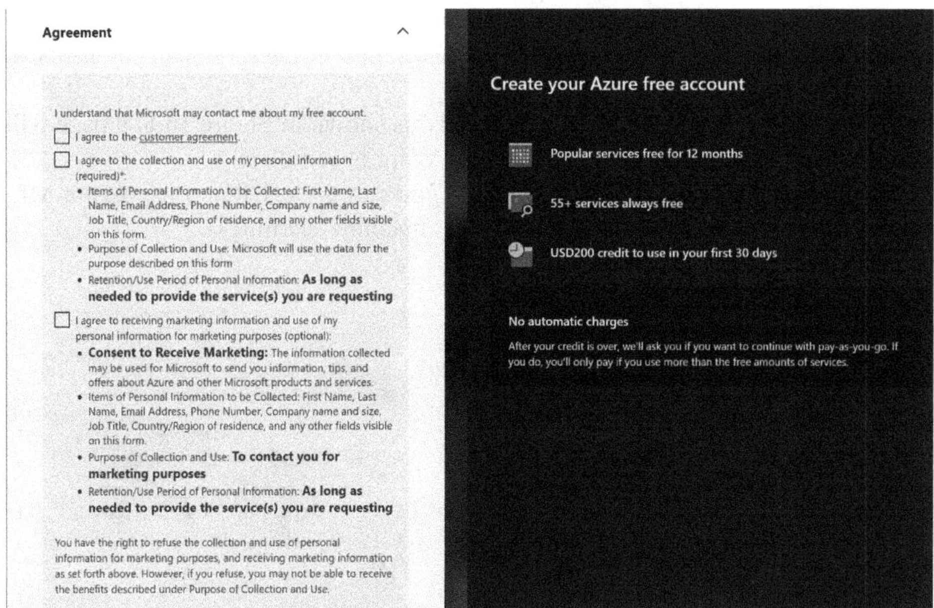

Figure 13.6 Azure contract details.

Proceed with the login by selecting “Sign in” located in the top right corner, as shown in Figure 13.9.

As shown in Figure 13.10, there are various login options, but proceed with logging in using the newly created ID.

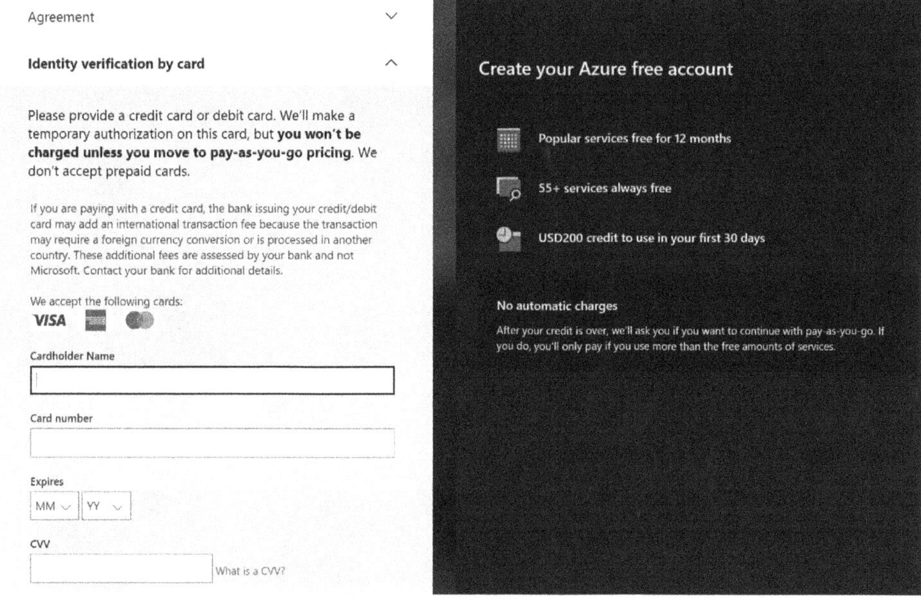

Figure 13.7 Card registration.

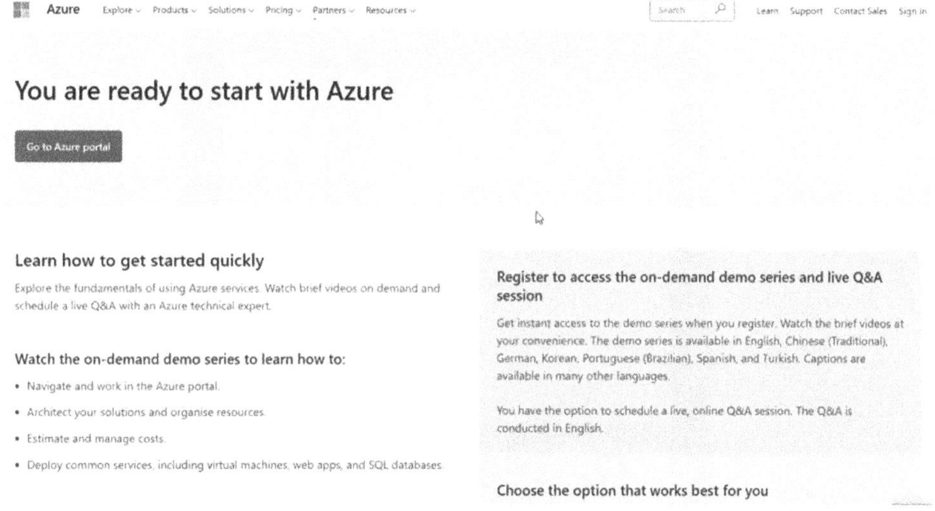

Figure 13.8 Azure registration completion page.

Figure 13.11 shows the main page of Azure ML Studio that appears after logging in. The first step is to create a Workspace, which is a comprehensive environment for preparing Azure ML. Within this environment, it is possible to collaborate with others or work alone to create ML models and group related tasks.

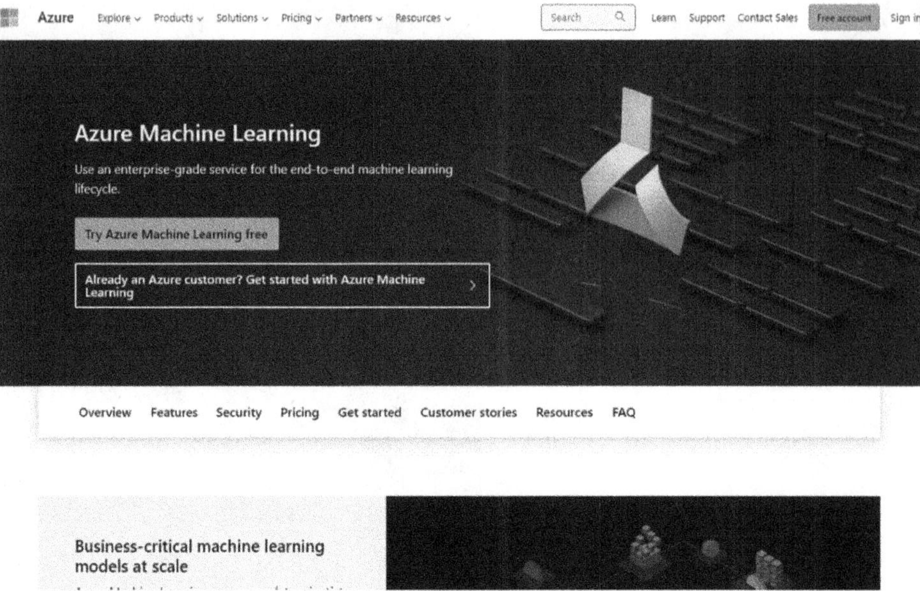

Figure 13.9 Azure ML homepage.

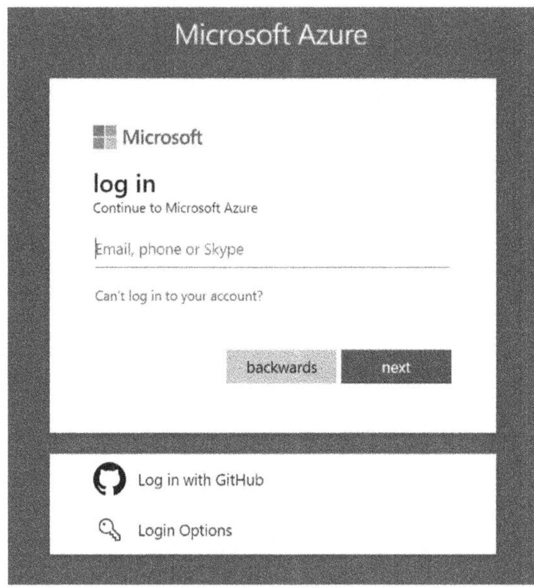

Figure 13.10 Azure ML login.

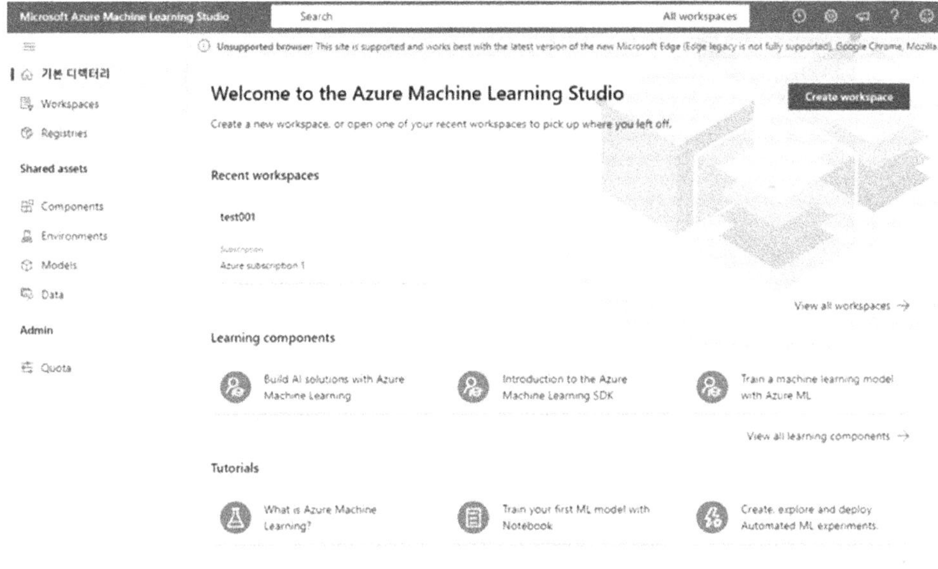

Figure 13.11 ML main screen.

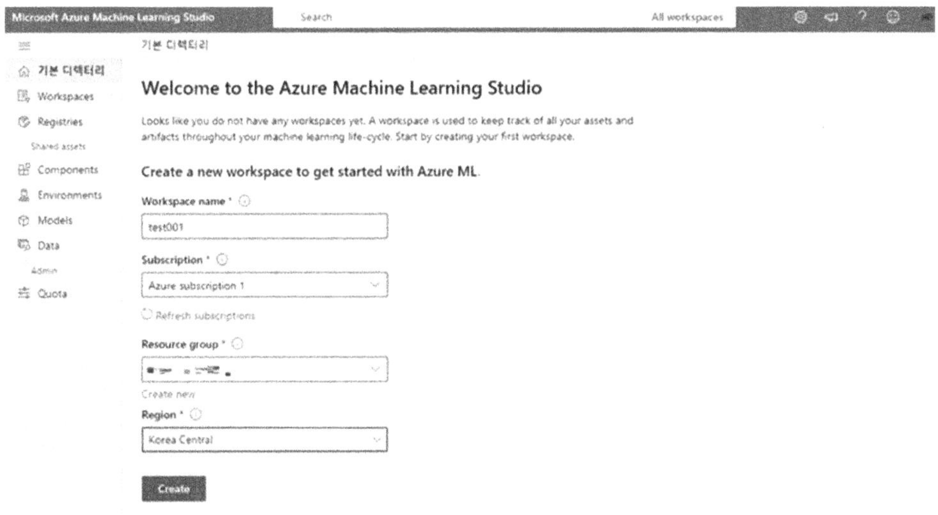

Figure 13.12 ML workspace.

The window for creating a workspace shown in Figure 13.12 appears when you first log in to Azure ML. After setting the name and region, create your own workspace by clicking "Create."

13.2 Practice Questions

Q1. Explain the advantages of Azure ML Studio.

Q2. Explain the algorithms supported by Azure ML Studio.

Q3. How does Azure ML Studio facilitate the creation of ML models for users without coding experience?

Q4. What steps are involved in the ML model development process within Azure ML Studio?

Q5. Describe the advantages of using Azure for ML projects, particularly with Azure ML Studio.

Chapter 14

Azure Automated Machine Learning

One of the features of Azure Machine Learning is Auto ML, which is an automated ML process that automates time-consuming and repetitive tasks in model development. Data scientists, analysts, and developers can build ML models with high scalability, efficiency, and productivity while maintaining model quality. Azure's automated ML is a No-Code automated ML model similar to AWS Sagemaker Studio described in the previous chapter. Let's learn about how to use Microsoft Azure with Titanic survival prediction data.

14.1 Azure Automated ML Titanic Prediction

Click on "Automated ML" as shown in Figure 14.1, and then select "New Automated ML job."

If you have selected a New Automated ML job, you will see the screen shown in Figure 14.2. Keep the Job name and New experiment name as the default settings, and then click "Next" to move on to the next step.

When you click the "Next" button, the screen shown in Figure 14.3 will appear. Click the "Create" button to load the dataset needed for ML.

Since there is no dataset to call here, first click on "Data" underneath Assets in the sidebar to go to the screen shown in Figure 14.4. Create a dataset. First, set the Name and Type according to the dataset you want to create.

As shown in Figure 14.5, you can choose a source to upload the data for the dataset you want to create. In this case, select "From local files" to use the data downloaded to your local drive.

If the previous workspace was created successfully, the workspace should be generated as shown in Figure 14.6. Select the default "workspaceblobstore" at the top.

As shown in Figure 14.7, click "Upload" to insert the training data, and then import the dataset to Azure by clicking "Next."

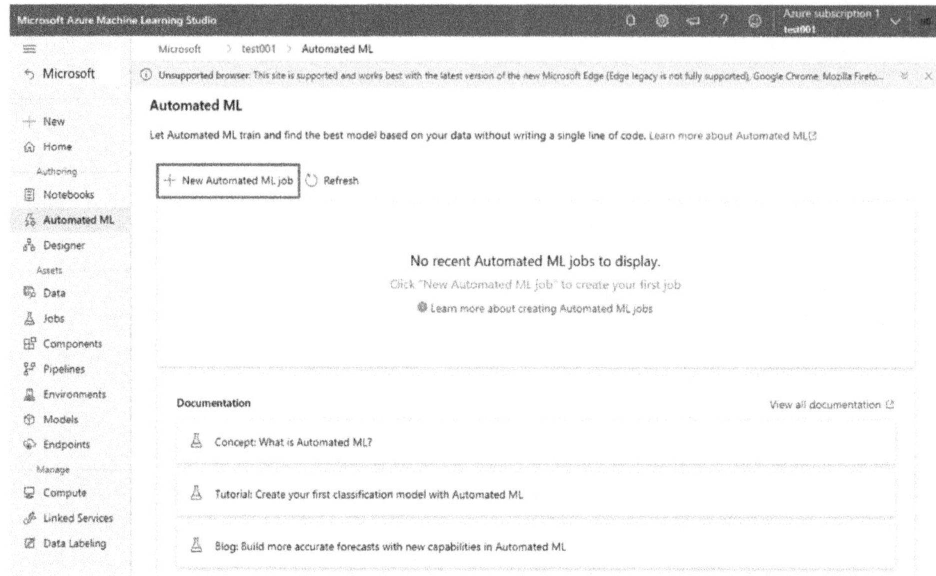

Figure 14.1 Automated ML start screen.

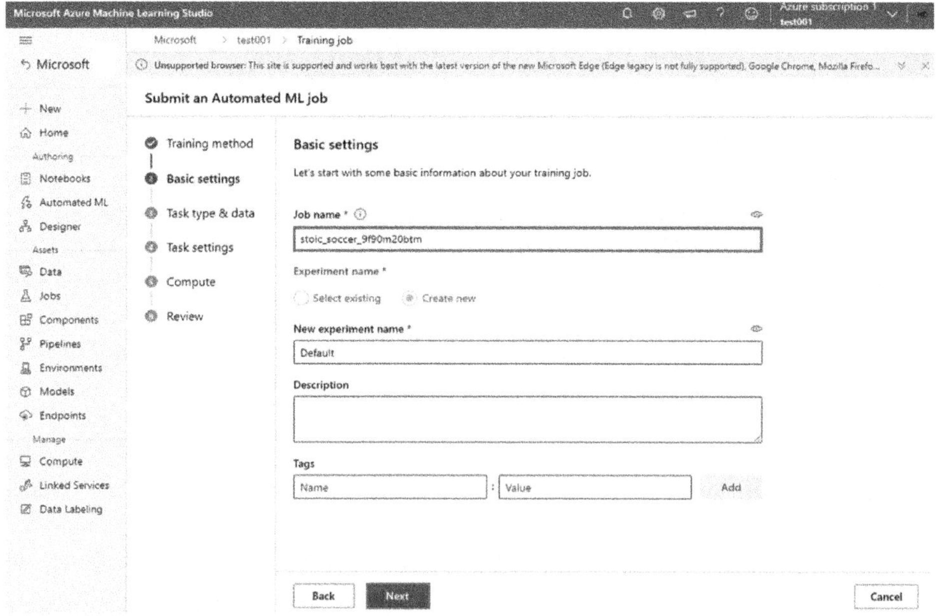

Figure 14.2 Basic setting.

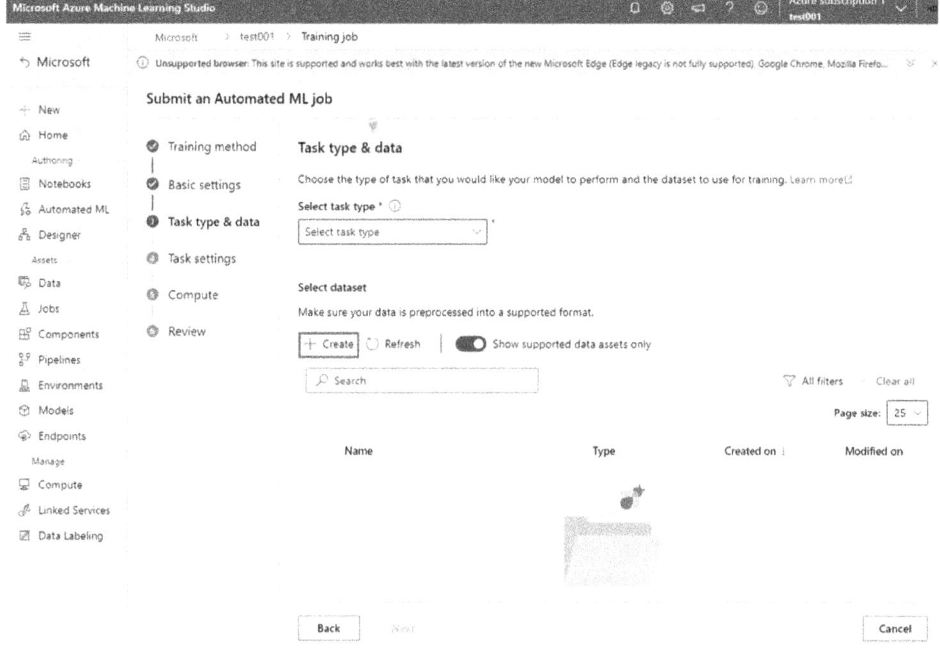

Figure 14.3 Dataset selection.

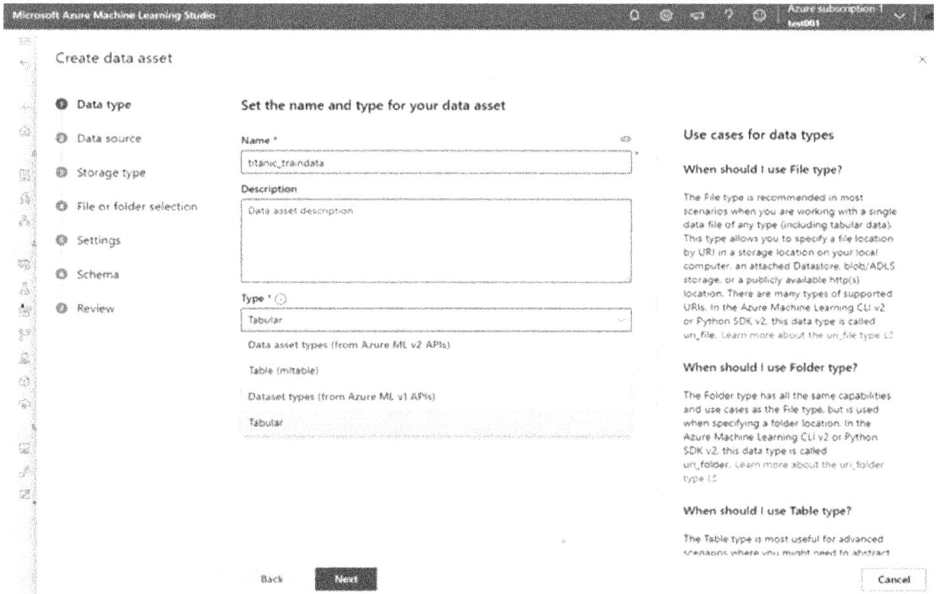

Figure 14.4 Data basic settings.

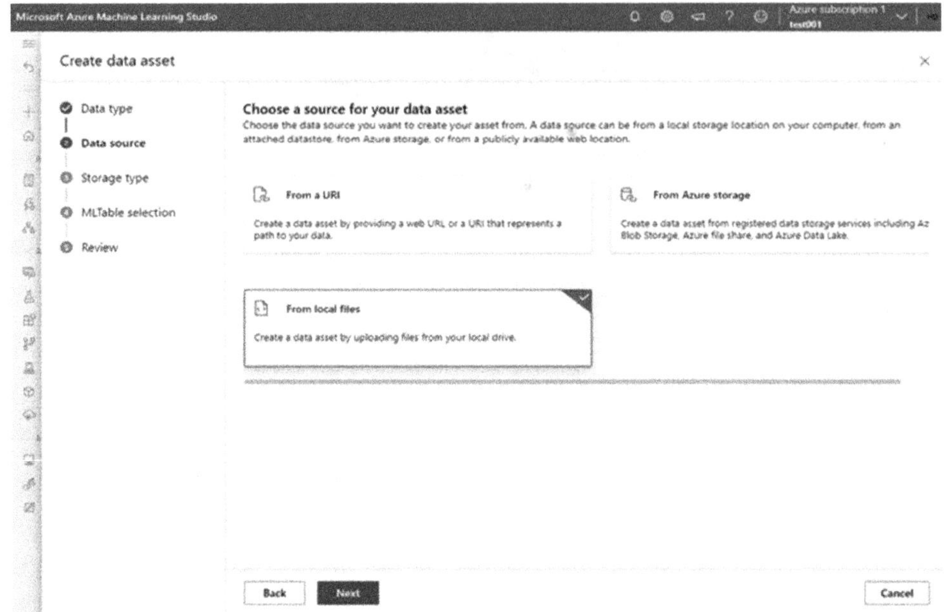

Figure 14.5 Choose data source.

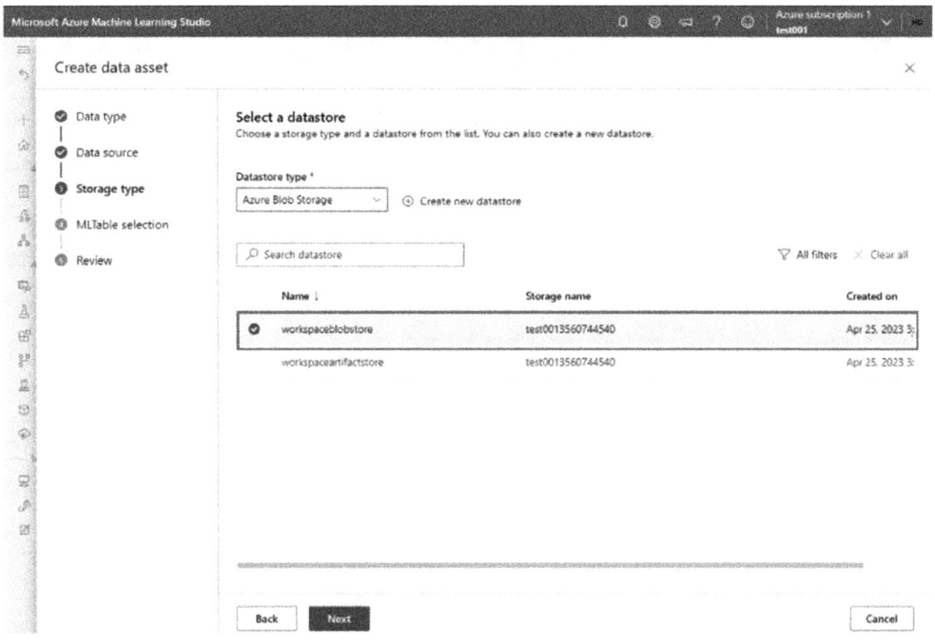

Figure 14.6 Selecting datastore.

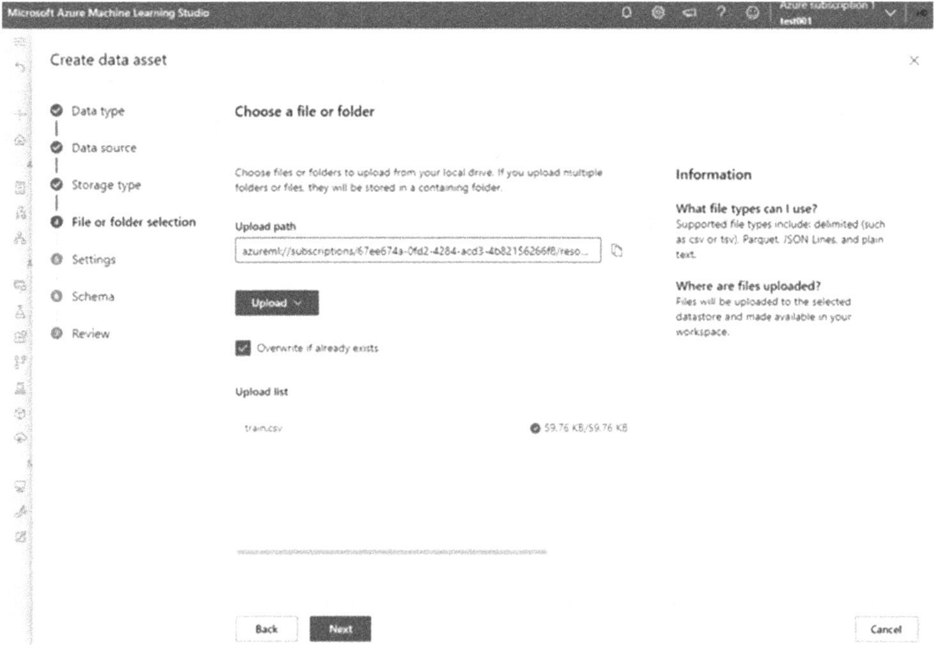

Figure 14.7 Dataset import.

If the dataset has been imported properly as shown in Figure 14.8, the preview screen shown below will appear. After confirming that the data has been imported correctly, proceed to the next step by clicking "Next."

Once the setup is complete, you need to configure the Schema as shown in Figures 14.9 and 14.10. Since the types are already set according to the recommended settings, there is no need for additional configuration. Exclude the data items Path, Passenger id, and Name from the Include settings, as shown in Figure 14.10, as they are unnecessary.

Through the Review process, as shown in Figure 14.11, you can perform a final check to ensure that your data is accurate, and then proceed by clicking "Create."

When the data is imported, the set data values will be displayed as shown in Figure 14.12. To proceed with the Titanic prediction model as a classification model, select the task type as "Classification" and choose the dataset. Then, proceed to the next step by clicking "Next."

To prepare for a ML job, a target column is necessary. As shown in Figure 14.13, we will select the "Survived" column to predict the survival of Titanic passengers and proceed. After setting the target column, we will configure the test dataset through "Select test dataset," as shown in Figure 14.14.

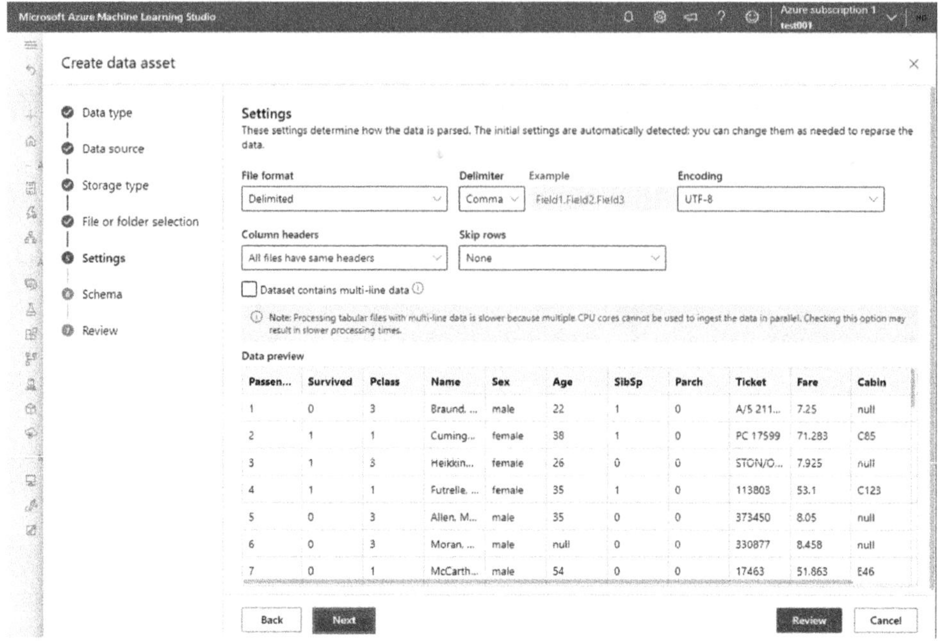

Figure 14.8 Data preview.

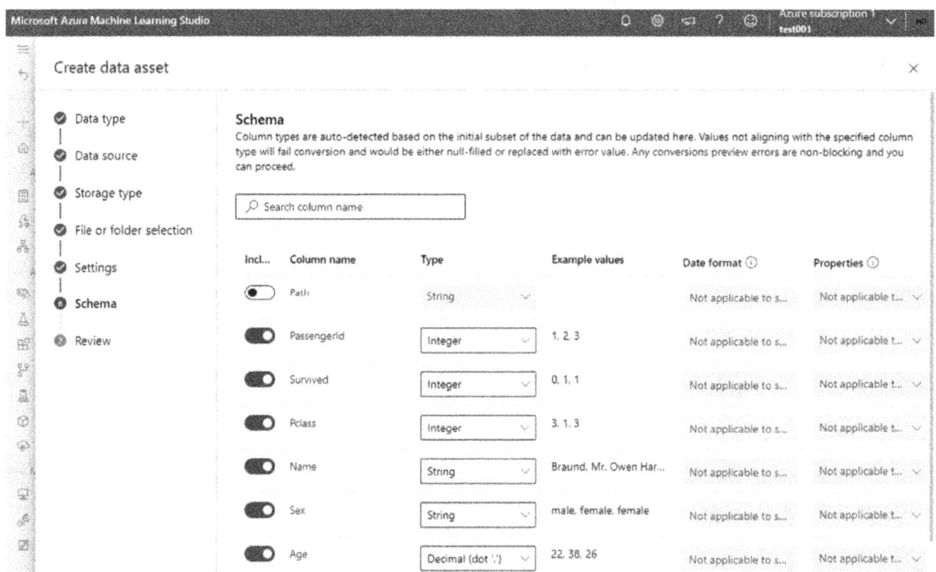

Figure 14.9 Dataset type setting 1.

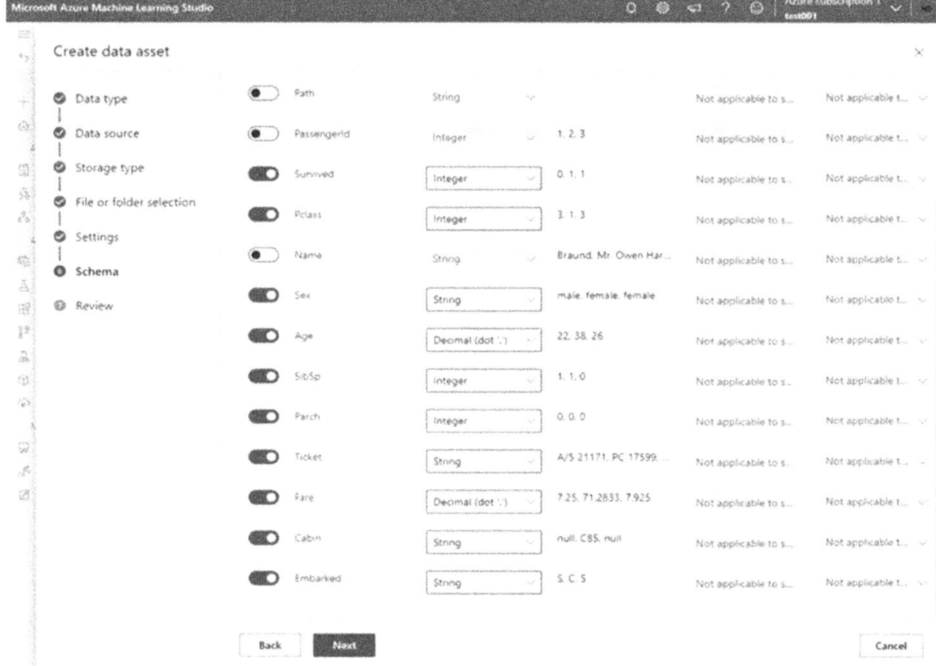

Figure 14.10 Dataset type setting 2.

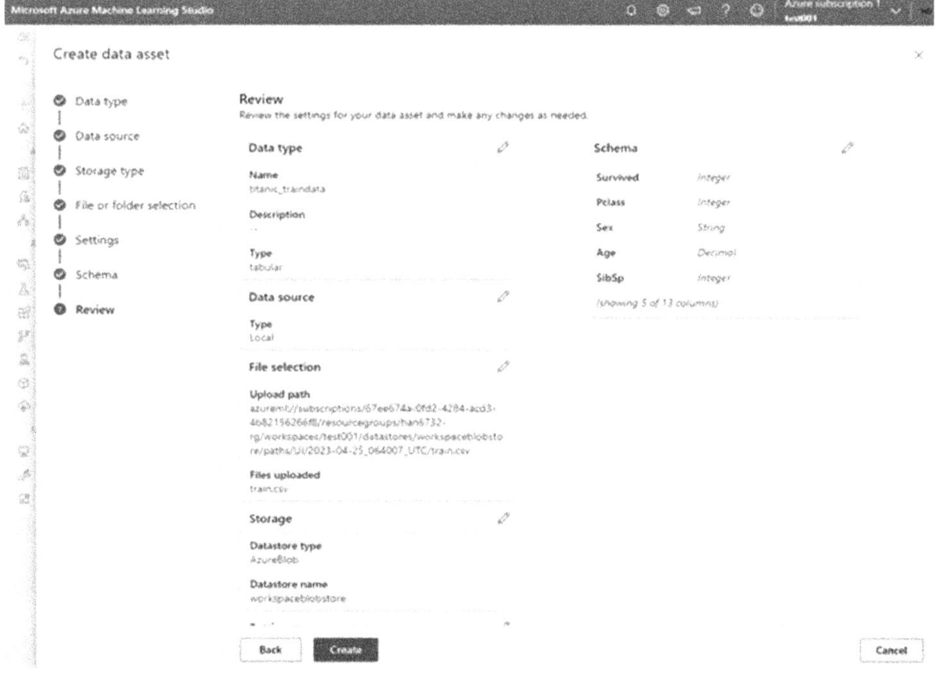

Figure 14.11 Data review.

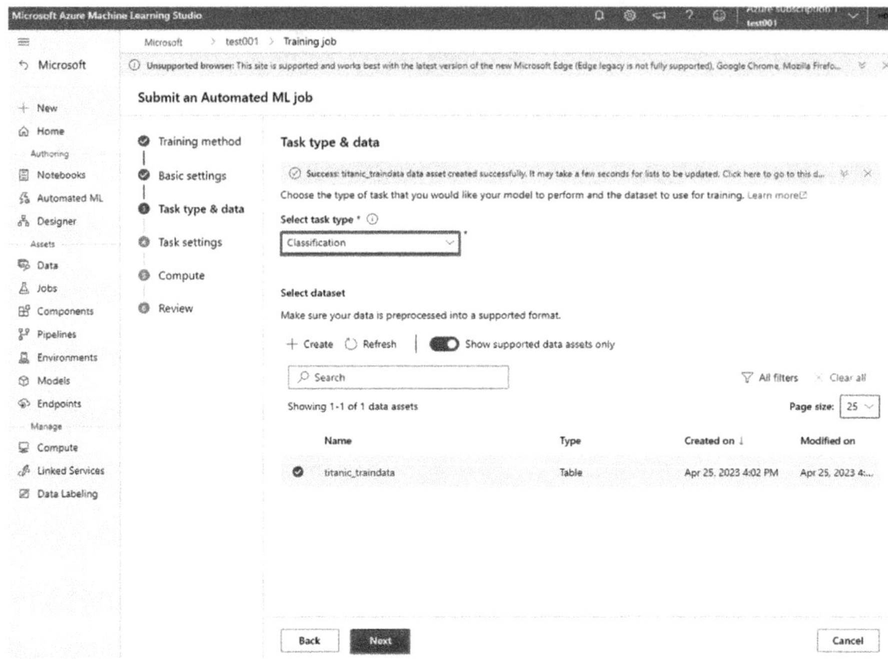

Figure 14.12 Task type setting.

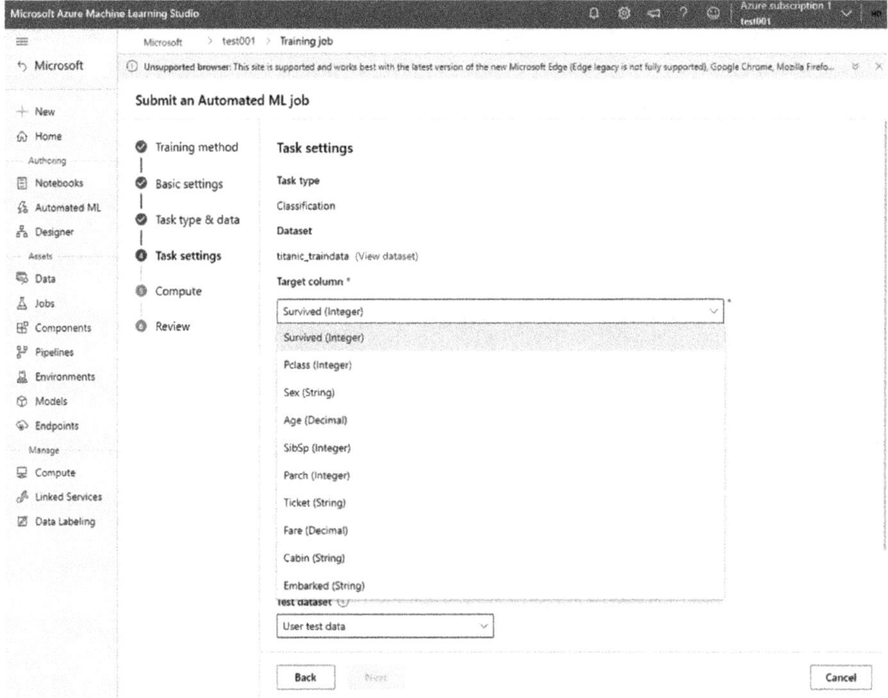

Figure 14.13 Target column setting.

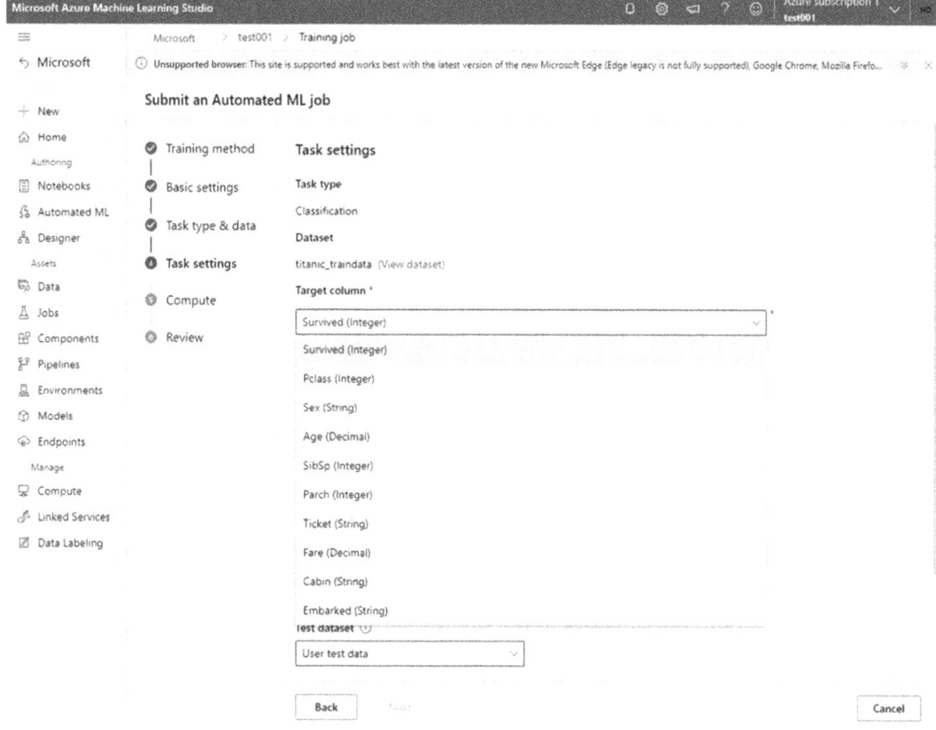

Figure 14.14 Select test dataset.

After completing the setting of the target column, set the test dataset through "Select test dataset" as shown in Figure 14.14. Then, set the dataset type in the same manner as shown previously in Figure 14.4.

In the data source step, set it up in the same way as shown previously in Figures 14.5–14.7.

Upload the test.csv file of the Titanic survivor prediction data, as shown in Figure 14.18.

As shown in Figure 14.19, the test dataset is automatically set, so you can proceed to the next step.

As with the train data, we exclude the rows of Path, Passengerid, and Name, as shown in Figure 14.20.

Figure 14.21 shows the final confirmation screen of the test dataset.

As shown in Figure 14.22 after uploading the test data, in Select test dataset, choose the "titanic_test data" at the top.

Figure 14.23 shows the step of selecting the type of computer instance. Azure-ML supports its own computing instances, allowing ML within Azure.

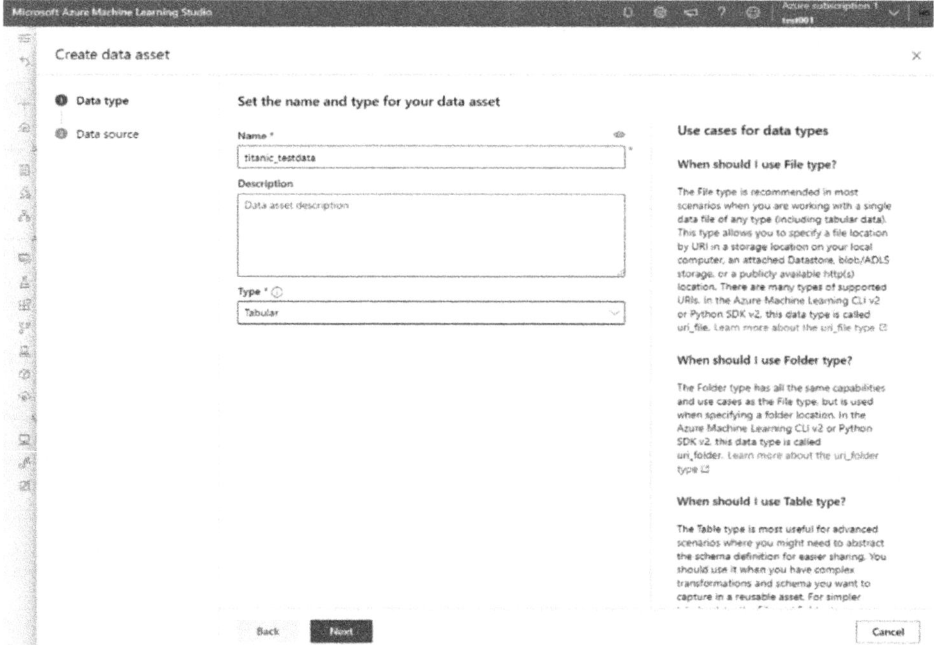

Figure 14.15 Test dataset 1.

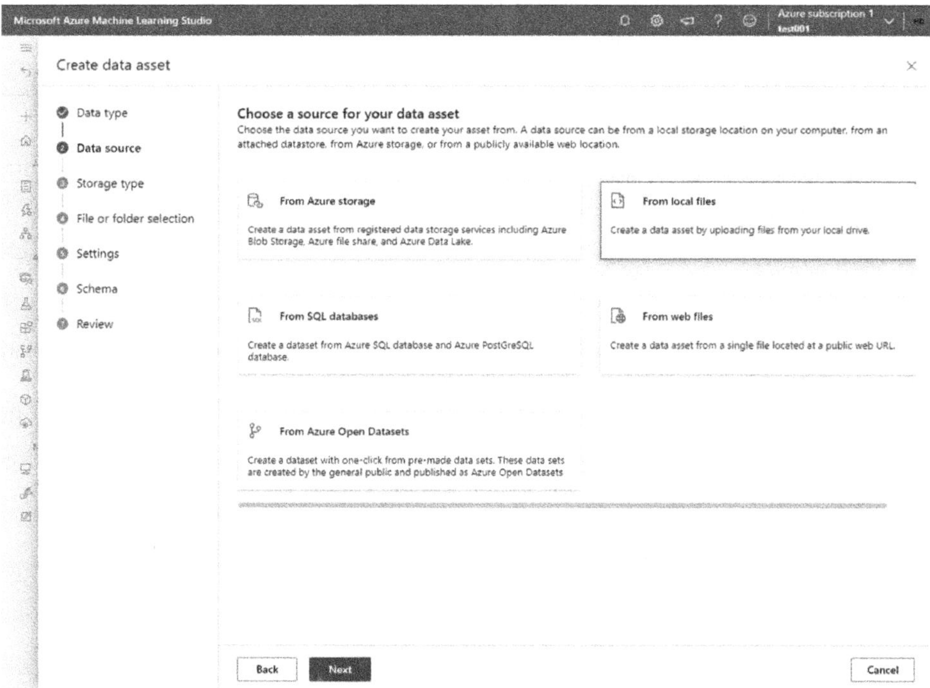

Figure 14.16 Test dataset 2.

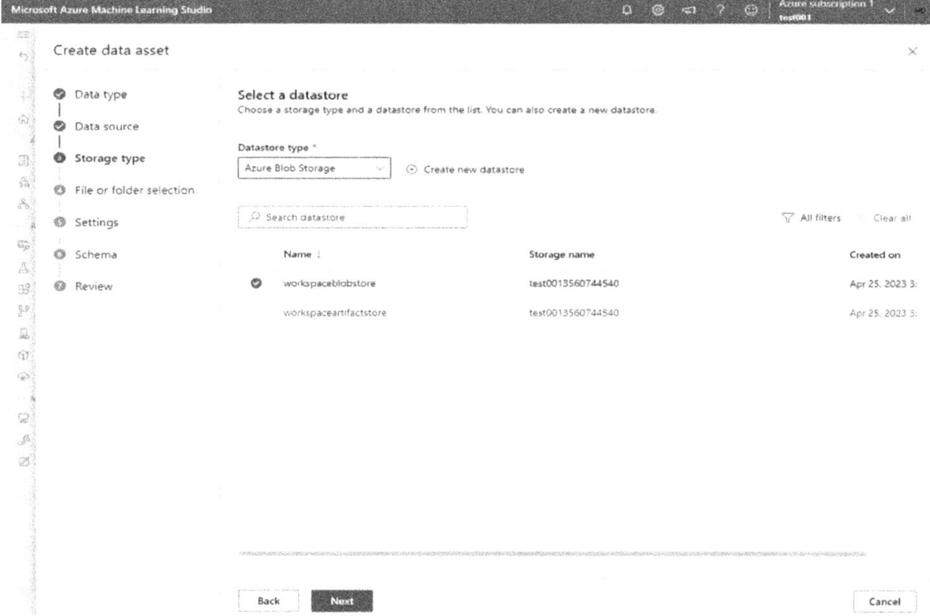

Figure 14.17 Test dataset 3.

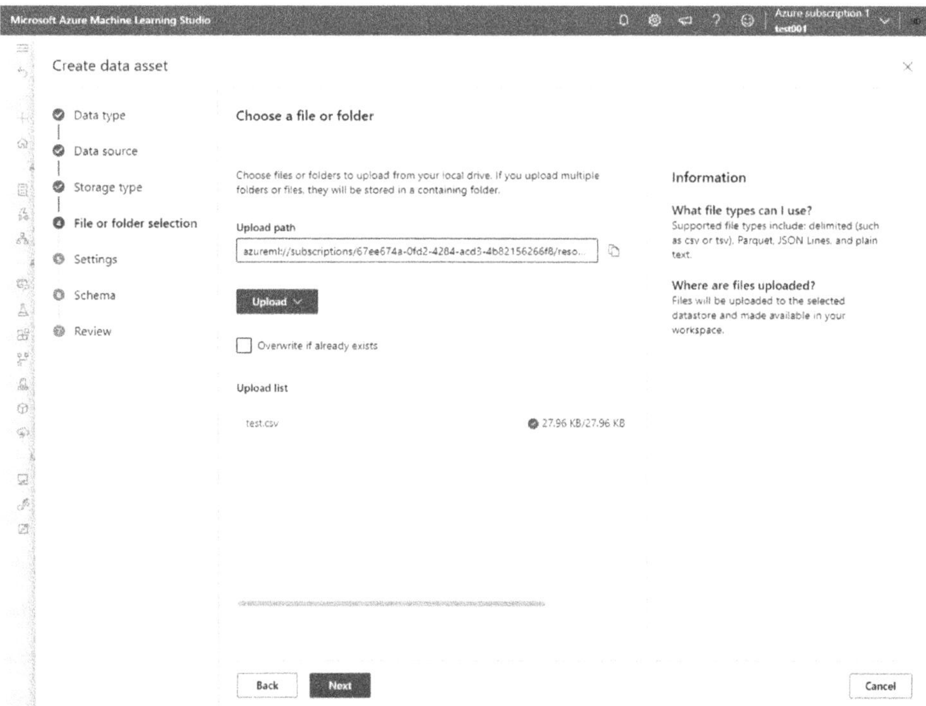

Figure 14.18 Test dataset 4.

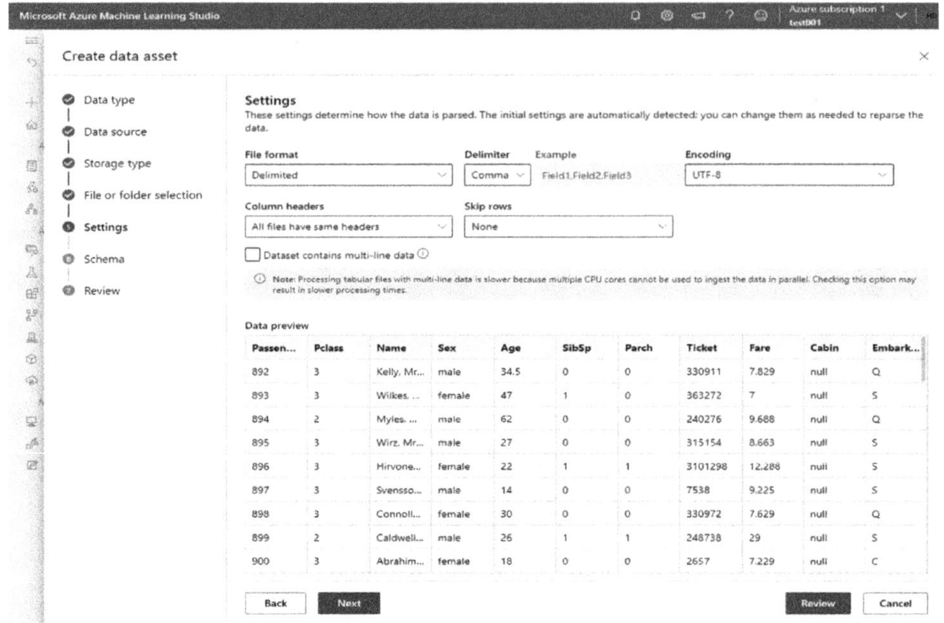

Figure 14.19 Test dataset 5.

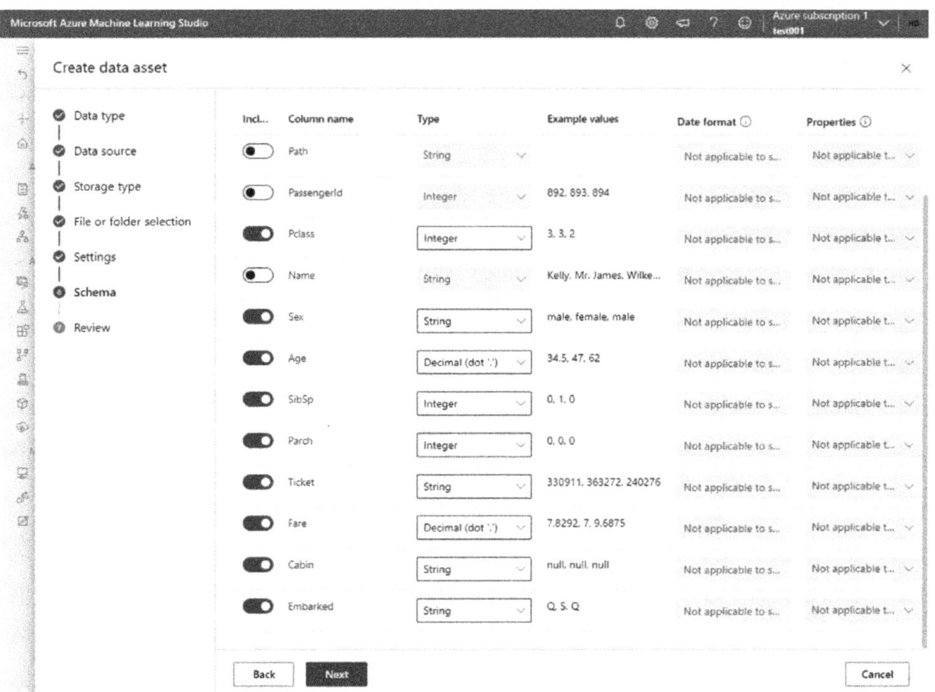

Figure 14.20 Test dataset 6.

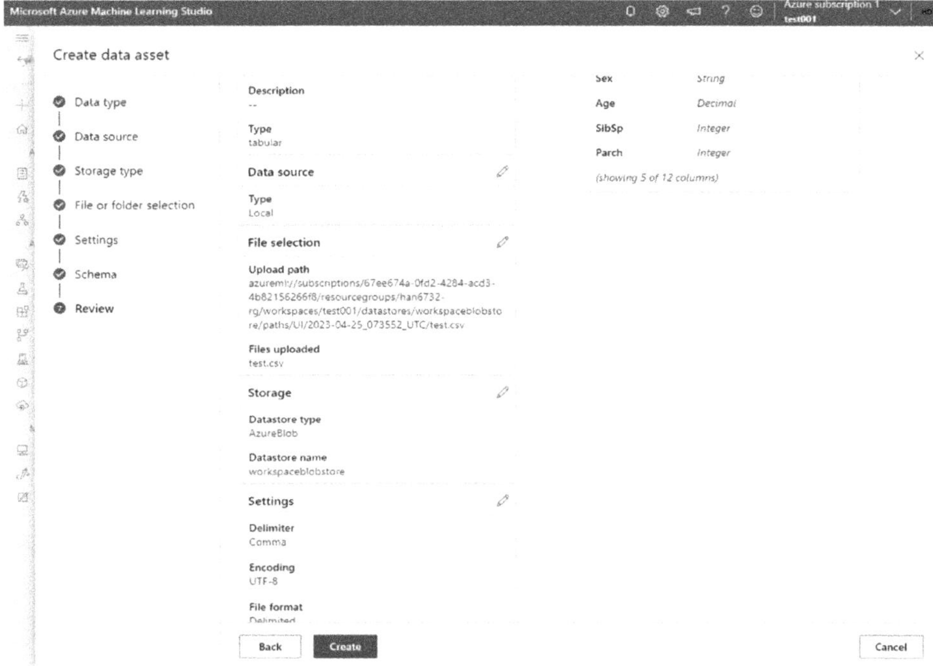

Figure 14.21 Test dataset 7.

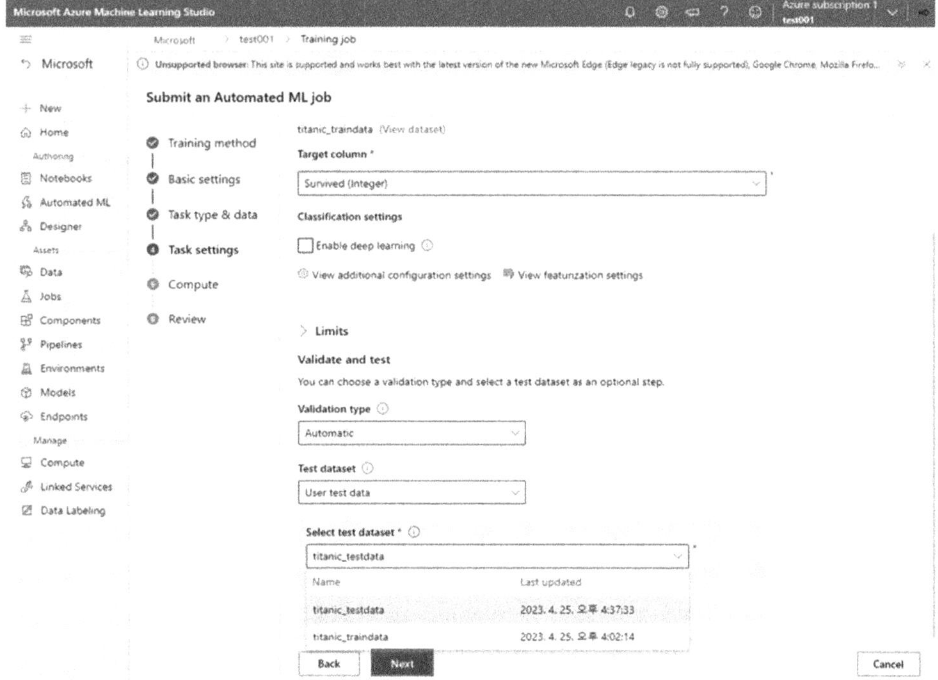

Figure 14.22 Selecting test data.

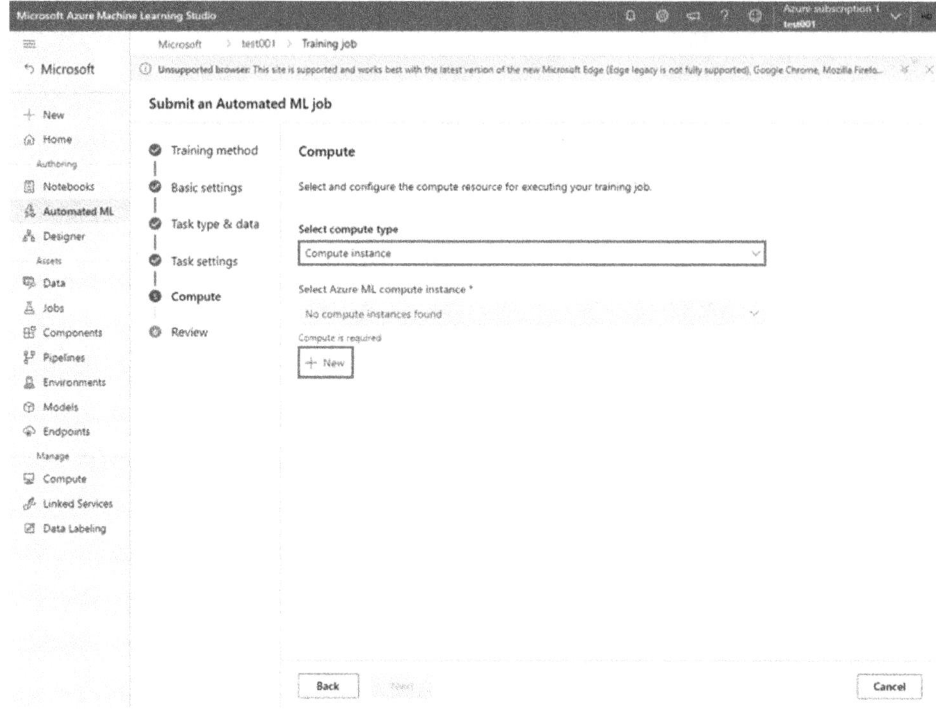

Figure 14.23 Compute type selection.

In Select compute type, select "Compute instance" and set up the computer environment to be trained by selecting "New."

As shown in Figure 14.24, Azure ML supports various PC environments. Set the Compute name and select "Standard_DS11_v2" from the recommended options on the CPU. Then, create the computing environment by clicking "Create."

After completing the setup, select the compute instance that was just set up and click "Next," as shown in Figure 14.25.

Figure 14.26 shows the final review and the last confirmation step before submission. If everything is set correctly, proceed by clicking "Submit training job."

Figure 14.27 shows the status page when running the execution of the Submit training job.

As shown in Figure 14.28, after execution is complete, users can see the results regarding which model was used by navigating to the "Models" tab on the Status page.

Figure 14.29 shows a page where you can check the models automatically generated by Azure. You can view detailed information of the utilized model by clicking "View explanation."

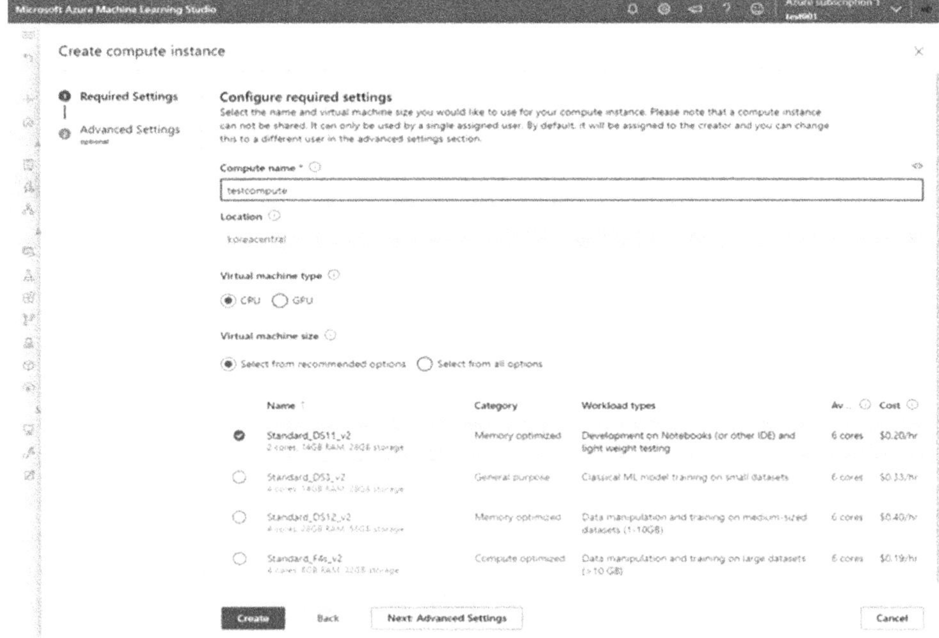

Figure 14.24 Compute environment setup and creation.

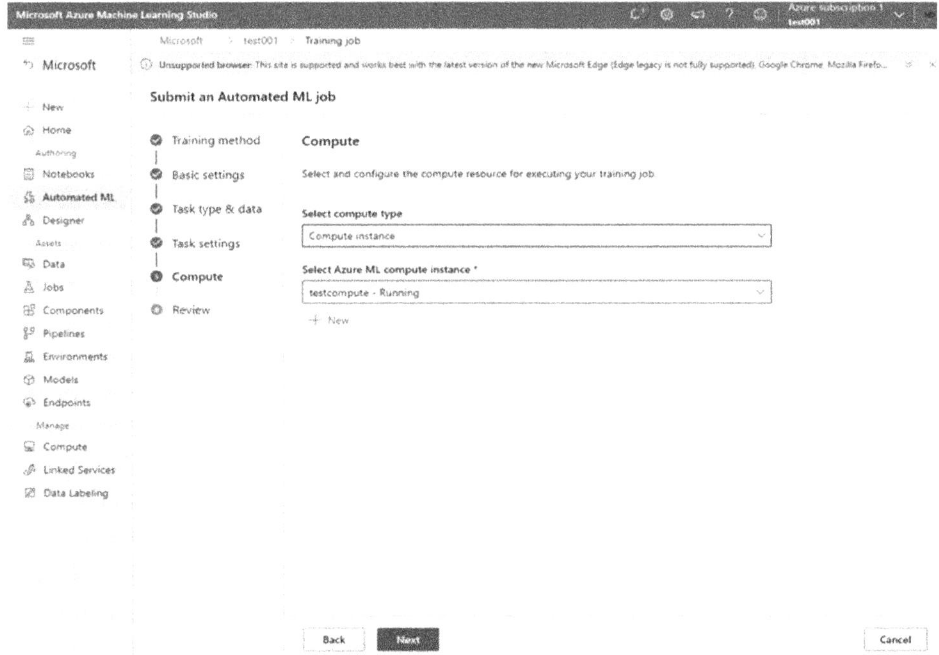

Figure 14.25 Selecting a compute instance.

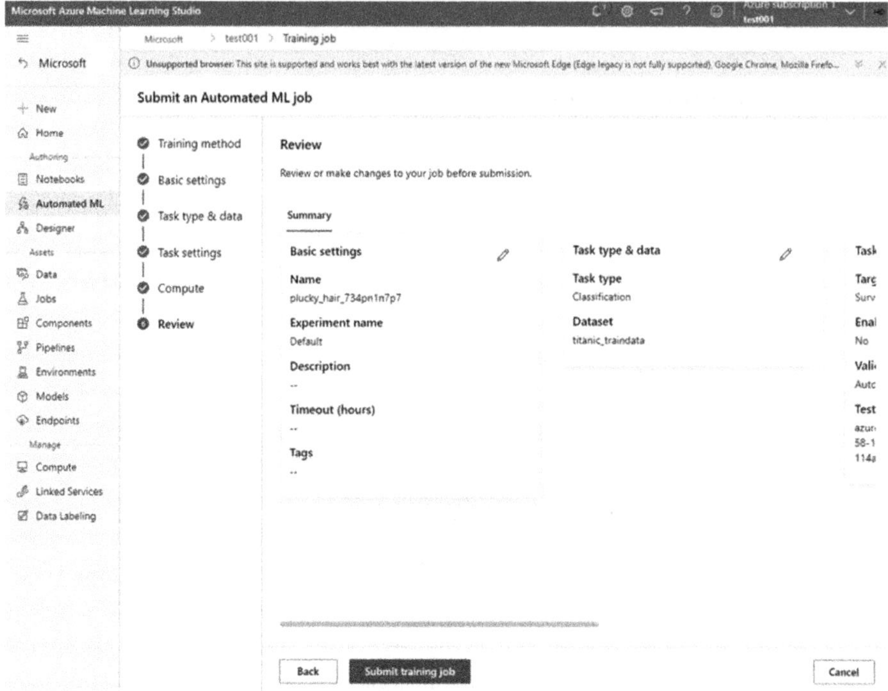

Figure 14.26 Training job review.

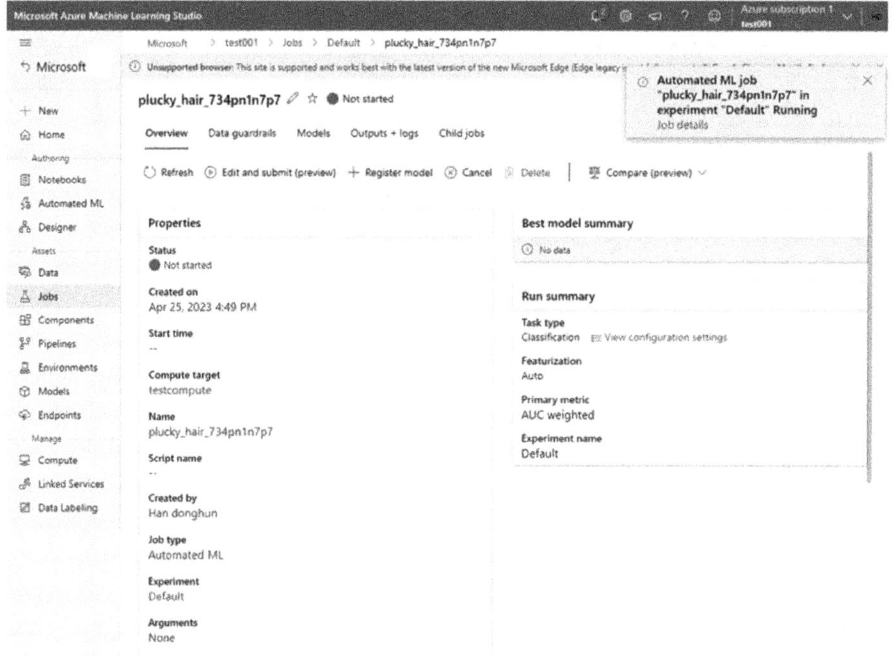

Figure 14.27 Training job running page.

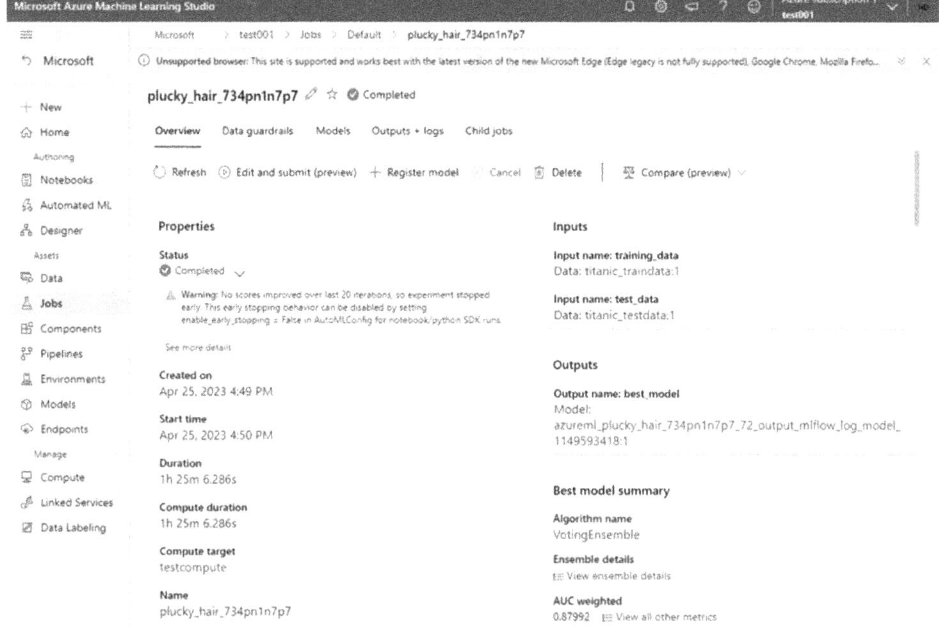

Figure 14.28 Training job completed.

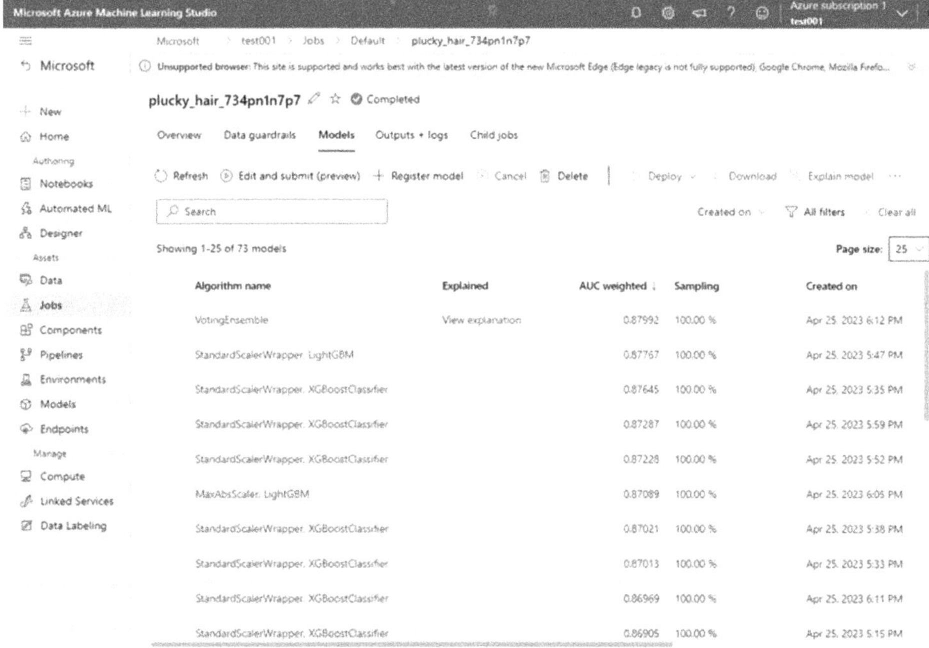

Figure 14.29 Models view page.

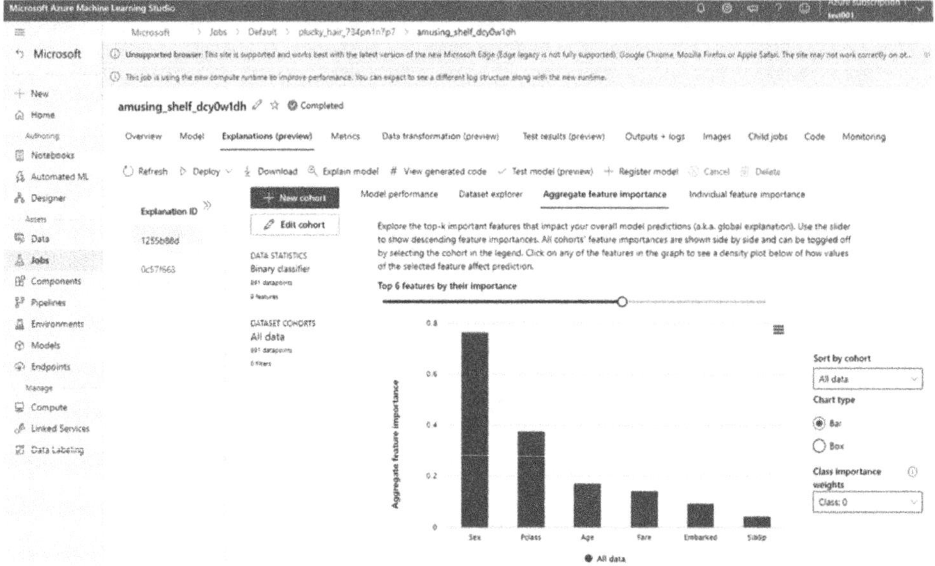

Figure 14.30 Aggregate feature importance.

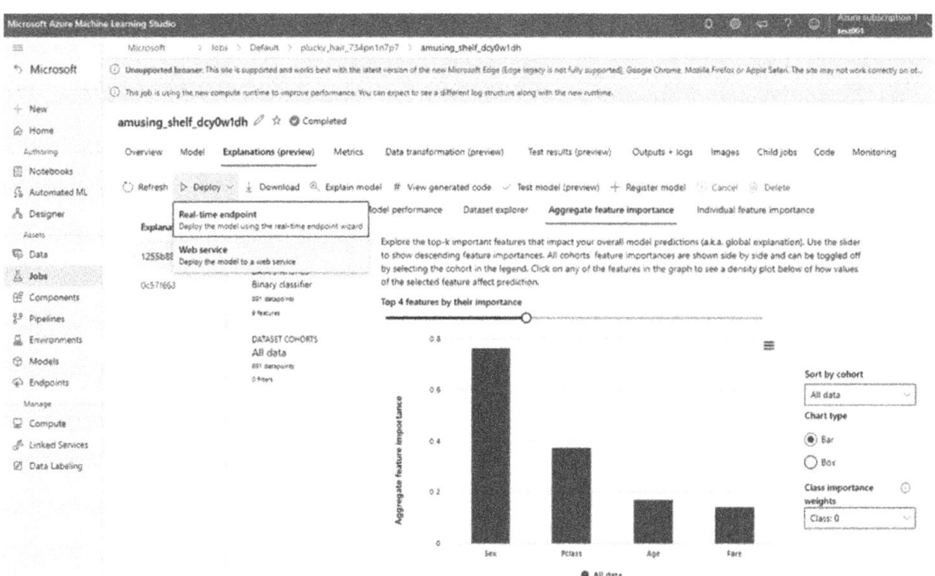

Figure 14.31 Select deploy model.

Figure 14.30 shows a graph illustrating which variable had the greatest impact in the detailed model explanation. According to the graph, it can be seen that Sex, Pclass, and Age had the greatest impact on survival, in that order.

Figure 14.31 shows the distribution of the created model. Of the two distribution methods, we will proceed by selecting “Web service.”

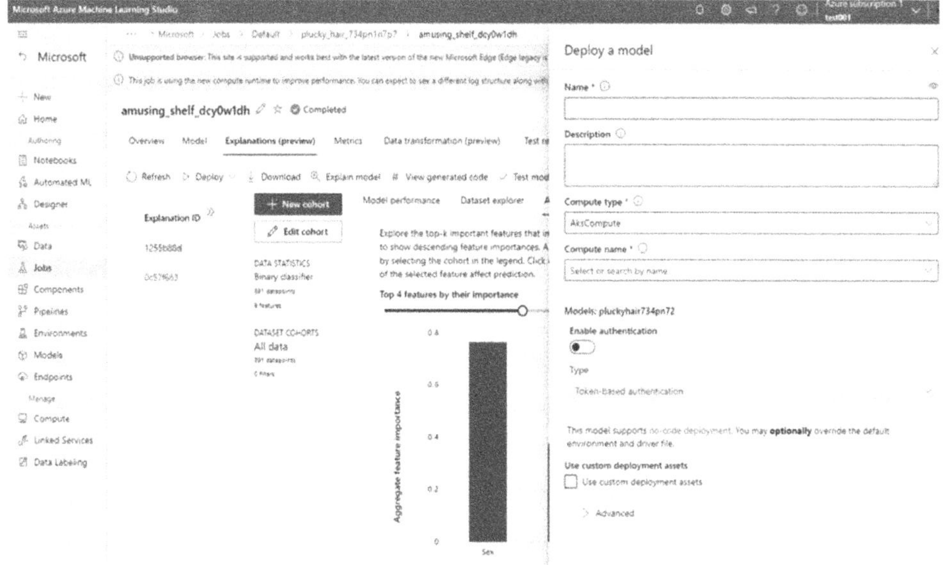

Figure 14.32 Deploy model setting.

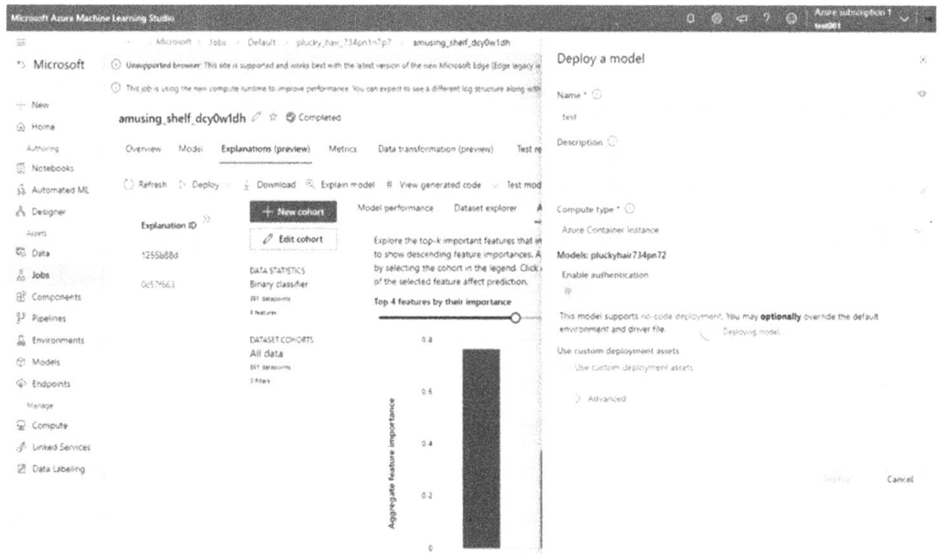

Figure 14.33 Model deploy.

As shown in Figure 14.32, set the Name and Compute type. Type an arbitrary name in the "Name" field, and proceed by clicking on "Azure Container Instance" under "Compute type."

After completing as shown in Figure 14.33, click "Deploy" to complete the creation.

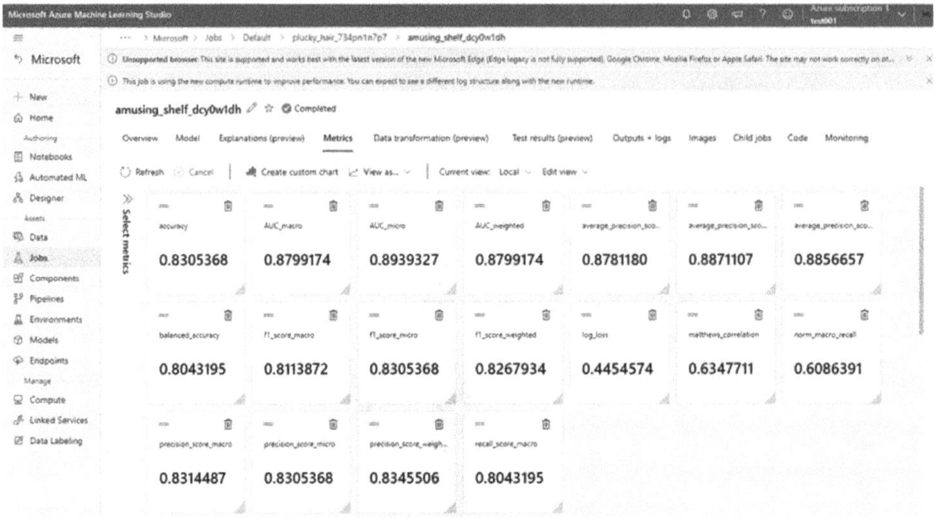

Figure 14.34 Model metrics.

As shown in Figure 14.34, you can check the metrics after deploying. Here, you can confirm various indicators of the used model, such as accuracy and AUC.

14.2 Practice Questions

Q1. How does Azure Automated ML streamline the ML model development process?

Q2. Describe the steps for setting up a Titanic survival prediction model in Azure Automated ML.

Q3. What benefits does Azure Automated ML provide for data scientists and developers?

Chapter 15

Azure Pipeline

Microsoft Azure Pipeline is one of the DevOps services provided by Microsoft Azure. It is a cloud-based service designed to automate and manage the software development and deployment processes. Azure Pipeline includes tools and services that enhances collaboration between software developers and operations teams and helps them to deliver applications quickly and reliably.

15.1 Azure Pipeline for Titanic Survivor Prediction

As shown in Figure 15.1, click "test001" under "Recent workspaces" to navigate to the user workspace.

Figure 15.2 represents the user's workspace screen.

As shown in Figure 15.3, click the [Designer] tab on the left side of the screen. Then, create a new pipeline.

Let's take a look at the process of building a pipeline using data, starting with Figure 15.4. Click on the (+) shown on the screen to import the data. Next, let's create a Data asset. First, set the data type.

Once the data type settings are complete, the next step is to select the data source (Figures 15.5 and 15.6). Users can choose the source of the imported data from Azure storage, local files, or web files.

Figure 15.7 shows the step of setting the storage type. Once you understand the storage types provided by Azure, you can set them according to the desired storage space type for each user.

Next is the process of selecting a file or folder. After setting the path in the upload path as shown in Figure 15.8, simply click "Upload."

If all stages are completed, click the "Create" tab below (Figures 15.9 and 15.10).

If the asset of the titanic_test data is finished in the pipeline, as shown in Figure 15.11, a Success indication will appear.

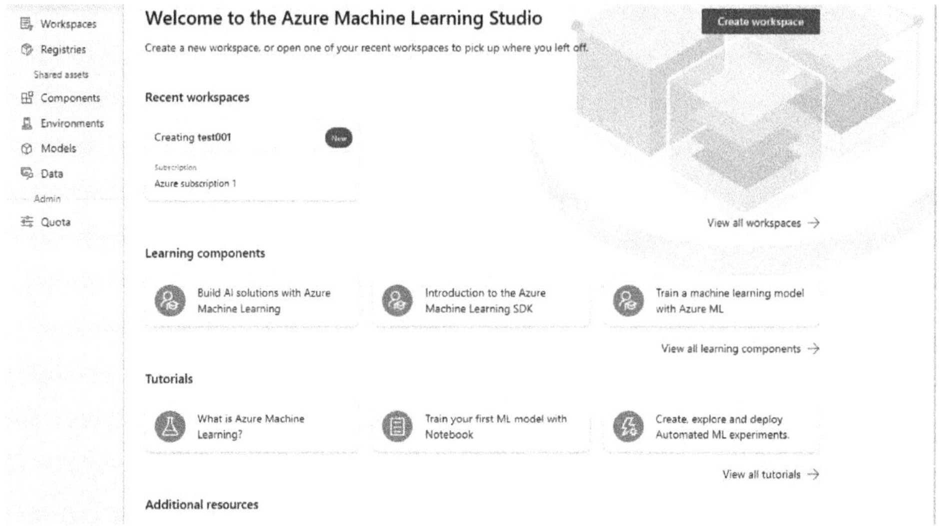

Figure 15.1 Microsoft Azure ML Studio home screen.

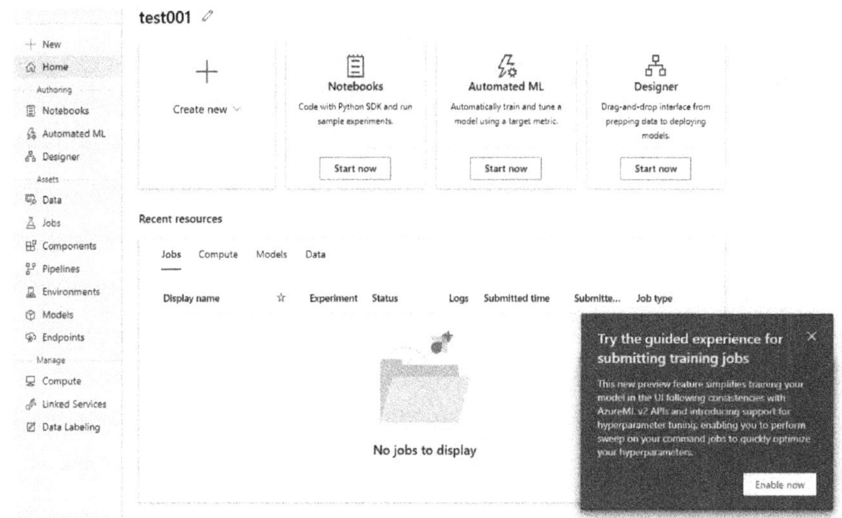

Figure 15.2 Microsoft Azure ML Studio user workspace.

Let's learn how to create a training dataset for Titanic. As shown in Figure 15.12, you can set the name and write a description for the data asset in the "Create Data asset." You can also set the type.

By clicking the "Next" tab shown in Figure 15.11 and setting the Data source, the screen shown in Figure 15.12 will appear.

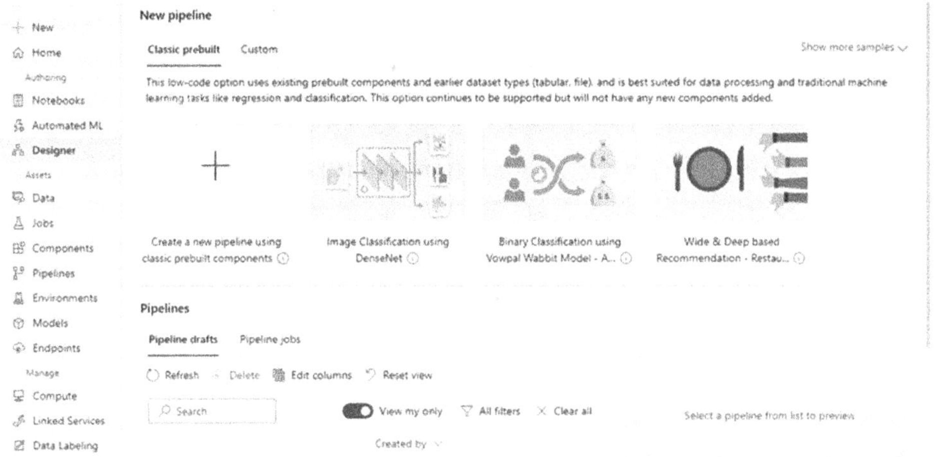

Figure 15.3 Pipeline creation screen.

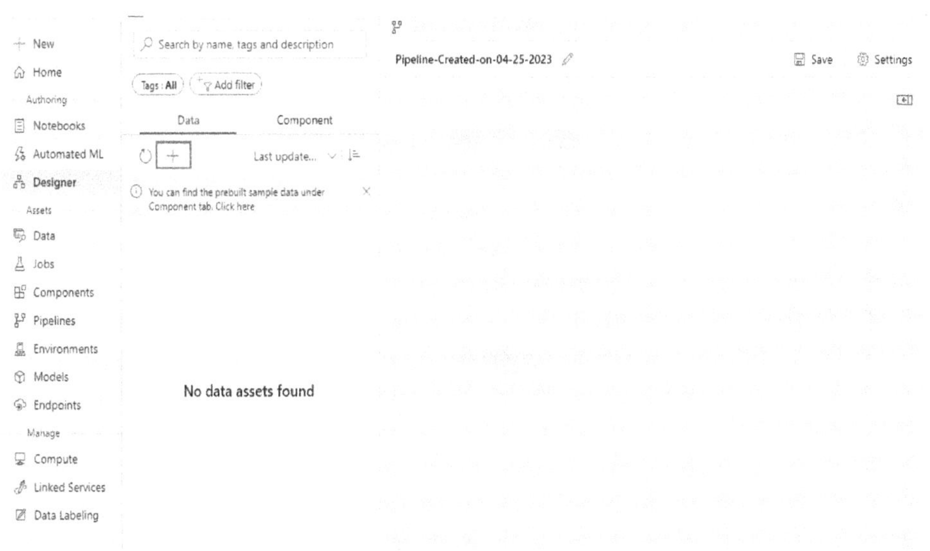

Figure 15.4 Microsoft Azure ML Studio user workspace.

As shown in Figure 15.13, clicking on titanic_train in the Data list will display the visualization field of the pipeline. Then, click on "Preview data" to examine the titanic_train data. Clicking on "Preview data" in Figure 15.13 will allow you to see the data at a glance, as shown in Figure 15.14 .

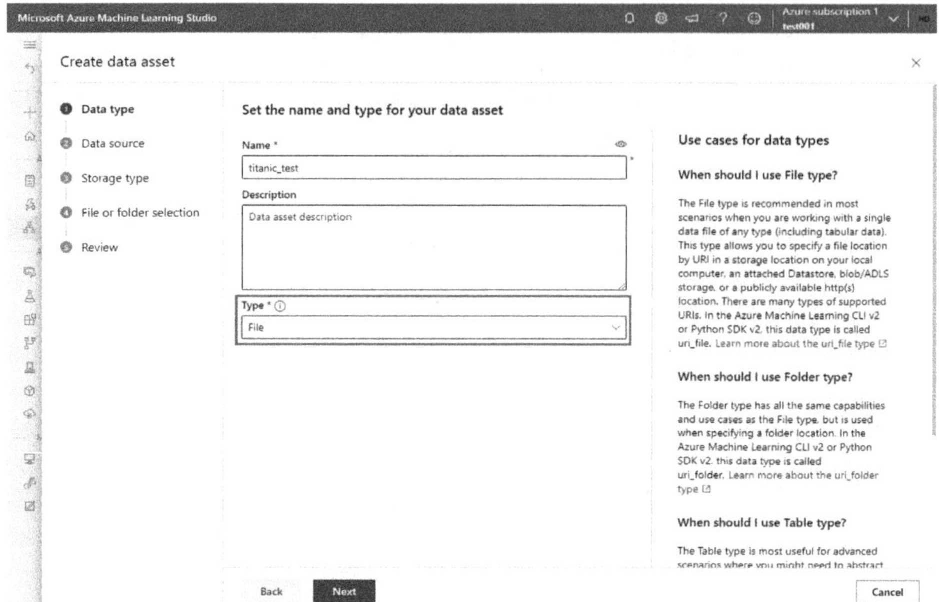

Figure 15.5 Data type.

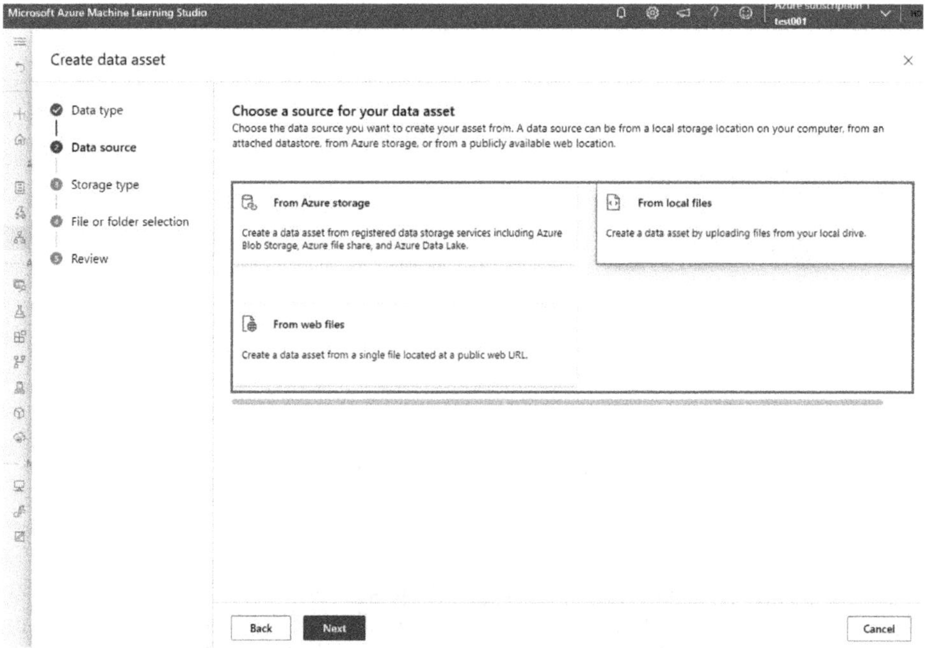

Figure 15.6 Data source.

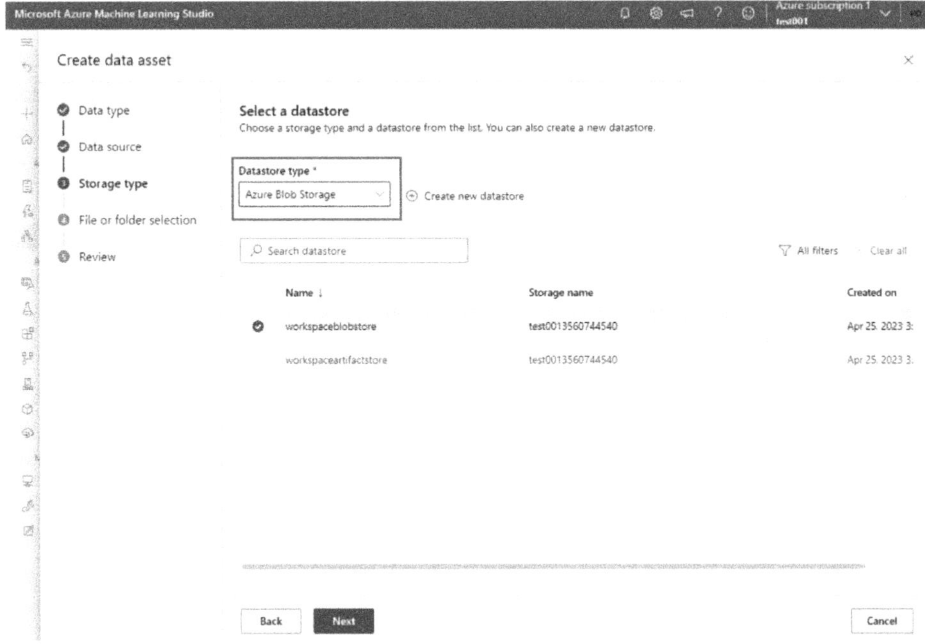

Figure 15.7 Storage type.

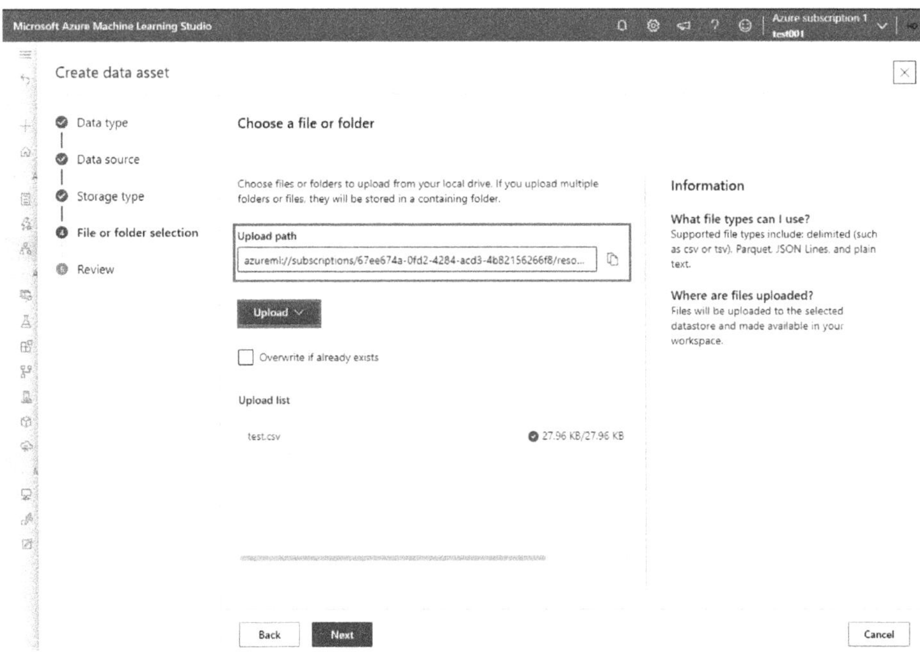

Figure 15.8 File or folder selection.

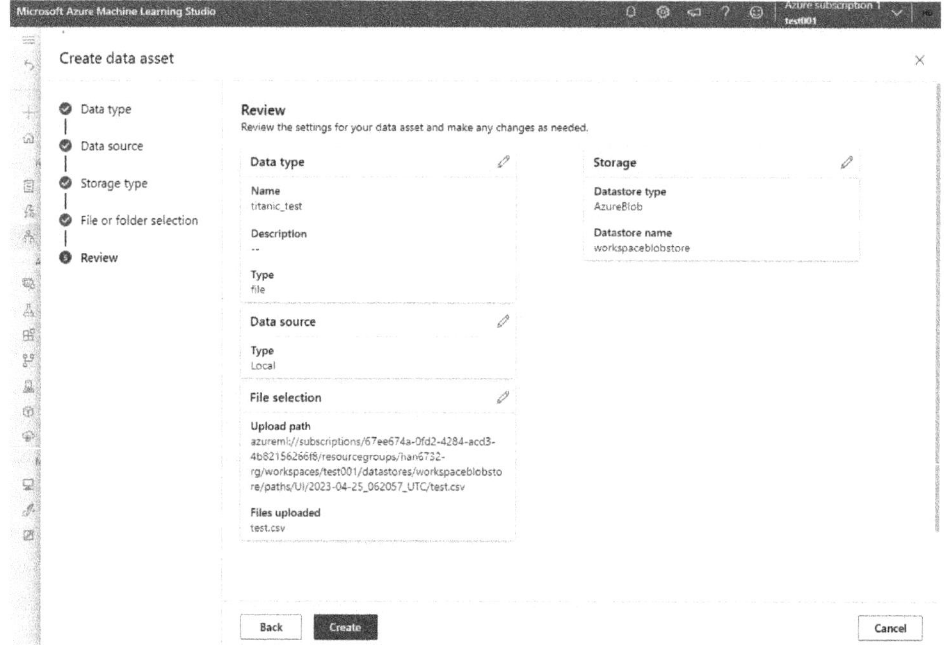

Figure 15.9 Creating review.

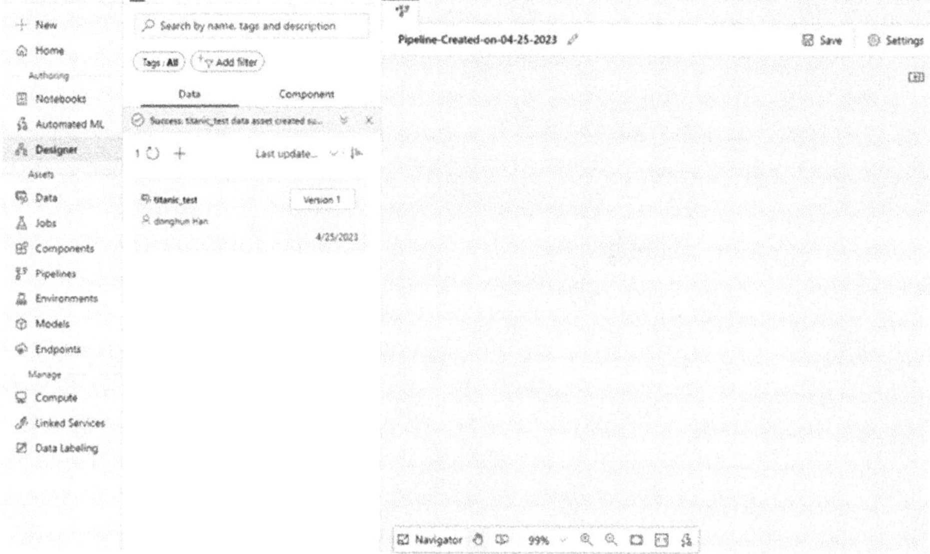

Figure 15.10 After create review.

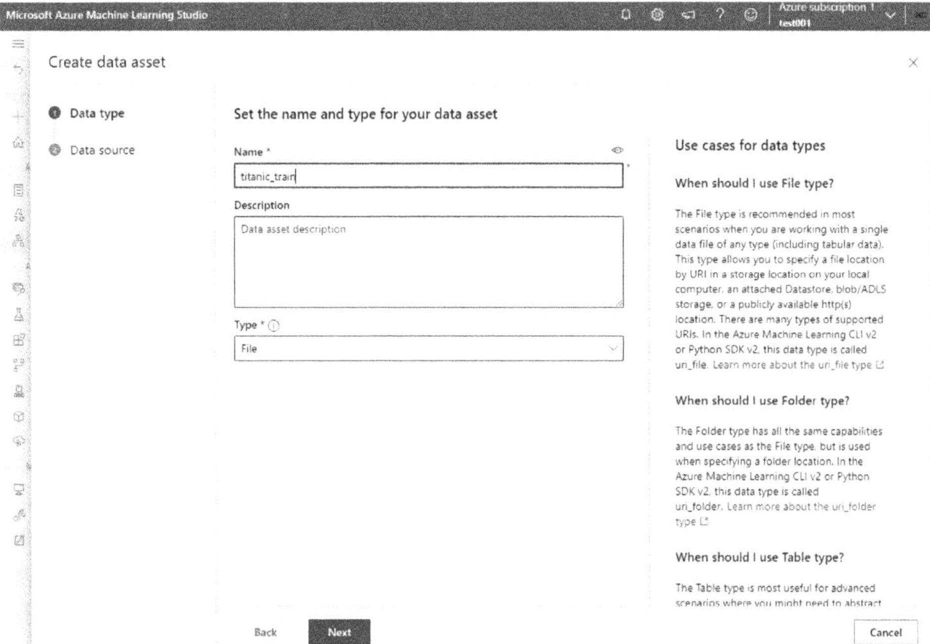

Figure 15.11 Create train data asset.

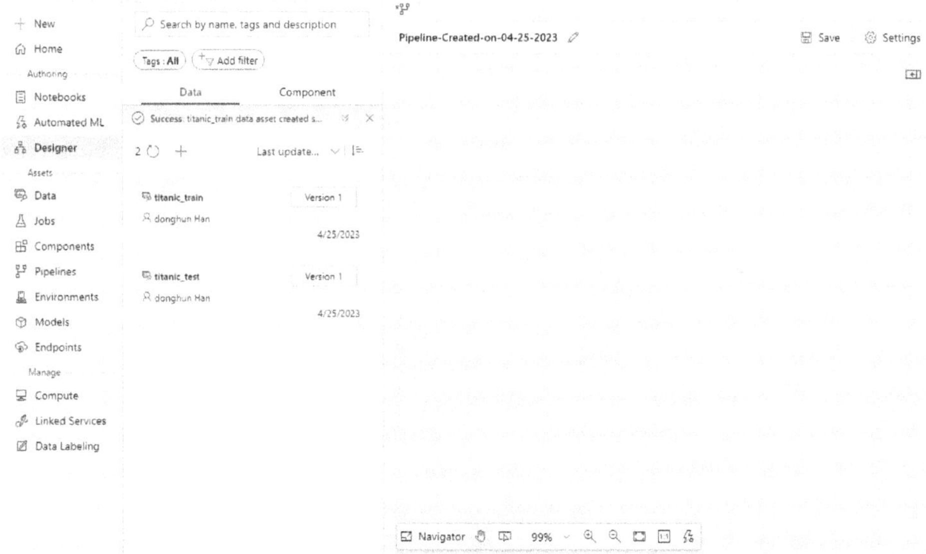

Figure 15.12 Confirm train data asset.

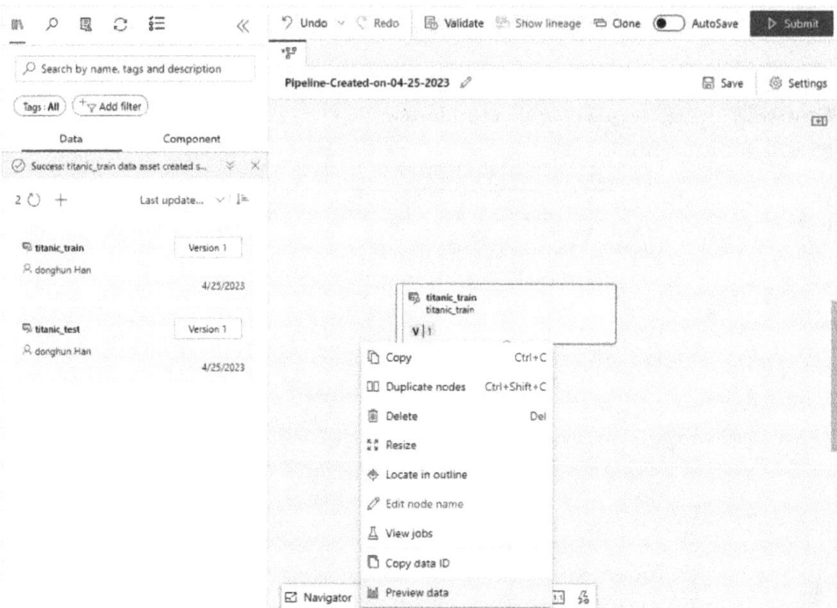

Figure 15.13 Pipeline dashboard.

DataOutput | DataOutput

Number of files: 1 (sampled)

Path	File Na...	Modifi...
/UI/202...	train.csv	2023-04.

Preview

Display as grid — With column headers

Passenge...	Survived	Pclass	Name	Sex	Age	SibSp	Parch	Ticket	Fare	Cabin	Embarked
1	0	3	Braund, M...	male	22	1	0	A/5 21171	7.25		S
2	1	1	Cumings, ...	female	38	1	0	PC 17599	71.2833	C85	C
3	1	3	Heikkinen,...	female	26	0	0	STON/O2. ...	7.925		S
4	1	1	Futrelle, M...	female	35	1	0	113803	53.1	C123	S
5	0	3	Allen, Mr. ...	male	35	0	0	373450	8.05		S
6	0	3	Moran, Mr...	male		0	0	330877	8.4583		Q
7	0	1	McCarthy, ...	male	54	0	0	17463	51.8625	E46	S
8	0	3	Palsson, M...	male	2	3	1	349909	21.075		S
9	1	3	Johnson, ...	female	27	0	2	347742	11.1333		S
10	1	2	Nasser, Mr...	female	14	1	0	237736	30.0708		C
11	1	3	Sandstrom...	female	4	1	1	PP 9549	16.7	G6	S
12	1	1	Bonnell, M...	female	58	0	0	113783	26.55	C103	S
13	0	3	Saunderco...	male	20	0	0	A/5. 2151	8.05		S

Figure 15.14 Preview data.

As shown in Figure 15.15, by searching for “select columns in dataset” in the top left corner, you can identify the components. After that, place “Select Columns in Dataset” onto the Pipeline Dashboard.

Connect the titanic_train dataset to “Select Columns in Dataset” as shown in Figure 15.16.

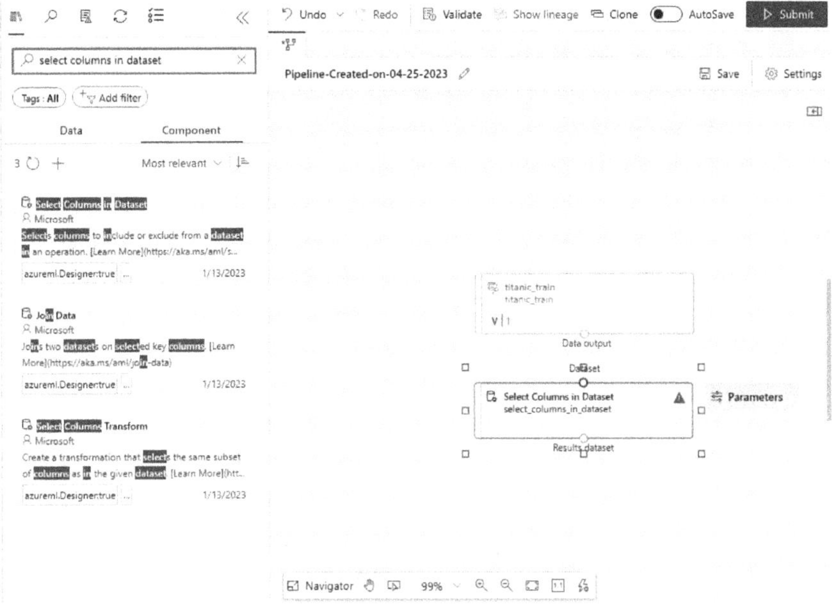

Figure 15.15 Select columns in dataset.

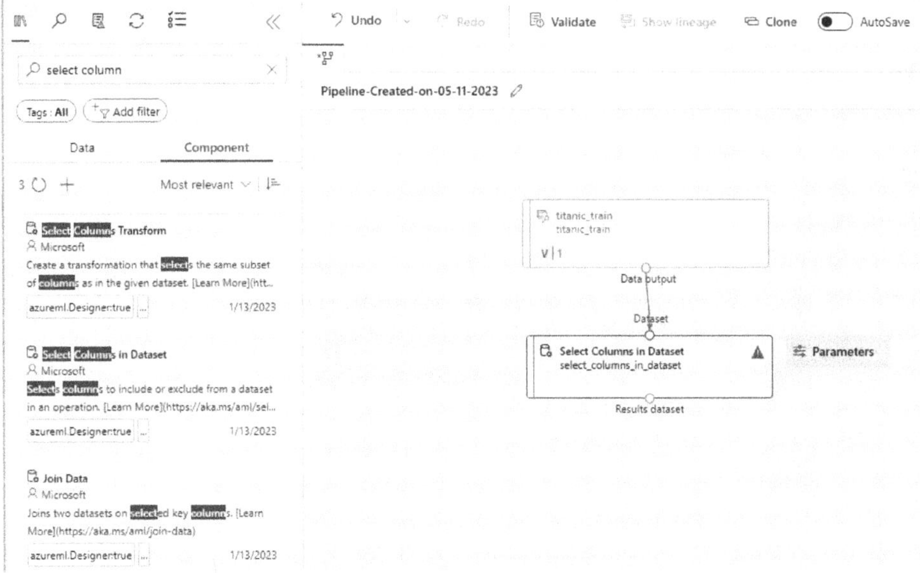

Figure 15.16 Connect dataset.

Select the desired columns in the dataset as shown in Figure 15.17.

Choose the columns "Survived," "Pclass," "Sex," and "Age," as shown in Figure 15.18.

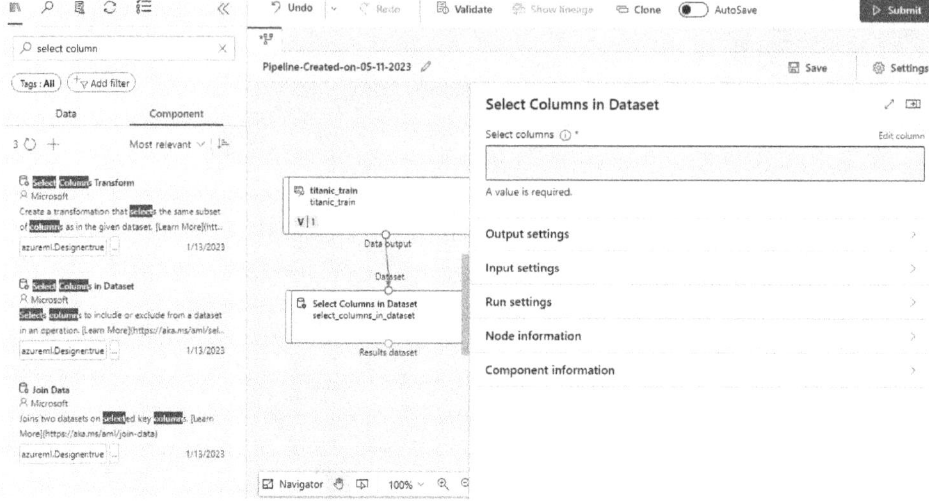

Figure 15.17 Select columns in dataset.

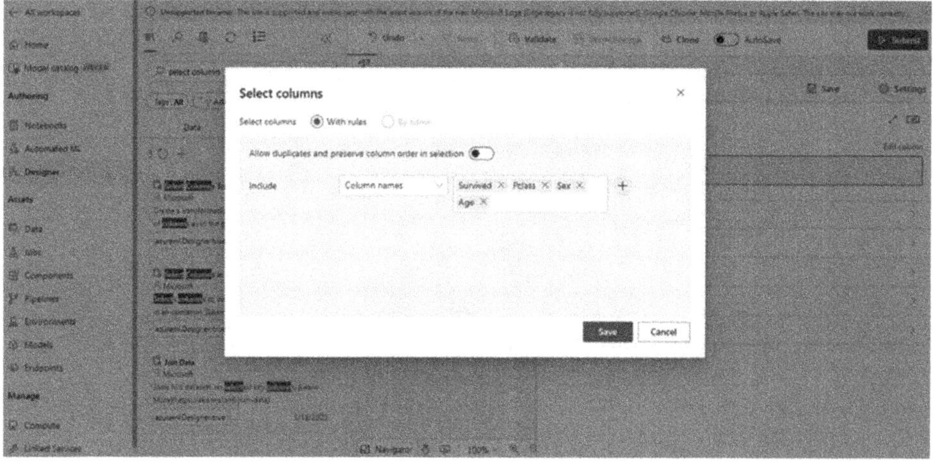

Figure 15.18 Select column.

If you search for “edit meta” in the search bar at the top left corner, you can find “Edit Metadata.” Afterward, upload “Edit Metadata” to the Pipeline Dashboard (Figure 15.19).

As shown in Figure 15.20, the Edit Metadata screen defines the metadata of the learning and prediction data, where you can set the data type, category, and field settings.

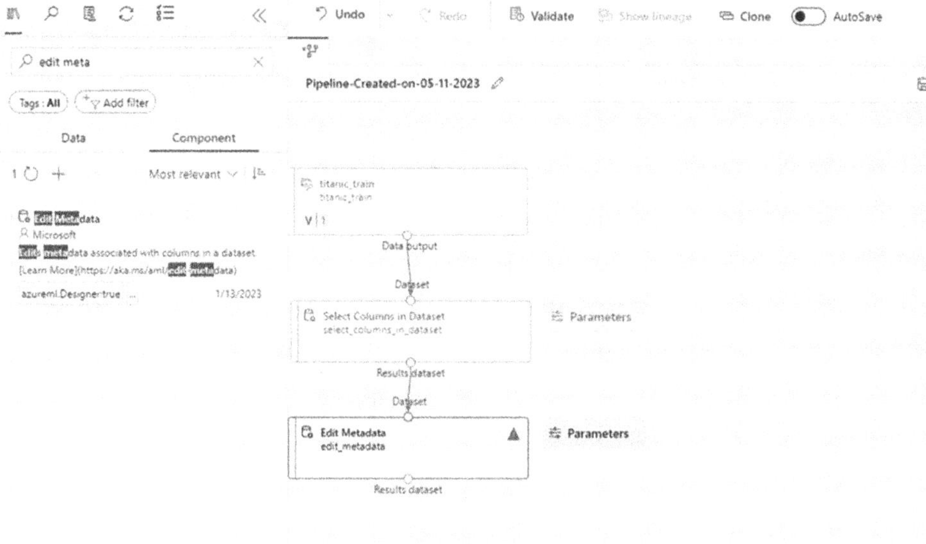

Figure 15.19 Edit metadata search.

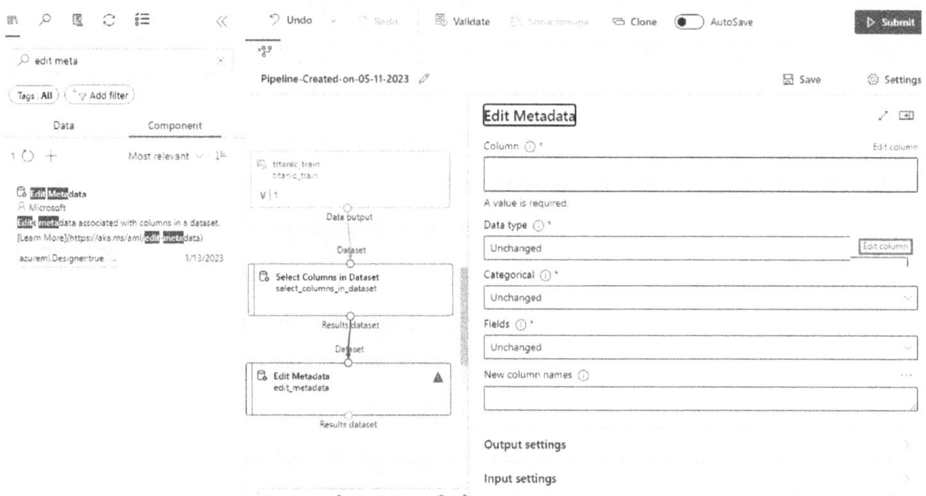

Figure 15.20 Edit metadata.

If you click "Edit column" on the right side here, the screen shown in Figure 15.21 will appear. You can set the column here, and select the "Survived" column. Once you have completed the settings, click "Save" below.

Figure 15.22 shows the step of setting the data type, and the data type is set as "Boolean."

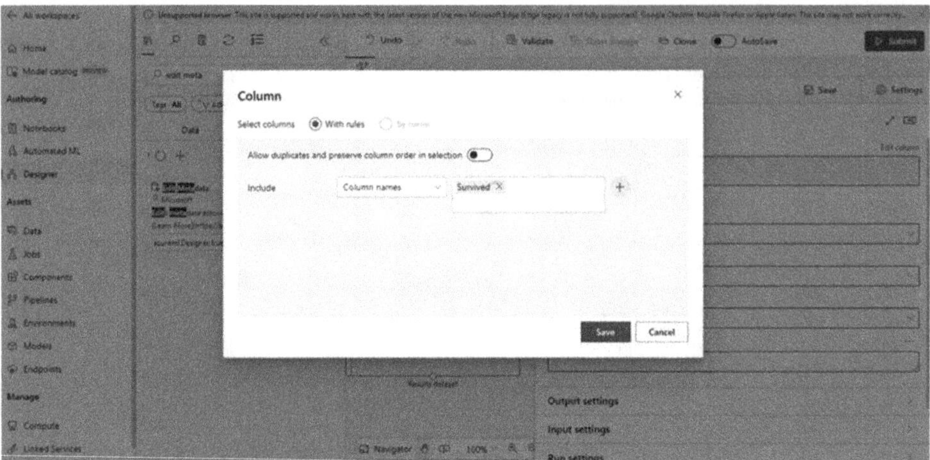

Figure 15.21 Setting the edit metadata column.

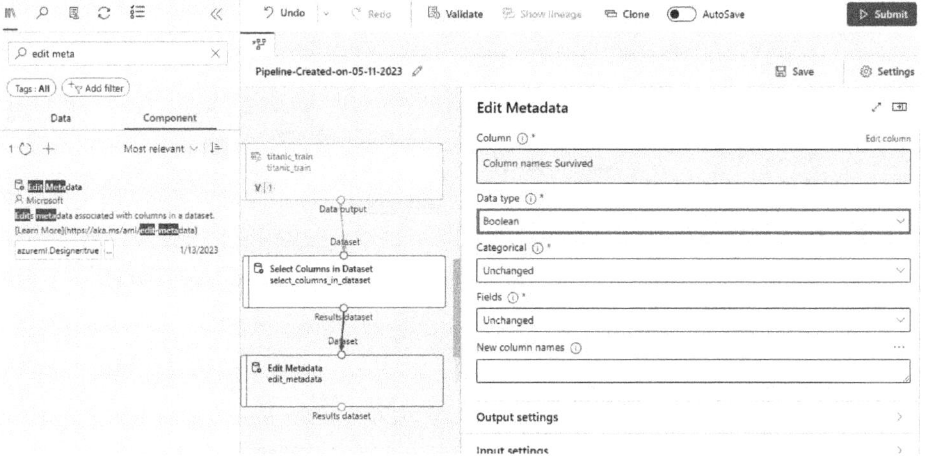

Figure 15.22 Set data type for edit metadata.

After completing the Edit Metadata step, search for “clean missing” in the search box at the top left corner, as shown in Figure 15.23. Select “Clean Missing Data” and upload it to the pipeline dashboard. This process is understood as the process of dealing with missing values in the data.

As shown in Figure 15.24, users can set Clean Missing Data on a per-item basis. After users confirm the following items, they can easily handle the missing values in the data. In the “Columns to be cleaned” section of Figure 15.24, if you look at the right side, click “Edit column.”

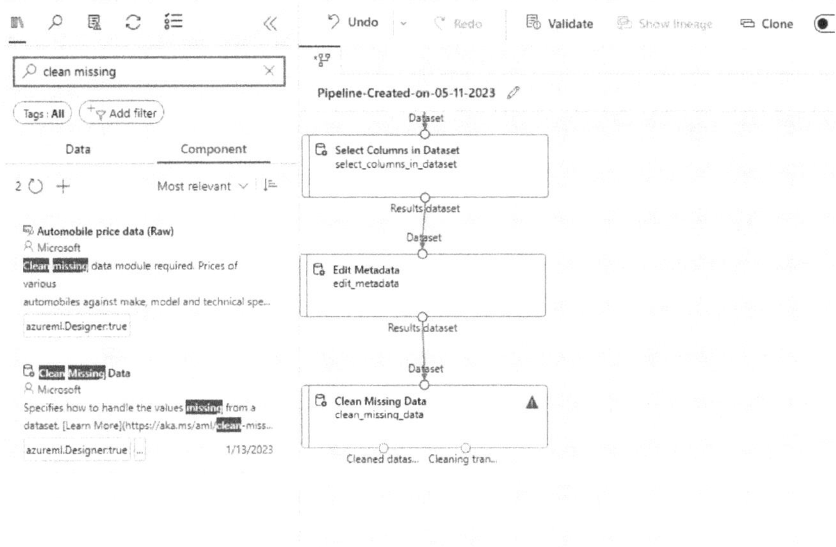

Figure 15.23 Clean missing data search.

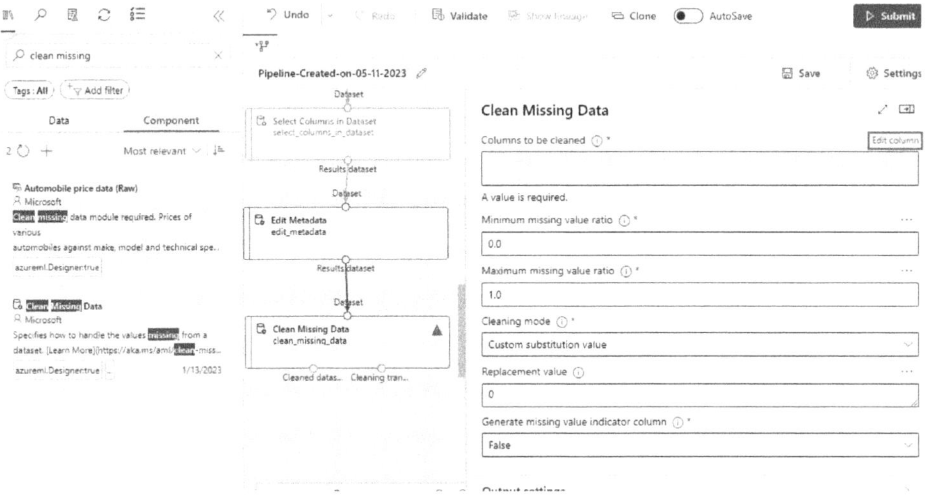

Figure 15.24 Clean missing data.

As shown in Figure 15.25, select “Age” for the column names and then click “Save.”

Afterwards, set Cleaning mode section to “Remove entire row,” as shown in Figure 15.26.

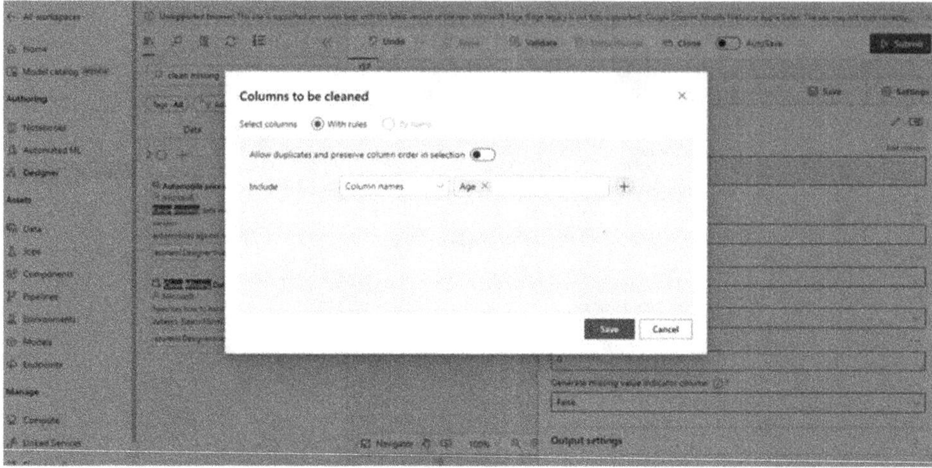

Figure 15.25 Setting up clean missing data column.

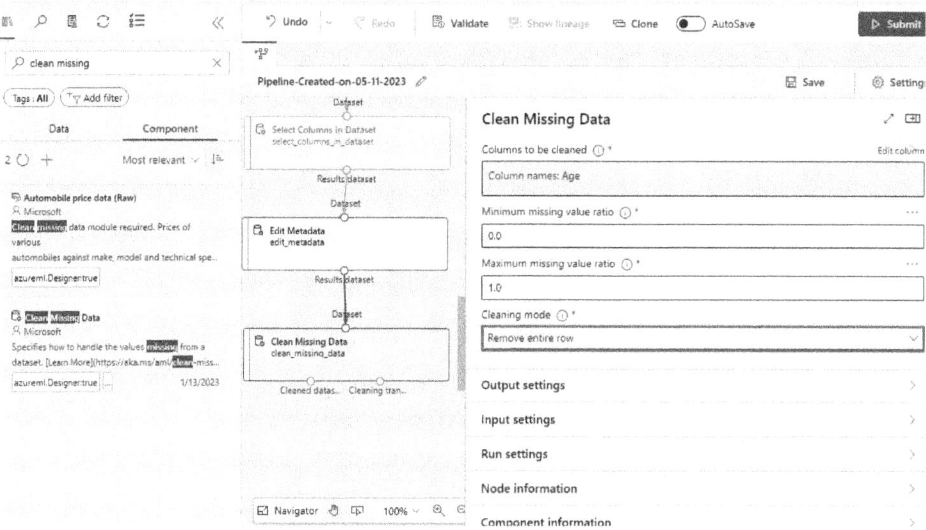

Figure 15.26 Setting of the cleaning mode.

As shown in Figure 15.27, if you search for "split" in the left-hand search bar, you can find the Split Data function. Upload "Split Data" to the pipeline dashboard.

As can be seen in Figure 15.28, users can set Split Data for five items.

As shown in Figure 15.29, search for "two class logistic" in the top-left search bar and upload "Two-Class Logistic Regression" to the pipeline dashboard.

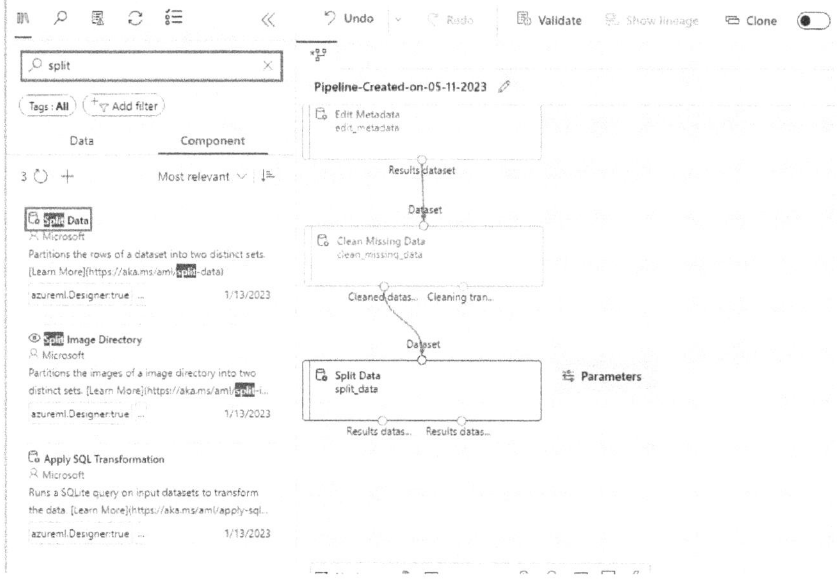

Figure 15.27 Split data search.

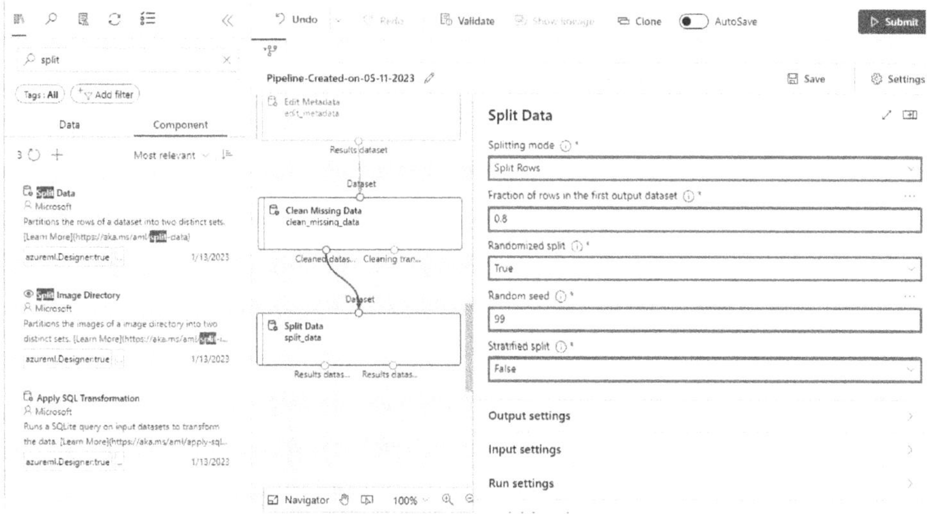

Figure 15.28 Split data setting.

As can be seen in Figure 15.30, after finding "train model" in the search box and uploading it, it has been connected to the Split Data and Two-Class Logistic Regression modules in the pipeline dashboard.

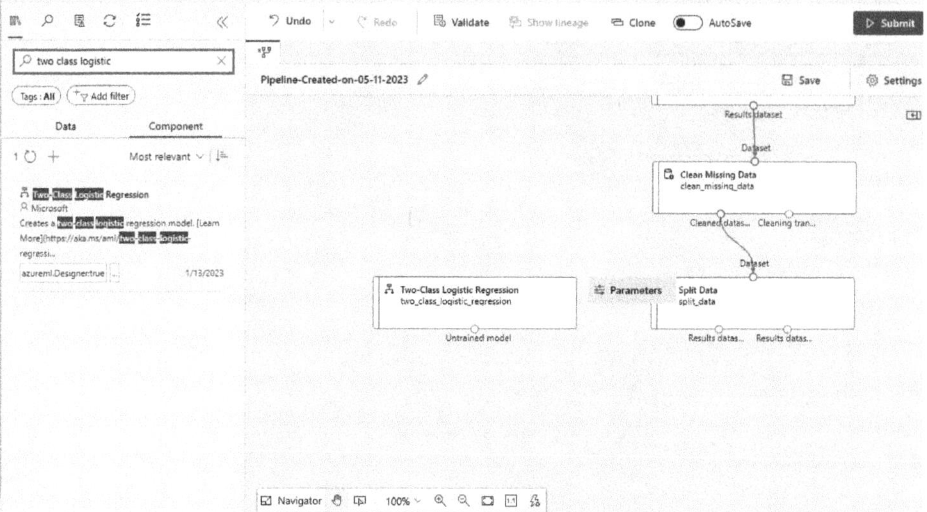

Figure 15.29 Two-class logistic.

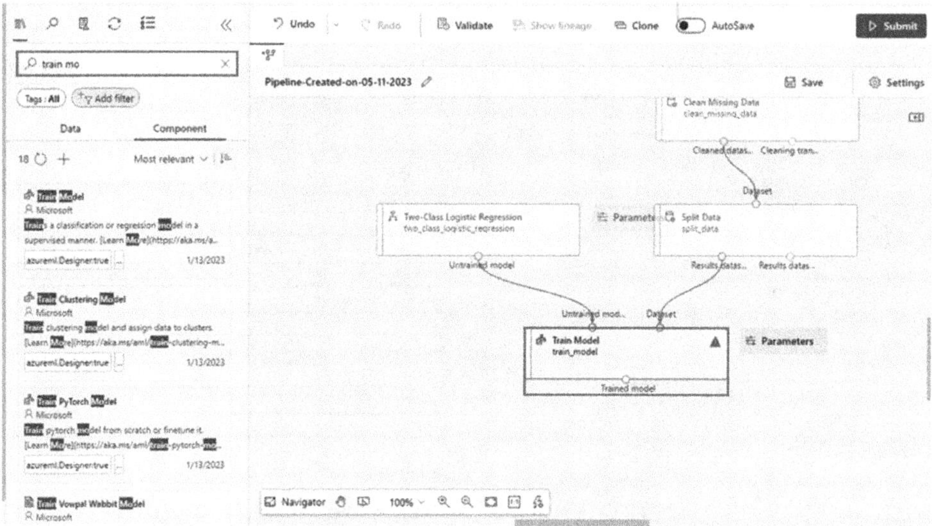

Figure 15.30 Train model.

As shown in Figure 15.31, after selecting "Train Model," you can see the "Edit Column" option in the upper right corner. Next, let's click on "Edit Column."

In the same screen as Figure 15.32, select "Survived" as the column name and click on the Save tab.

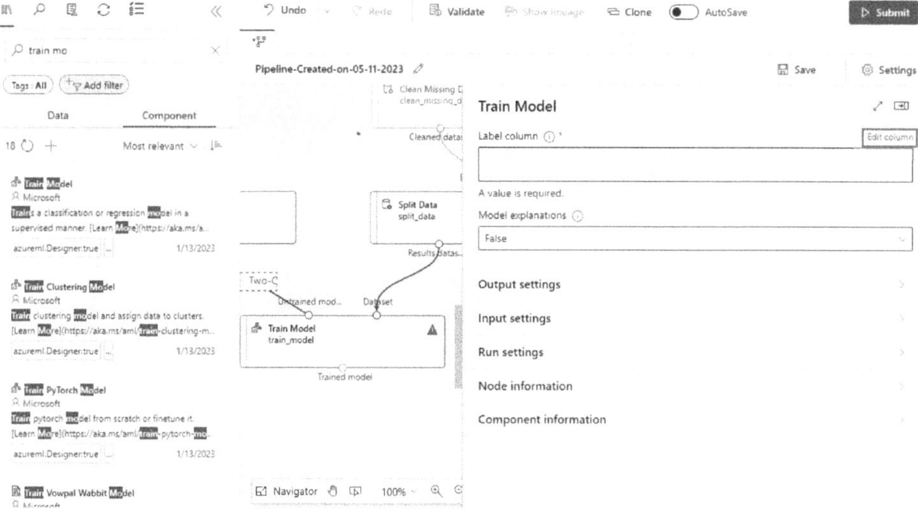

Figure 15.31 Train model settings.

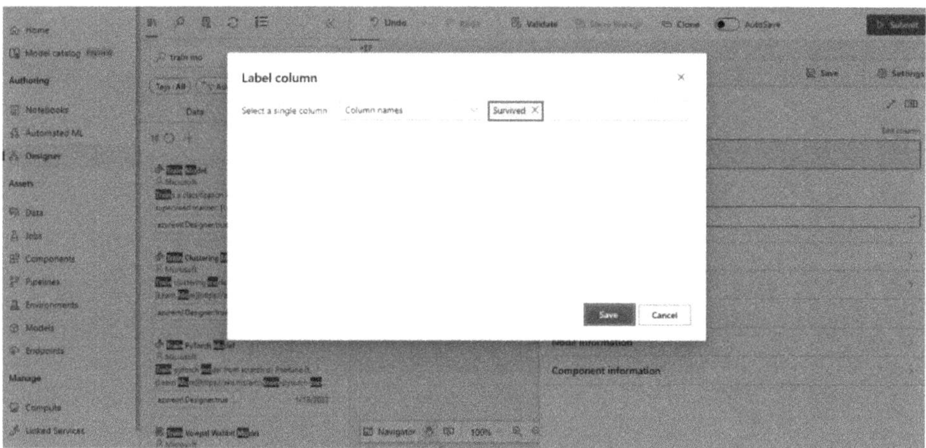

Figure 15.32 Train model column configuration.

As shown in Figure 15.33, search for "evaluate model" in the top left search bar and upload the "Evaluate Model" module to the pipeline dashboard.

As shown in the Evaluate Model setup in Figure 15.34, click on the "Output settings" section and select "Use default output settings workspaceblobstore." Then, click on "Run settings."

The reasons why "invalid" might appear in the Default compute, as shown in Figure 15.35, is because there is currently no pipeline compute instance. Therefore, click "Create Azure ML compute instance" below to create an instance.

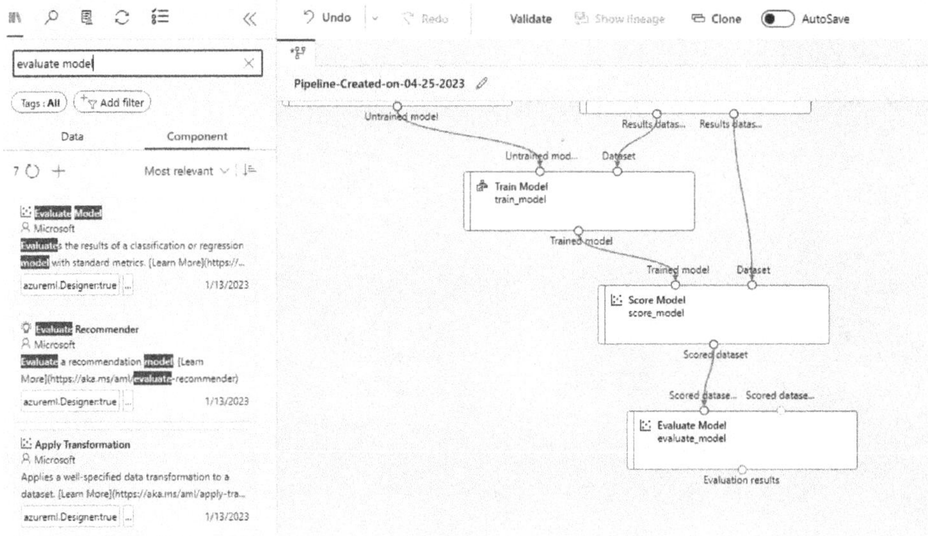

Figure 15.33 Evaluate model search.

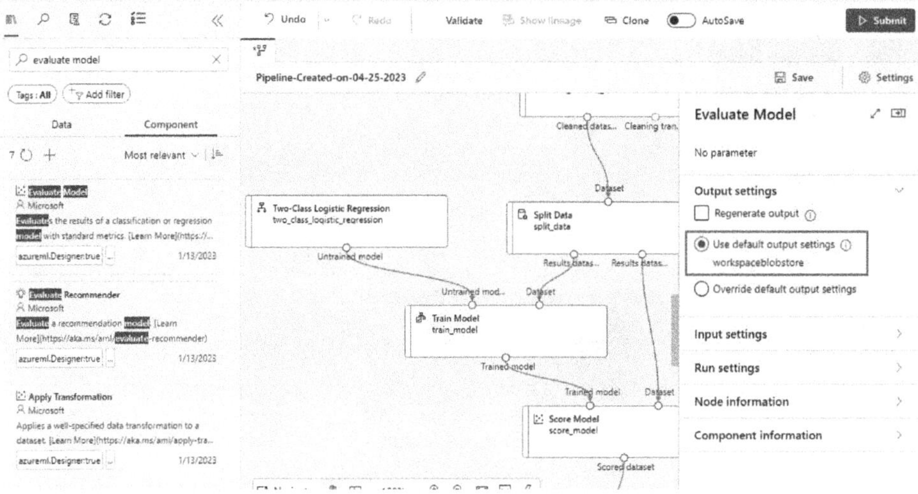

Figure 15.34 Evaluate model settings.

Figure 15.36 shows the process of creating a compute instance. First, in the “Required Settings” step, enter the compute name and set the virtual machine type, then select the size.

Set the compute type and select the Azure ML compute instance as shown in Figure 15.37.

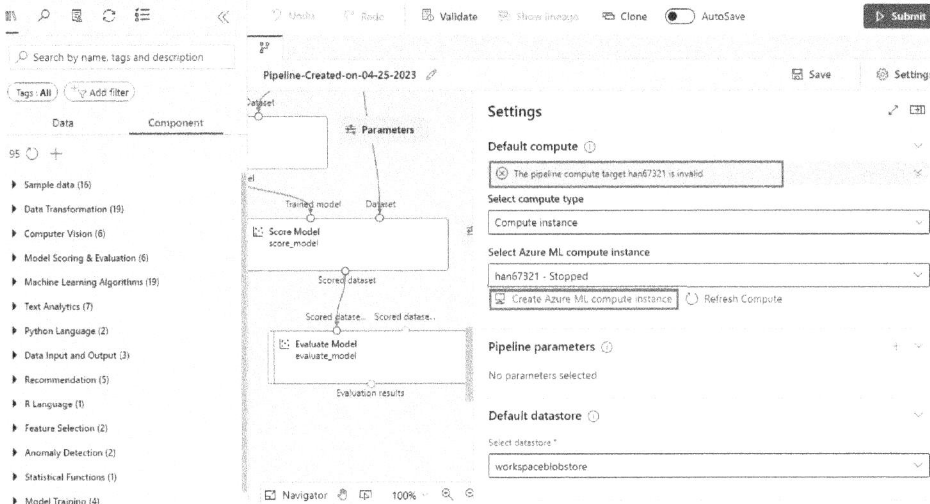

Figure 15.35 Evaluate model settings.

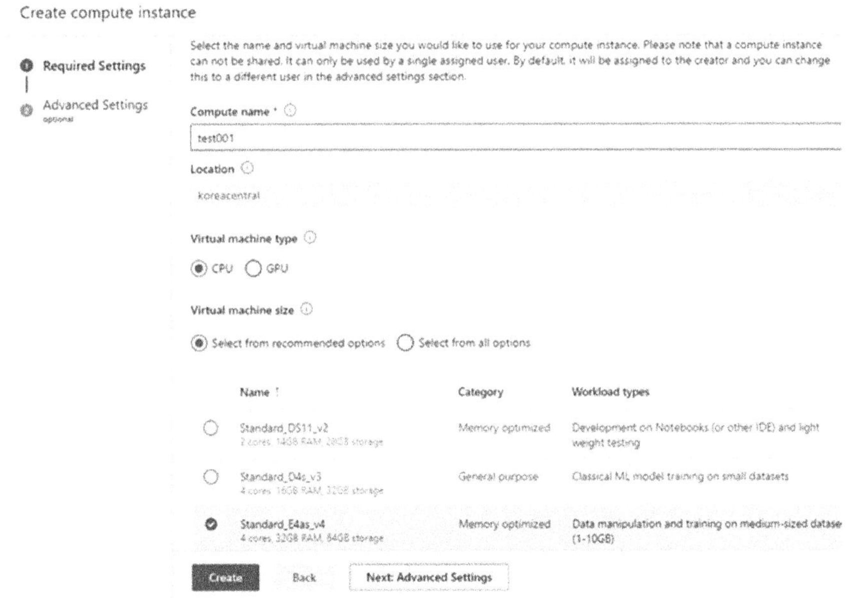

Figure 15.36 Creating a compute instance.

As shown in Figure 15.38, once the Evaluate Model configuration is complete, upload it to the pipeline dashboard. After completing all of the above steps, click [Submit] in the upper right corner.

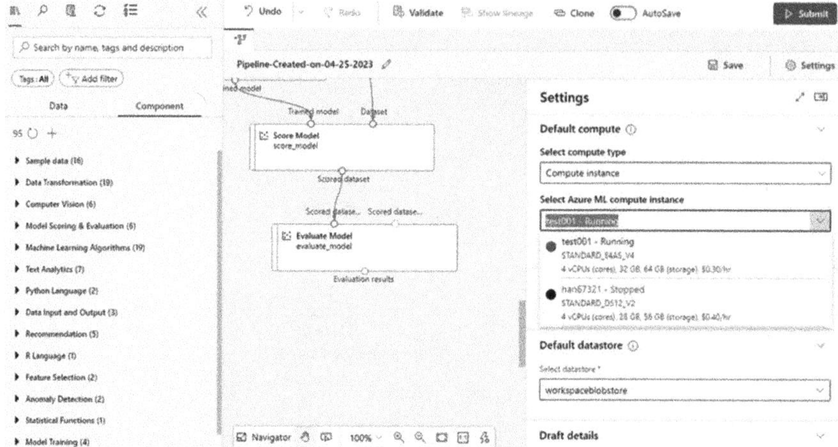

Figure 15.37 Compute instance settings.

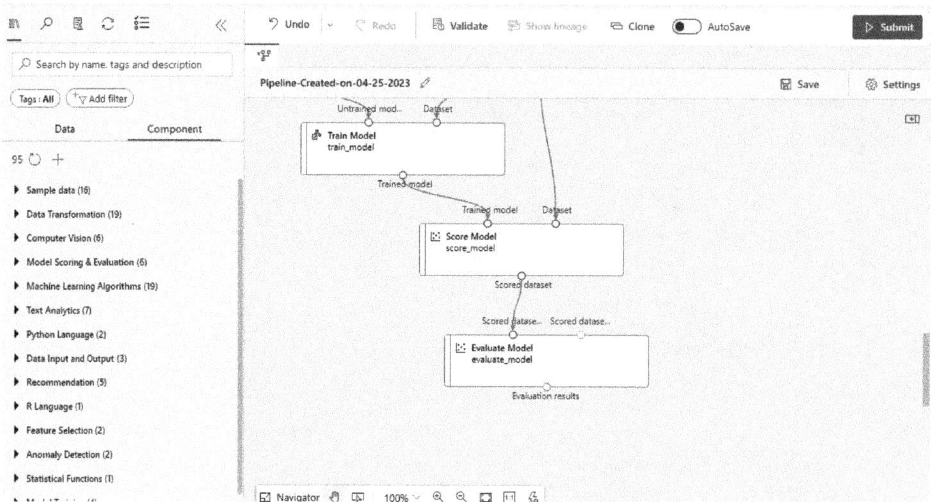

Figure 15.38 Evaluate model upload.

A set up window, as shown in Figure 15.39, will appear. Enter the appropriate content for each item and click "Submit" at the bottom.

Figure 15.40 shows the screen where the pipeline job process is executed, and the job will be completed in approximately 1 minute.

Figure 15.41 is the result of running all the modules created in the Pipeline Dashboard.

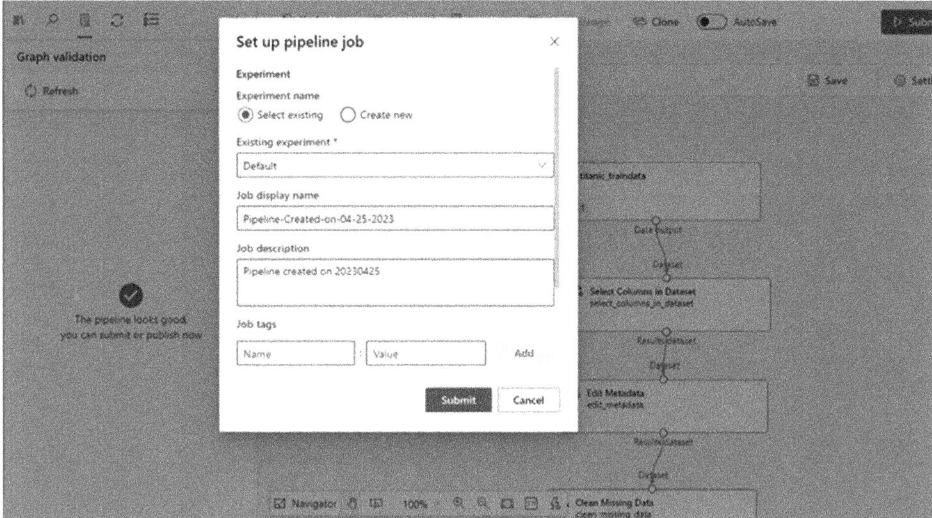

Figure 15.39 Setting up a pipeline job.

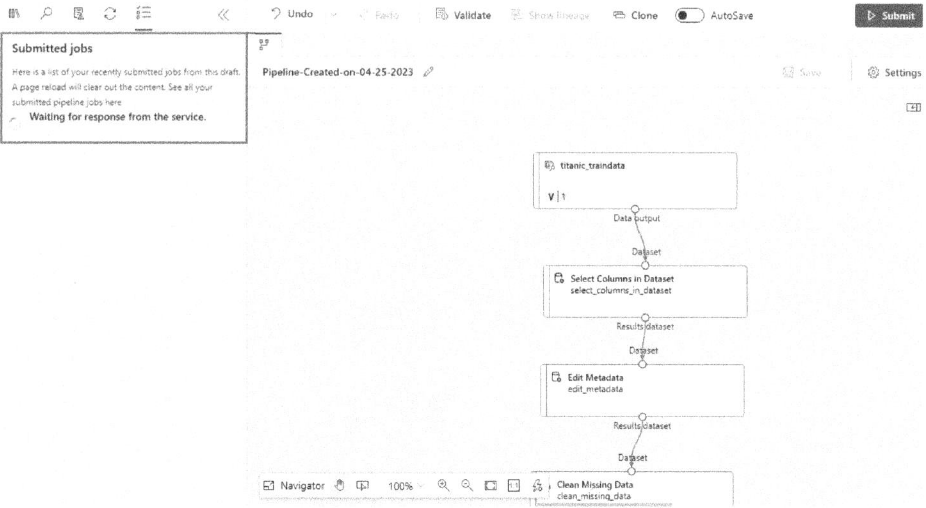

Figure 15.40 Set up pipeline job.

In Figure 15.42, you can confirm the results of the model by clicking “Preview data” in the Score Model, and then clicking “Scored dataset.”

When you click on the scored dataset here, the result will be shown as in Figure 15.43.

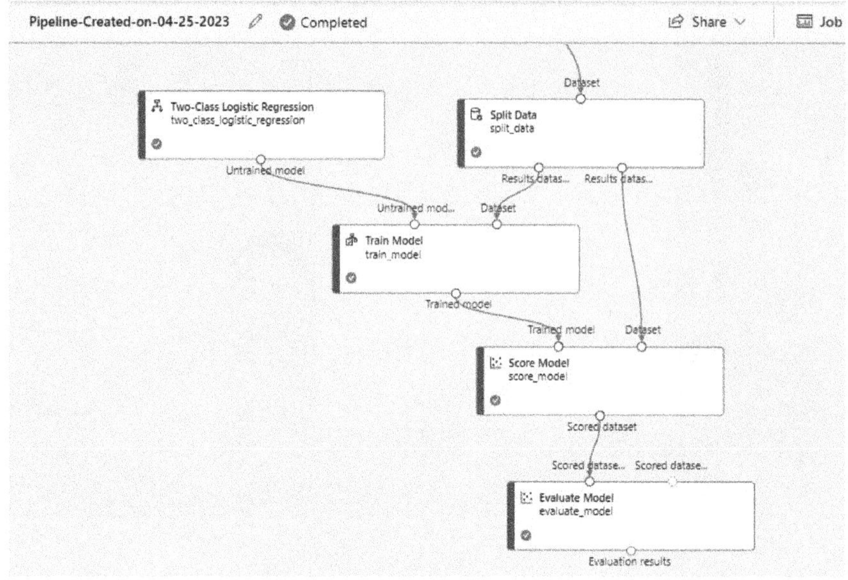

Figure 15.41 Pipeline job results.

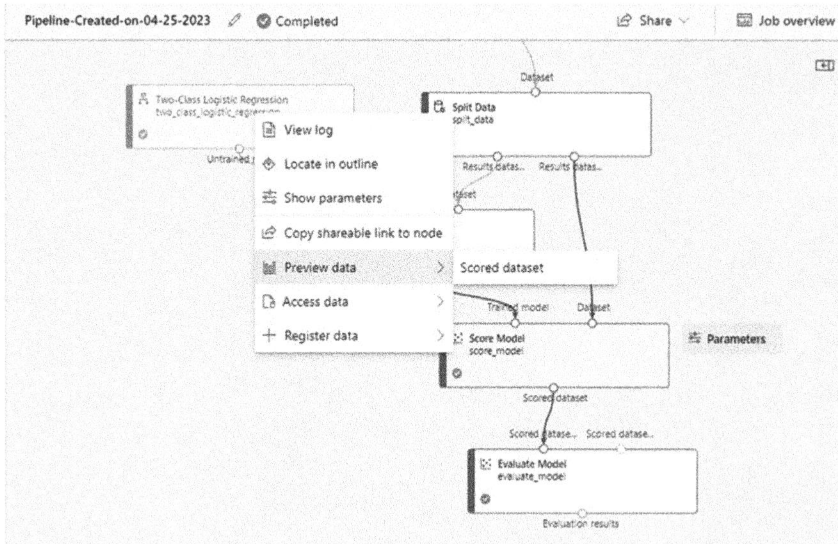

Figure 15.42 Score model.

Through Figure 15.44, you can learn how to check the evaluation metrics of the model. Click on "Evaluation results" in the Preview data.

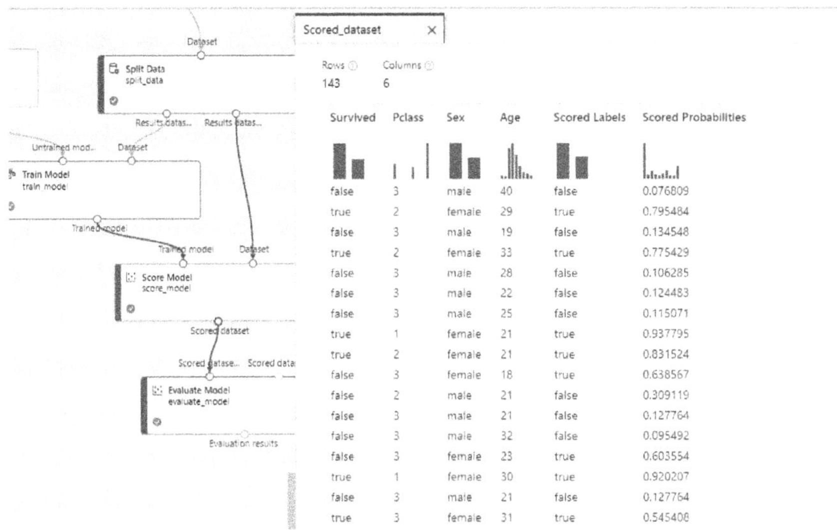

Survived	Pclass	Sex	Age	Scored Labels	Scored Probabilities
false	3	male	40	false	0.076809
true	2	female	29	true	0.795484
false	3	male	19	false	0.134548
true	2	female	33	true	0.775429
false	3	male	28	false	0.106285
false	3	male	22	false	0.124483
false	3	male	25	false	0.115071
true	1	female	21	true	0.937795
true	2	female	21	true	0.831524
false	3	female	18	true	0.638567
false	2	male	21	false	0.309119
false	3	male	21	false	0.127764
false	3	male	32	false	0.095492
false	3	female	23	true	0.603554
true	1	female	30	true	0.920207
false	3	male	21	false	0.127764
true	3	female	31	true	0.545408

Figure 15.43 Score model results.

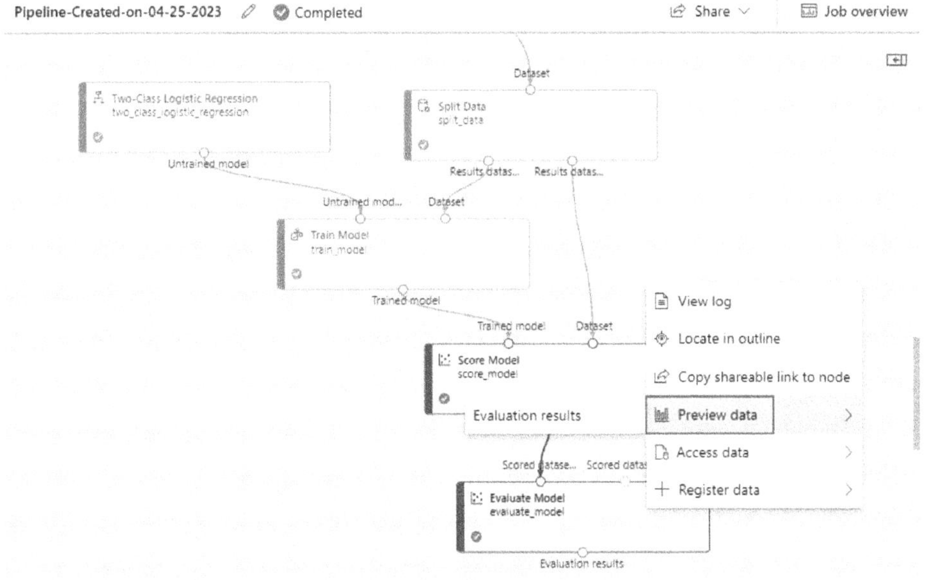

Figure 15.44 Evaluate model.

Figure 15.45 shows the evaluation metrics of the model, visualized and presented to the user. Next, let's explore how to publish the generated pipeline.

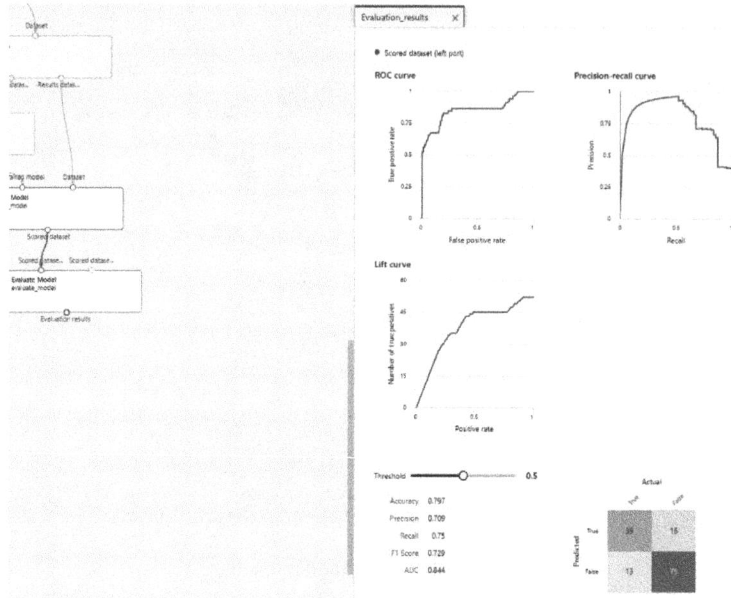

Figure 15.45 Evaluate model evaluation metrics.

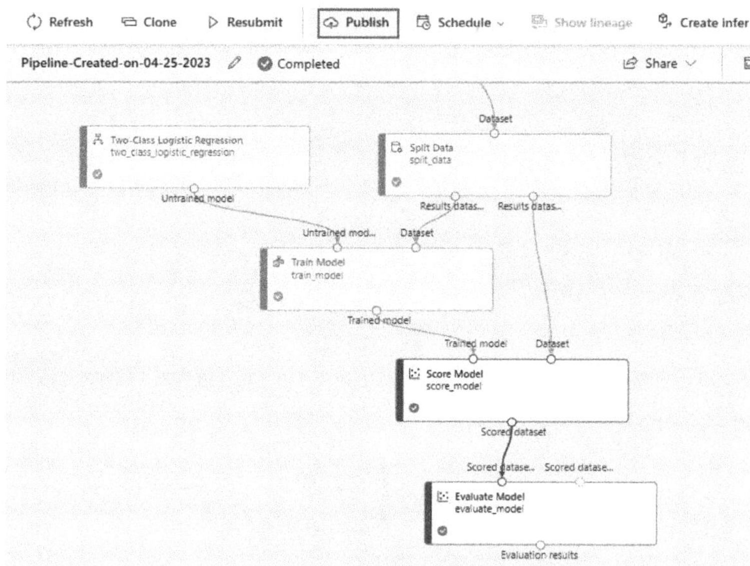

Figure 15.46 Published pipeline.

Let's click on the "Publish" tab shown in Figure 15.46.

To share a pipeline as shown in Figure 15.47, the following window appears. Select "Create new" under PipelineEndpoint, and then enter the appropriate content for each item. Once done, click "Publish" at the bottom right.

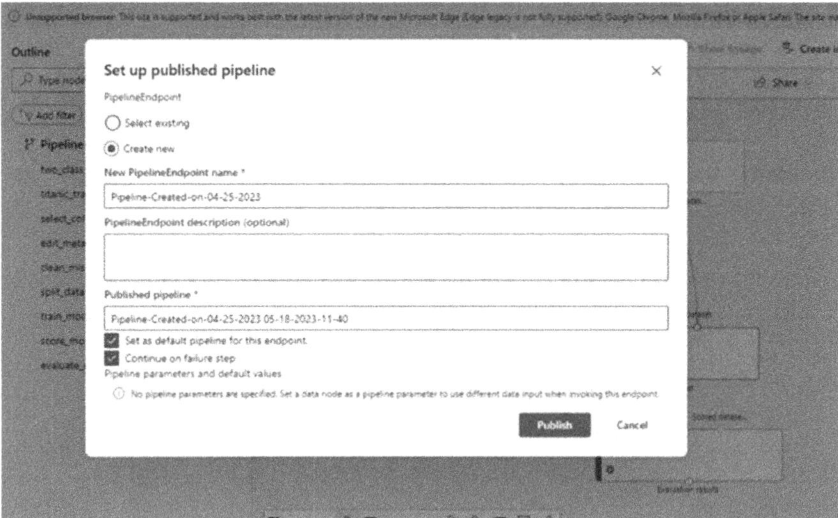

Figure 15.47 Published pipeline settings.

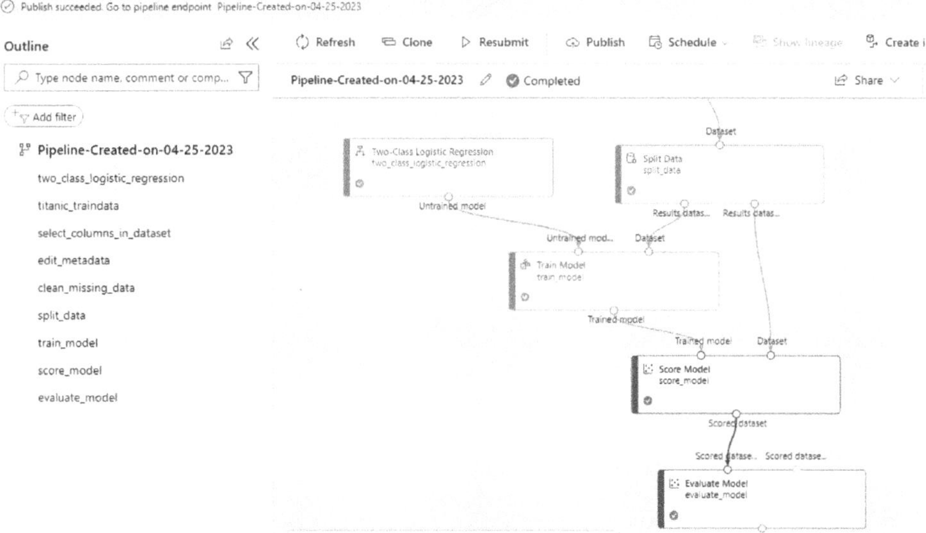

Figure 15.48 Published pipeline completed.

As shown in Figure 15.48, once the pipeline is created and sharing is completed, the message "Publish succeeded" will appear.

15.2 Practice Questions

Q1. How does Microsoft Azure Pipeline enhance the software development and deployment process for machine learning projects?

Q2. Describe the steps involved in creating a pipeline for the Titanic survivor prediction using Azure ML Studio.

Q3. What are the key benefits of using Azure Pipeline in machine learning model development and deployment?

Part 3

BI Solutions

Chapter 16

Amazon QuickSight

Amazon QuickSight is a cloud-based business intelligence (BI) tool by AWS. This tool allows users to connect and combine various data sources to analyze data and present the results in the form of reports or dashboards. Released in 2015, QuickSight is compatible with various data sources of AWS as well as big data, spreadsheets, SaaS data, B2B data, and more. QuickSight supports enterprise security requirements, global availability, and built-in redundancy. BI developers can use this tool to easily visualize data and create dashboards to understand business trends and insights. Analysis starts with a single dataset, and users can register multiple datasets and generate various visualizations for each dataset. These visualizations can be placed on specific sheets, and multiple sheets can be created.

The completed sheets can be published as dashboards and can be shared with team members or sent as periodic report emails. The main advantages of Quick sight are as follows (Figures 16.1 and 16.2):

- Can perform collaborative analysis without the need to install applications.
- Various data can be combined into one analysis.
- Analysis can be posted and shared on a dashboard.
- Efficient analysis of service (purchasing rate, MAU, data visualization on advertising effectiveness, etc.)

Serverless auto scaling

QuickSight is serverless and can automatically scale to tens of thousands of users, without any infrastructure to manage or capacity to plan for. You no longer need to apply software patches or upgrade hardware to meet increasing demand.

Amazon Q in QuickSight

The new Amazon Q dashboard authoring capabilities in QuickSight make it easier than ever before to analyze data and build interactive dashboards. Amazon Q in QuickSight makes it easier for business users to use existing dashboards to better inform business stakeholders, distill key insights, and simplify decision-making.

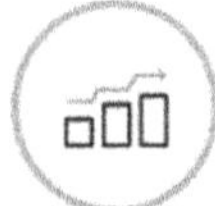

Broad data source support

QuickSight allows you to directly connect to and import data from various cloud and on-premises data sources. These include software-as-a-service (SaaS) applications such as Salesforce, Square, ServiceNow, Twitter, GitHub, and Jira; third-party databases such as Teradata, MySQL, Postgres, and SQL Server; native AWS services such as Amazon Redshift, Amazon Athena, Amazon Simple Storage Service (Amazon S3), Amazon Relational Database Service (Amazon RDS), and Amazon Aurora; and virtual private cloud (VPC) subnets. You can also upload various file types including Excel, CSV, and JSON.

SPICE (Super-fast, Parallel, In-memory Calculation Engine)

With SPICE, the QuickSight in-memory calculation engine, you achieve blazing fast performance at scale. SPICE automatically replicates data for high availability, allowing thousands of users to simultaneously perform fast, interactive analysis, while shielding your underlying data infrastructure, which saves you time and resources.

Support for billion-row datasets

SPICE now supports datasets up to a billion rows, so you can access and analyze very large datasets quickly. Billion-row dataset support is now available to all Amazon QuickSight Enterprise Edition customers.

Global collaboration and multitenancy

As an AWS service with customers all over the world, QuickSight has been designed and built as a global product from the beginning. The QuickSight application is localized in 10 major languages, including English, German, Spanish, French, Italian, Portuguese, Japanese, Korean, Simplified Chinese, and Traditional Chinese. QuickSight is also available across multiple AWS Regions, including US East (Ohio), US East (N. Virginia), US West (Oregon), Asia Pacific (Singapore), Asia Pacific (Sydney), Asia Pacific (Tokyo), and Europe (Ireland).

Built-in security and compliance

QuickSight provides built-in security features so that you can distribute dashboards and insights more securely to tens of thousands of users. In addition to the multi-Region availability and built-in redundancy, you can use QuickSight to manage your users and content through a comprehensive set of security features, including role-based access controls (RBACs), Active Directory integration, AWS CloudTrail auditing, single sign-on (such as AWS Identity and Access Management [IAM] and third parties), VPC subnets, and data backup. QuickSight is also FedRAMP-, HIPAA-, PCI DSS-, ISO-, and SOC-eligible to help you meet any industry-specific or regulatory requirements.

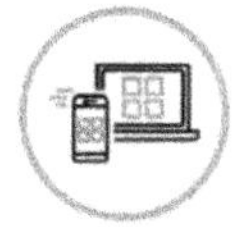

Mobile app support

Amazon QuickSight Mobile for iOS and Android helps you securely get insights from your data from anywhere. Favorite, browse, and interact with all your dashboards in a straightforward mobile-optimized experience. You can explore your data by drilling down and filtering, stay ahead of the curve through forecasting, get email alerts when unexpected changes happen in your data, and share those insights with colleagues. QuickSight Mobile is available as a free download for all QuickSight users from the App Store and Google Play Store.

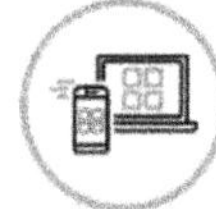

Extensive API capabilities

QuickSight now offers expanded API capabilities that make it easier to create and manage users, datasets, and assets. QuickSight APIs provide programmatic access to BI assets such as dashboards, analysis, and reports. This API access to the underlying data models allows these BI assets to be treated as software source code and managed in your DevOps pipelines. You can use these APIs to programmatically migrate from legacy BI solutions to QuickSight, thereby accelerating the BI transition to the cloud.

Figure 16.1 QuickSight key tasks.

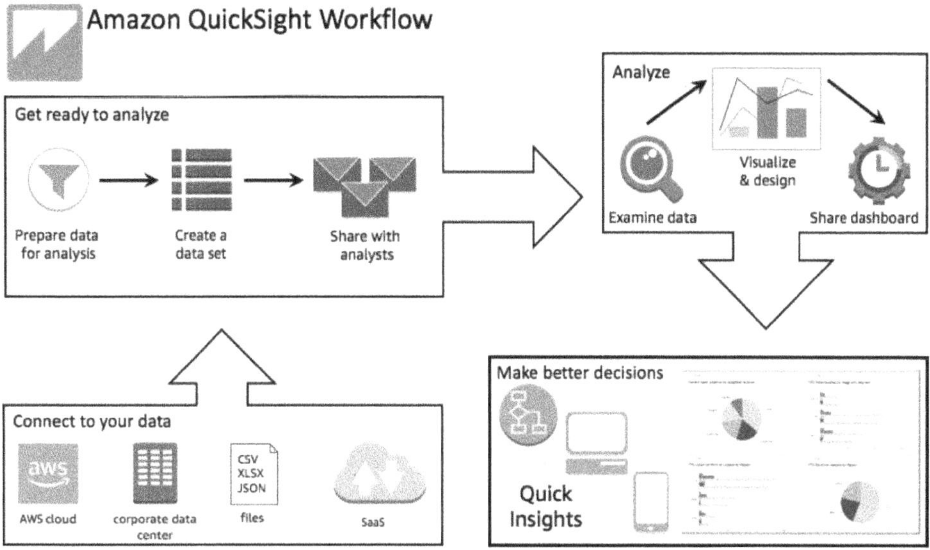

Figure 16.2 QuickSight workflow.

16.1 QuickSight Functions

The key feature of QuickSight is its ability to integrate with various data sources. For example, it can integrate with various databases and cloud storage solutions, including Amazon Redshift, Amazon RDS, Amazon S3, and Amazon Athena. Additionally, it enables easy visualization of data using visualization tools. Users can create dashboards and reports in the form of charts, graphs, and tables using an intuitive interface. They can export and share these dashboards and reports with other users in various formats. QuickSight also offers high performance and scalability, allowing analysis of large datasets and enabling multiple users to access the data simultaneously. Users can economically utilize the service by paying only for what they use. With these features, QuickSight serves as an effective tool for data visualization and BI.

16.1.1 *Dataset*

Amazon QuickSight supports multiple data sources, and QuickSight datasets are data imported into QuickSight for visualization. They can be built from one or more data sources, and can be temporarily saved in QuickSight or directly queried from the data source.

16.1.2 *Visualization and insight*

QuickSight analysis is composed of screen components called widgets that include data visualization, and BI developers can add these two different types of components. Visualization generally refers to representing a dataset graphically using plot types and charts. Other types of visualizations include simple or pivot tables, and geographic spatial visualizations. Insights are components that provide insights into data using common analysis or machine learning functions.

16.1.3 *DashboardS*

Once the analysis is developed and ready to be published, the BI developer can export it to the QuickSight dashboard. The dashboard is typically accessed by business users, and the BI developer can also add customized controls and interactivity to the dashboard. Users must view and interact with the dashboards and reports created by the BI developer.

The user types of Quicksight are as follows (Figure 16.3):

- **READER:** Has read-only access to QuickSight content.
- **AUTHOR:** BI developer who creates and publishes dashboards and reports. They have full access privileges to QuickSight content and can perform analysis, visualization, and editing of dashboards.

	READER	AUTHOR	ADMIN
ACCESS QUICKSIGHT	V	V	V
RECEIVE ALERTS	V	V	V
USE DASHBOARDS	V	V	V
CREATE DATASETS		V	V
CREATE ANALYSIS		V	V
SHARE ANALYSIS		V	V
MANAGE USERS			V
MANAGE QUICKSIGHT			V
MANAGE STORAGE			V

Figure 16.3 QuickSight user type.

- **ADMIN:** An administrator is a user who manages QuickSight accounts and their resources. They create and manage user accounts, set up data sources and permissions, and can monitor usage and activity.

16.1.4 *Business intelligence*

BI is the use of data by organizations to make strategic decisions, with increasing numbers of organizations becoming data-centric. BI tools help organizations transform data into practical insights, allowing users to analyze data and visualize the results in reports or dashboards. These reports or dashboards can be provided to decision-makers who want to understand the business situation.

16.1.5 *Data warehouse*

A data warehouse is a storage repository and an important component of the BI process, and the stored data is typically structured. Data from the existing operational data storage is collected and centralized in the data warehouse. The data warehouse is designed to optimize analysis queries for large amounts of data, and the results are usually calculated after aggregating multiple rows from one or more tables. BI applications aggregate and visualize data using analysis queries. Using a data warehouse to provide data to BI applications is a common architecture approach. Cloud-based data warehouses include Snowflake and Google BigQuery, and in 2012, AWS released Amazon Redshift, a cloud-native, fully managed data warehouse service. It is now integrated with most BI tools, including Amazon QuickSight, and is widely used by many companies across various industries.

16.1.6 *Data lake*

A data lake is a centralized storage repository that stores various data in its original format and can be used for various purposes such as large-scale analysis, predictive analysis, and visualizations. The data is stored in the lake in its original state and can be transformed and utilized for analysis and processing as needed. Cloud-based data lakes use cloud object storage to store data. However, one of the main issues with data lakes is that there are no restrictions on the schema or format of the data, so without proper governance, it can quickly deteriorate into a simple data storage repository. To address this problem, AWS provides various services that support data governance. AWS Glue catalog is one of these services and is a core component of AWS Glue. This catalog is a managed data catalog compatible with Apache Hive, allowing big data applications to explore and process data using metadata. The AWS Glue catalog stores and manages metadata for various data storage repositories such as Amazon S3, Redshift, Aurora, and DynamoDB.

16.2 QuickSight Start

Go to the AWS homepage address (https://aws.amazon.com/ko/). When you access the homepage, the screen shown in Figure 16.4 will appear. If you haven't logged in, click on the login tab in the upper right corner and log in.

When you log in, you will be directed to the AWS Console home screen as shown in Figure 16.5.

To use AWS QuickSight, search for QuickSight in the [Search] window as shown in Figure 16.5. When you search, you will see the QuickSight service provided by AWS as shown in Figure 16.6. Click on QuickSight to run it.

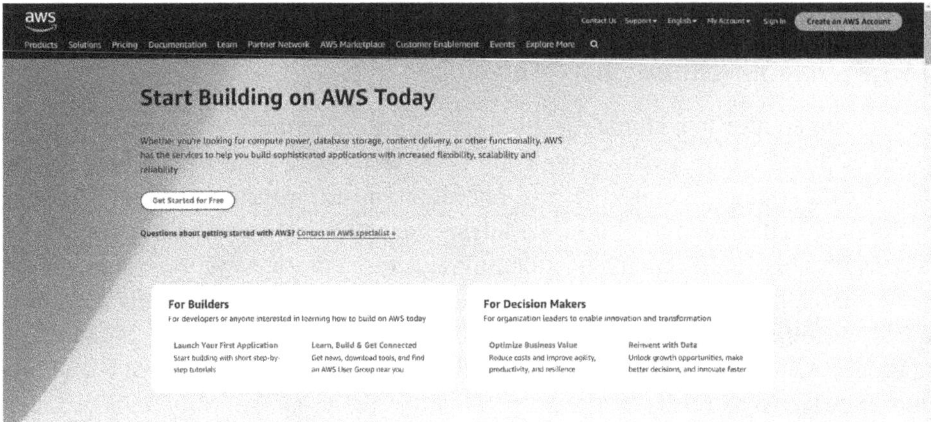

Figure 16.4 AWS homepage.

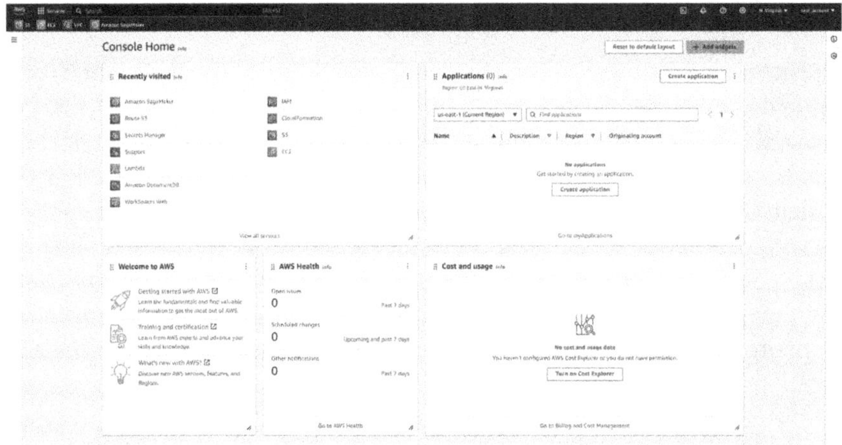

Figure 16.5 AWS Console homepage.

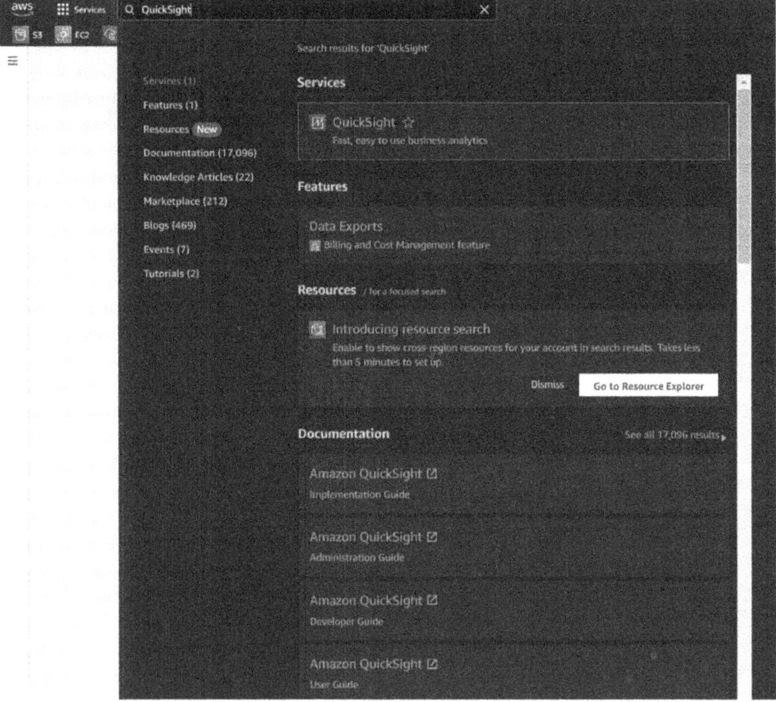

Figure 16.6 QuickSight search screen.

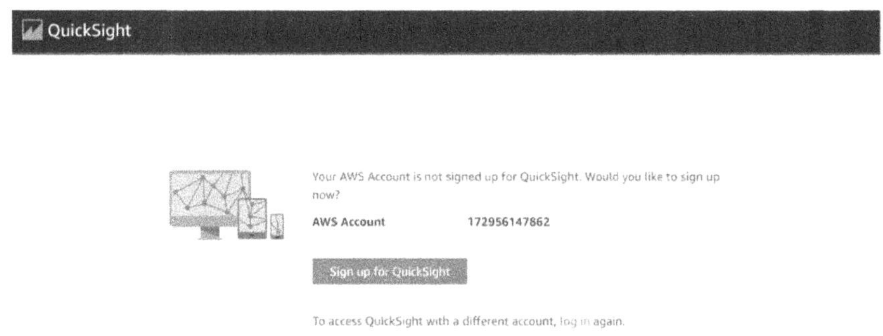

Figure 16.7 QuickSight sign up.

When executed, the screen shown in Figure 16.7 will appear. To use QuickSight, you must first sign up. Therefore, starting from the next chapter, we will take a closer look at the registration process.

16.3 Creating a QuickSight Account

As shown in Figure 16.7, by clicking on the "Sign up for QuickSight" button in the middle, you can sign up for QuickSight. There are two ways to sign up: by selecting Enterprise or Enterprise + Q when signing up.

When you click on "Sign up for QuickSight" in Figure 16.7, the screen shown in Figure 16.8 will appear. When starting QuickSight for the first time, select Enterprise to create an account. If you need more services for QuickSight later, you can select Enterprise + Q to create an account. Once you have made your selection, click "Continue" tab at the bottom right to proceed to the next step.

Figure 16.8 shows the step for configuring the Enterprise settings, QuickSight region settings, and access permissions for QuickSight and AWS services. First, for the Enterprise settings, select "Use IAM federated identities & QuickSight-managed users" for the Authentication method. QuickSight region settings allow users to

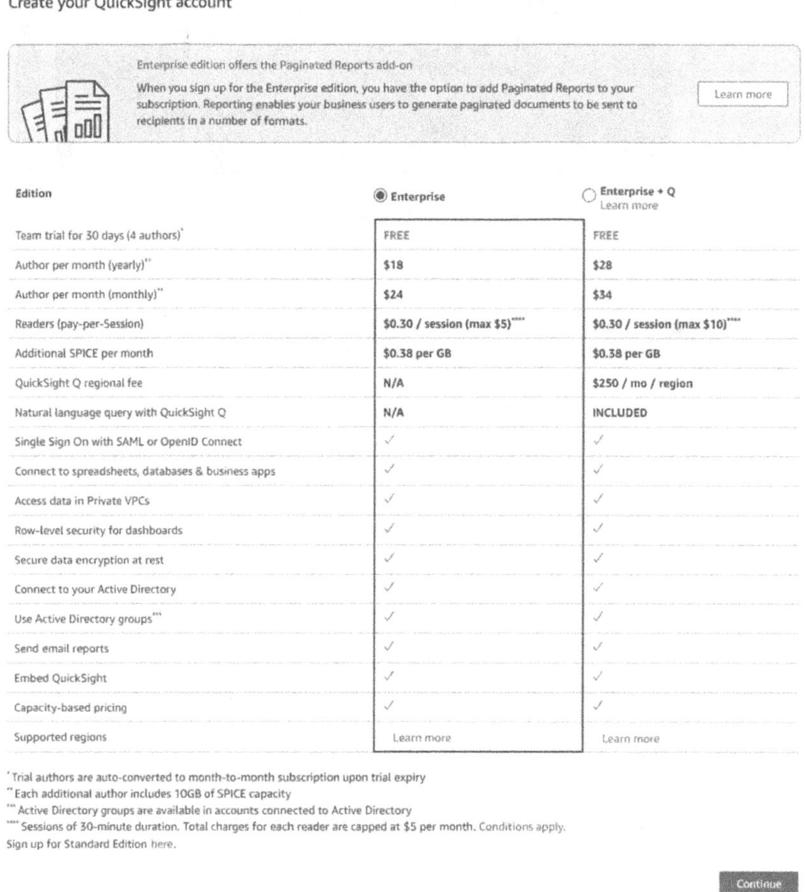

Create your QuickSight account
Enterprise edition offers the Paginated Reports add-on
When you sign up for the Enterprise edition, you have the option to add Paginated Reports to your subscription. Reporting enables your business users to generate paginated documents to be sent to recipients in a number of formats.
Learn more
Edition
Enterprise
Enterprise + Q
Learn more
Team trial for 30 days (4 authors)*
FREE
FREE
Author per month (yearly)**
$18
$28
Author per month (monthly)**
$24
$34
Readers (pay-per-Session)
$0.30 / session (max $5)****
$0.30 / session (max $10)****
Additional SPICE per month
$0.38 per GB
$0.38 per GB
QuickSight Q regional fee
N/A
$250 / mo / region
Natural language query with QuickSight Q
N/A
INCLUDED
Single Sign On with SAML or OpenID Connect
Connect to spreadsheets, databases & business apps
Access data in Private VPCs
Row-level security for dashboards
Secure data encryption at rest
Connect to your Active Directory
Use Active Directory groups***
Send email reports
Embed QuickSight
Capacity-based pricing
Supported regions
Learn more
Learn more
*Trial authors are auto-converted to month-to-month subscription upon trial expiry
**Each additional author includes 10GB of SPICE capacity
***Active Directory groups are available in accounts connected to Active Directory
****Sessions of 30-minute duration. Total charges for each reader are capped at $5 per month. Conditions apply.
Sign up for Standard Edition here.
Continue

Figure 16.8 QuickSight membership registration.

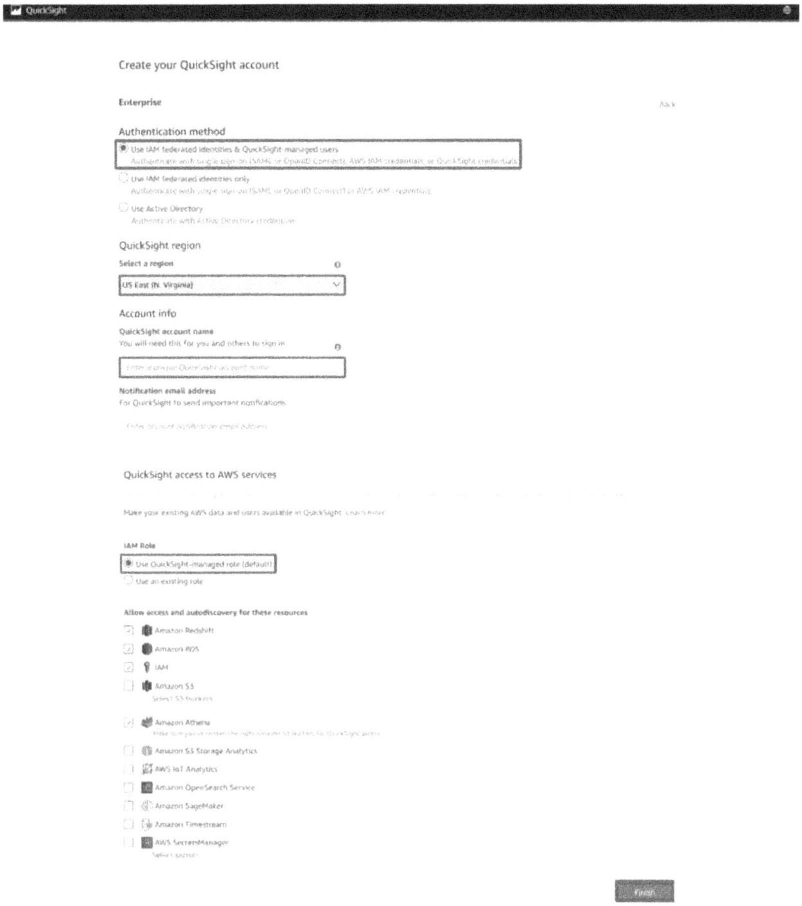

Figure 16.9 QuickSight membership registration process.

select their desired region. Account Info is the step where users input their account information, such as the account name to be used in QuickSight service and the email address to receive notifications when using the QuickSight account.

Finally, in the QuickSight access to AWS Services step, there are two ways to configure the IAM Role. Configuration should be done according to the user's environment, but for default configuration, select the "Use QuickSight-Managed role (default)" provided by AWS. In the next step, for integration between the services provided by AWS and QuickSight, automatic recommendations are given to configure access settings for four services. However, if the user wants to allow access and integration with other services, they can select additional services. After completing all the steps, click the "Finish" button to finalize.

When you click the "Finish" button in Figure 16.9, the screen shown in the following Figure 16.10 will appear. Afterward, click on the "Go to Amazon QuickSight" button to learn about QuickSight.

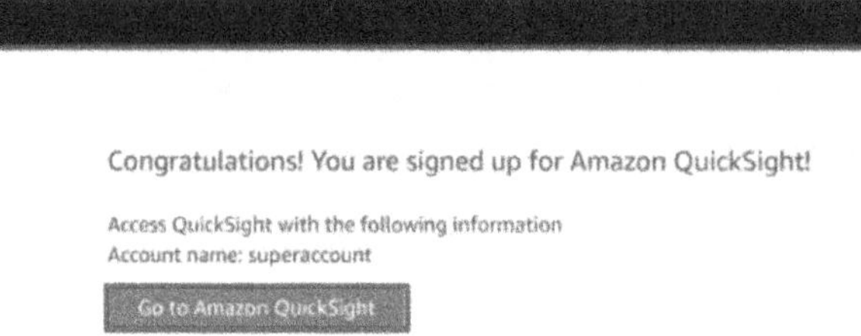

Figure 16.10 QuickSight membership registration completed.

16.4 QuickSight Analysis Creation

Let's click on the [New Analysis] tab on the top right of the Amazon QuickSight start page, as shown in Figure 16.11, and go to the dataset page (Figure 16.12).

Select the Business Review sample dataset on the dataset page, and then choose to generate analysis.

16.4.1 *Create visual chart objects*

Select a visual object type and drag the desired fields into the field collection to create a visual object. To create a chart object, click the "Add" tab at the top left corner in Figure 16.13.

Select the visual object type and create a visual object by dragging the desired fields into a field collection. To add a visual chart object, click the "Add" tab in the top left corner as shown in Figure 16.13.

As shown in Figure 16.14, select "Visualize" on the application toolbar of the analysis page, and then click "Add Visual." Choose the scatter plot icon in the visual types pane.

Let's select the scatterplot icon in the visual types pane, as shown in Figure 16.15.

Next, in the Fields list window, select the fields to add to the field web panel. Choose "Desktop uniques" to fill the X-axis field set and choose "Mobile uniques" to fill the Y-axis field set. Select "Date" to fill the "Group/Color" field set, and then a scatter plot using these fields will be generated, as shown in Figure 16.16.

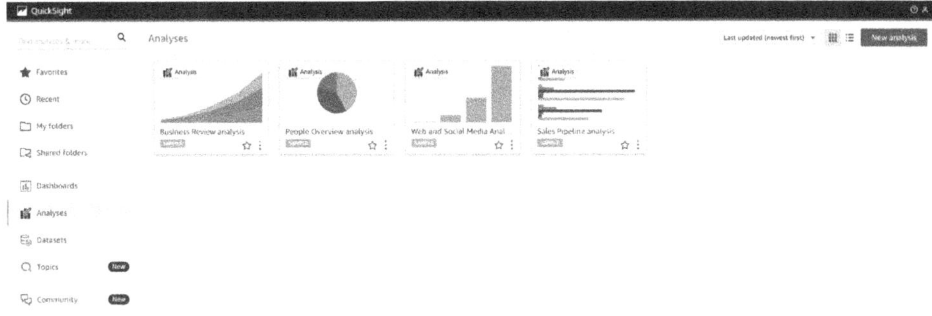

Figure 16.11 QuickSight home page.

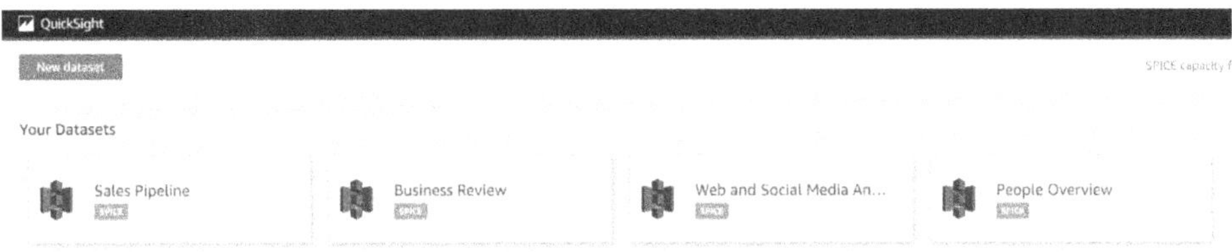

Figure 16.12 QuickSight dataset screen.

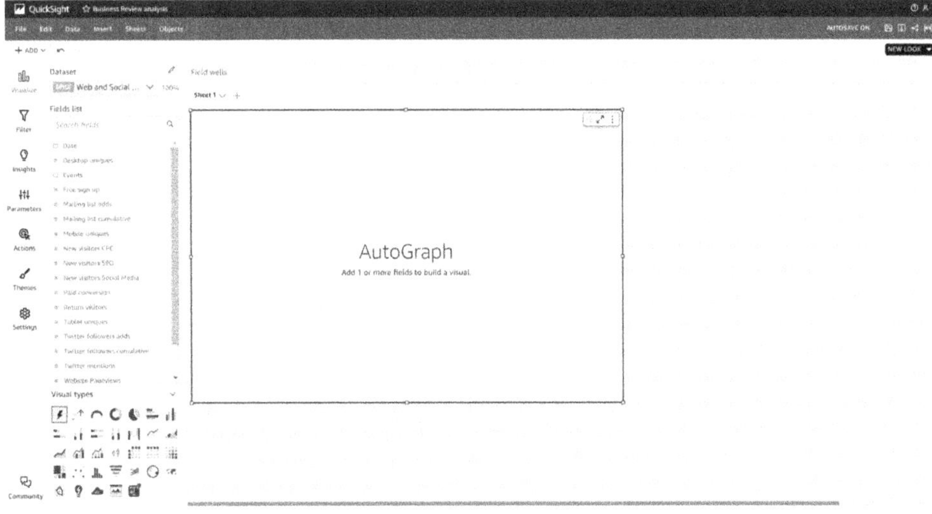

Figure 16.13 Business review analysis.

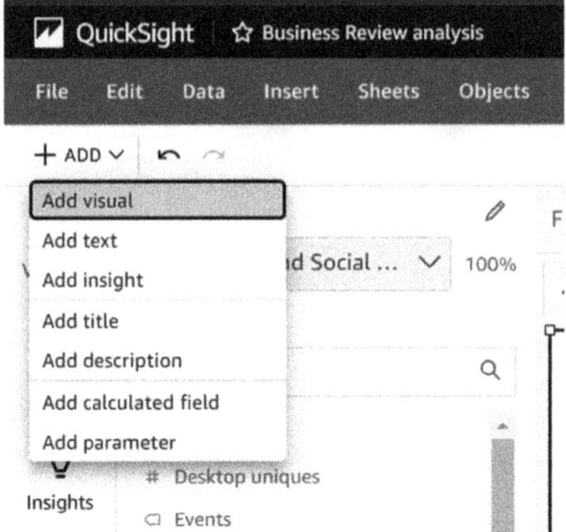

Figure 16.14 Adding visuals.

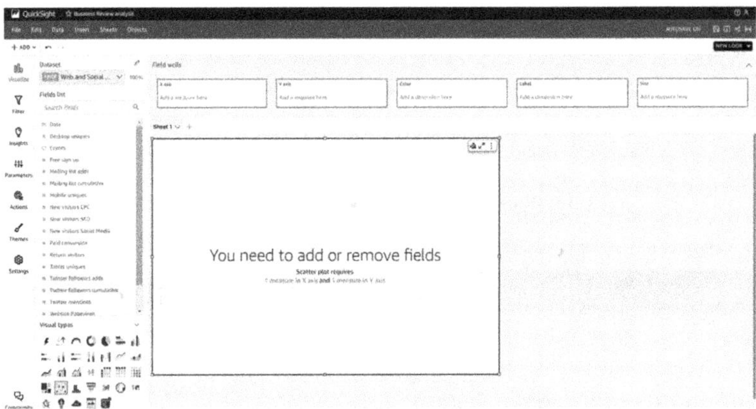

Figure 16.15 Visual object types.

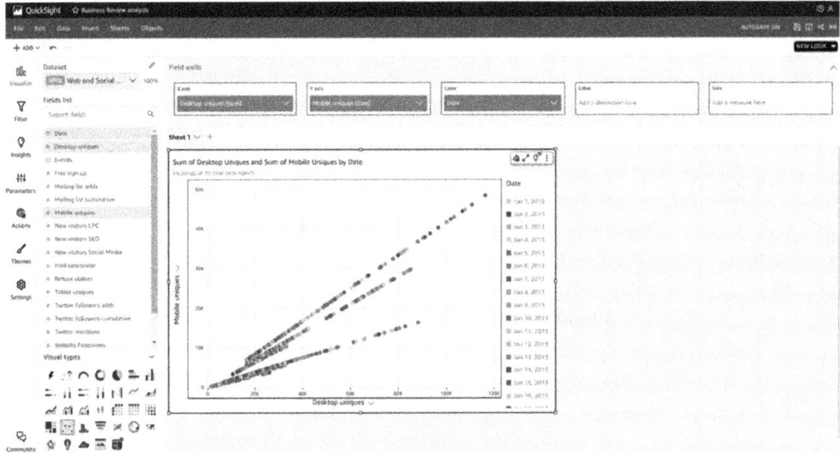

Figure 16.16 Scatter plot graph.

16.5 QuickSight Visualization

16.5.1 *Line chart visualization*

Let's display additional measurement values by date and modify the line chart visually by changing the color. To modify the line chart, select linear chart visual in the analysis. Add a different measure to the visual. In the Fields list pane, select the "New visitors SEO" field. This measure will be added to the Value field well, and the line chart will be updated to represent the measure (Figures 16.17–16.19).

Change the color of the line used to indicate the "Return visitors" measure. Select the line in the chart representing "Return visitors." To do this, select the end of the line, not the middle. After selecting "Color Return visitors," choose the red icon in the color picker.

In the X-axis field, carefully select the date field, then choose aggregation, and finally select the month.

16.5.2 *Modifying visual chart objects*

Let's learn how to modify visual charts by changing the level of data detail. To modify a visual chart object, select the scatter plot visual in the analysis. Then, select the "Group/Color" field collection, choose "Aggregate," and finally select "Month."

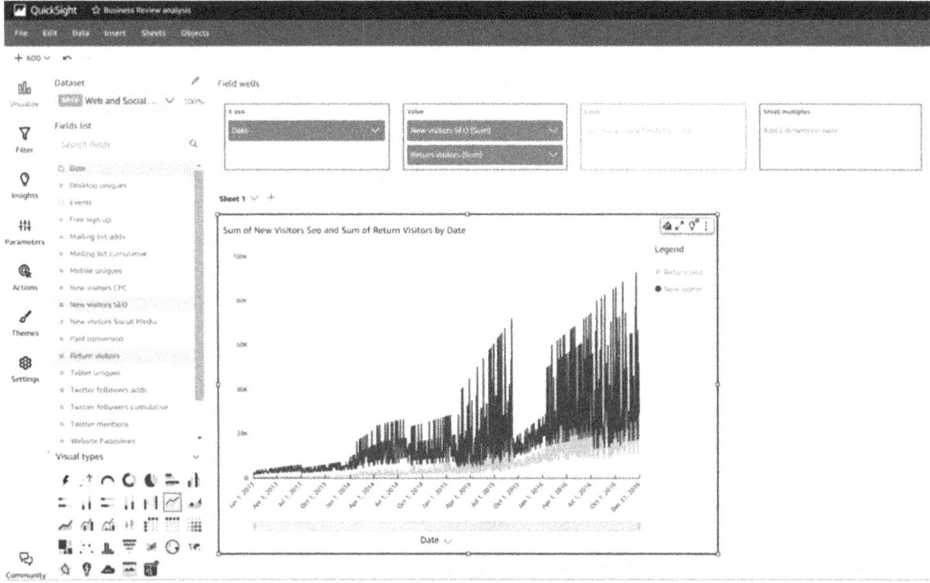

Figure 16.17 Linear chart visualization.

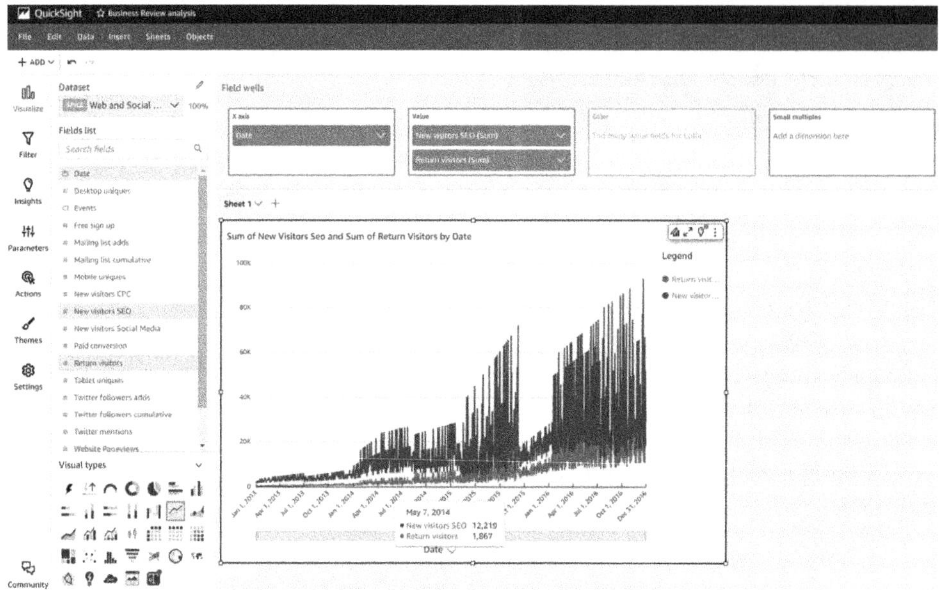

Figure 16.18 Linear chart visualization results.

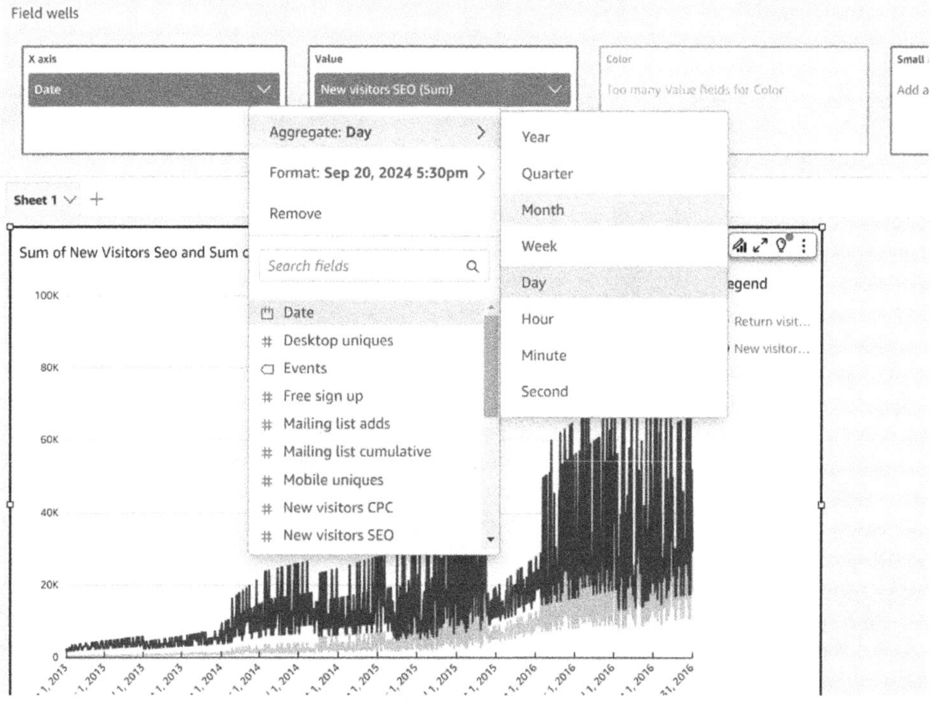

Figure 16.19 Linear chart date configuration.

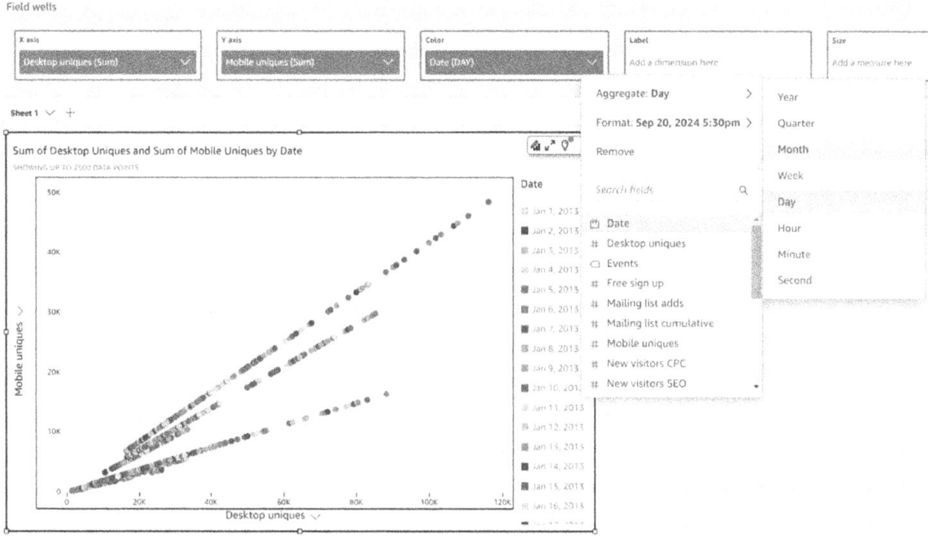

Figure 16.20 Scatter chart visual.

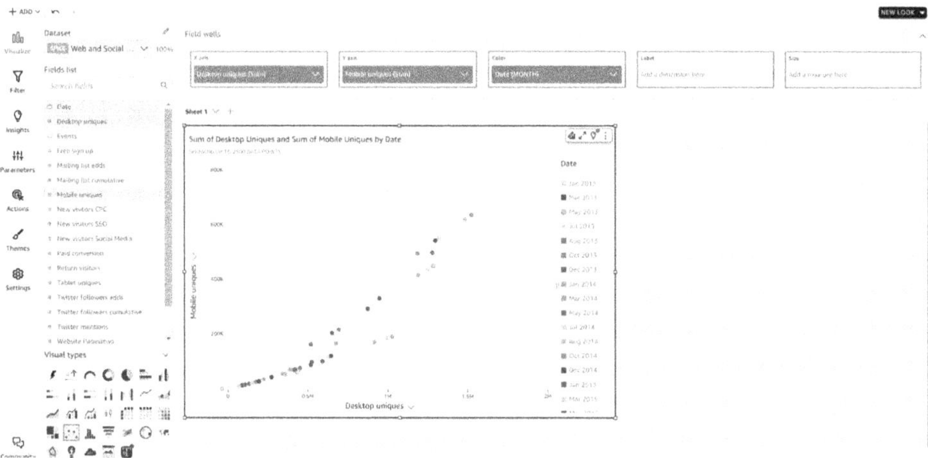

Figure 16.21 Scatter chart visual 2.

The scatter plot is updated to display monthly measures instead of the default yearly measures (Figures 16.20 and 16.21).

16.5.3 *Visual layout change*

Let's learn about how to change the visual layout, add filters, and modify both visual elements. Change the size and position of visuals, and modify them by adding and applying filters.

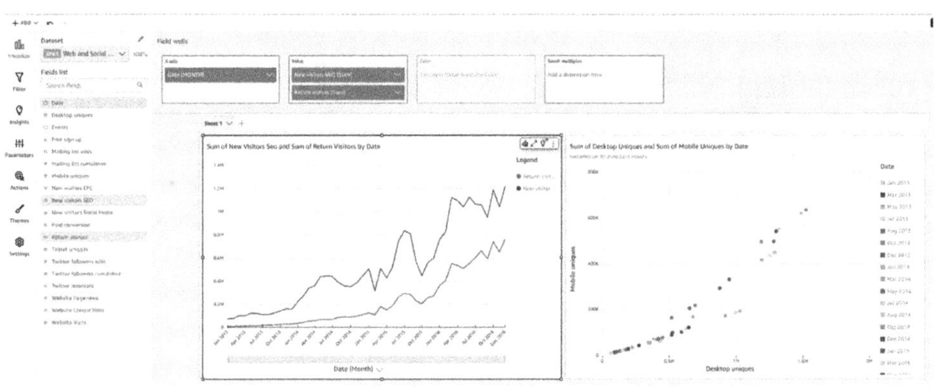

Figure 16.22 Chart merge.

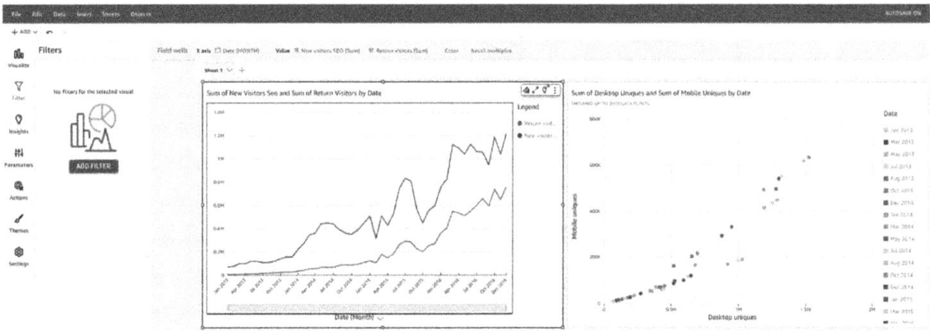

Figure 16.23 Chart filter added.

- Modifying two visuals.
- Select "Line Chart Visual" in the analysis.
- Select the size adjustment handle at the bottom right corner of the visual object and move it upwards to the top left until the visual object is reduced to half its previous size in both horizontal and vertical directions.
- Repeat this procedure in scatter plot visual.
- Select the mobile handle on the scatter plot chart and drag the linear chart to the right to move it.

As shown in Figure 16.22, add and apply filters to modify both visuals. Once you have added filters to the two charts, select the scatter plot visual for analysis. On the left, select the filter and then select the plus icon in the filter panel, and then select the date field to use as the filtering criteria.

As shown in Figure 16.23, click on the "Add Filter" tab and select the new filter to expand. Then, in the filter type section of the Filter Edit window, select

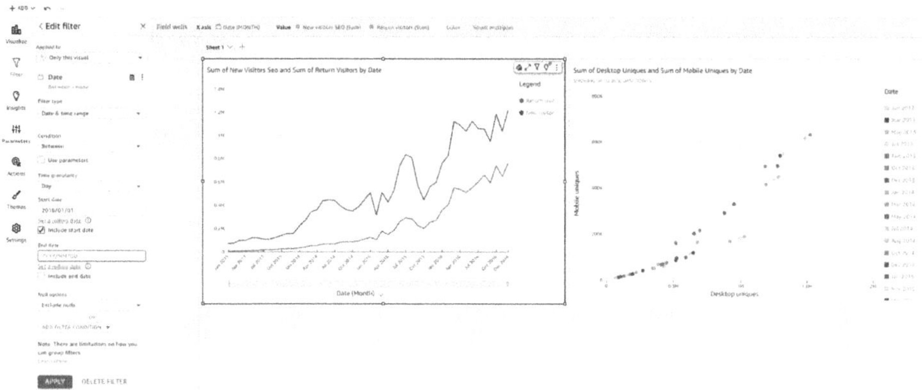

Figure 16.24 Scatter chart filter edit.

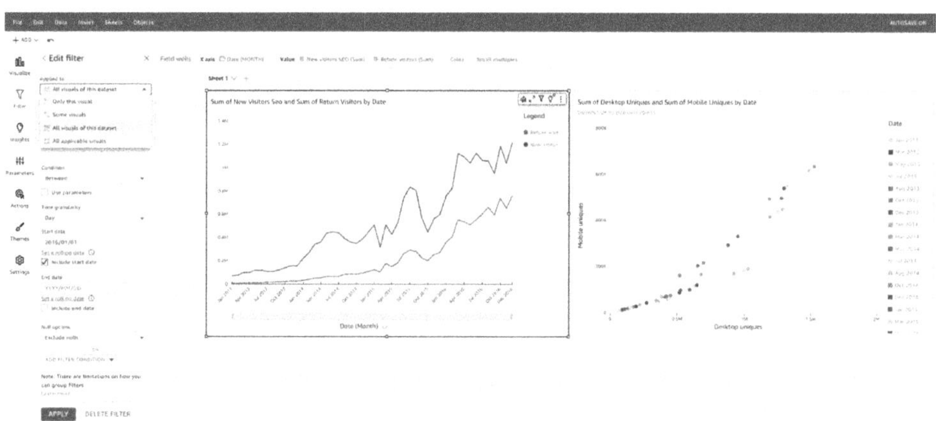

Figure 16.25 Linear chart filter editing.

the comparison type. After that, enter the start date value as 2014/01/01. When selecting the date, choose the year as 2014, and choose January as the month.

Apply filters to the line chart. Re-select the date filter from the left filter panel and then select all visuals of this dataset (Figure 16.24).

Figure 16.25 shows the result of editing visuals to change the layout. The chart was edited by applying chart filters.

16.6 QuickSight Dashboard

In this chapter, let's learn how to create a dashboard using QuickSight analysis. First, select "Share" from the application toolbar in the top right corner of the analysis, and then choose "Publish dashboard" (Figures 16.26 and 16.27).

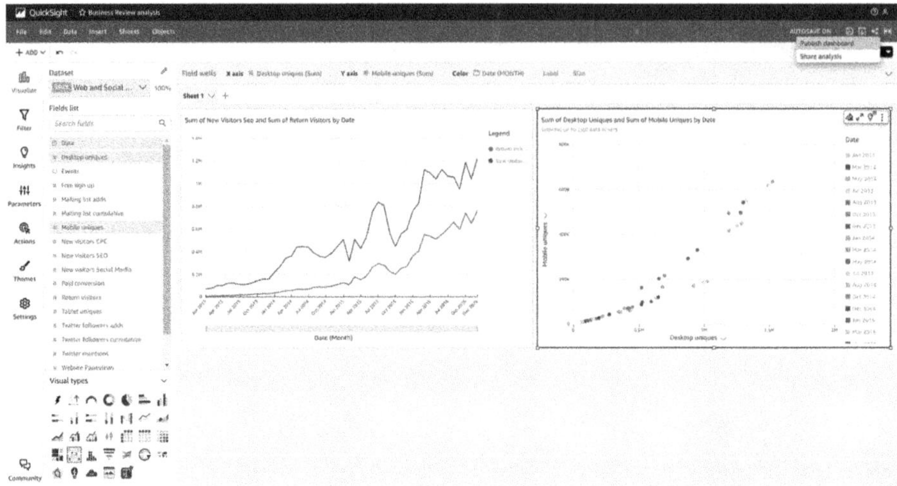

Figure 16.26 Dashboard posting.

Publish a dashboard ×

Publish new dashboard as

New test1

Replace an existing dashboard

ALL SHEETS SELECTED ▾

Advanced publish options ∨

Publish dashboard

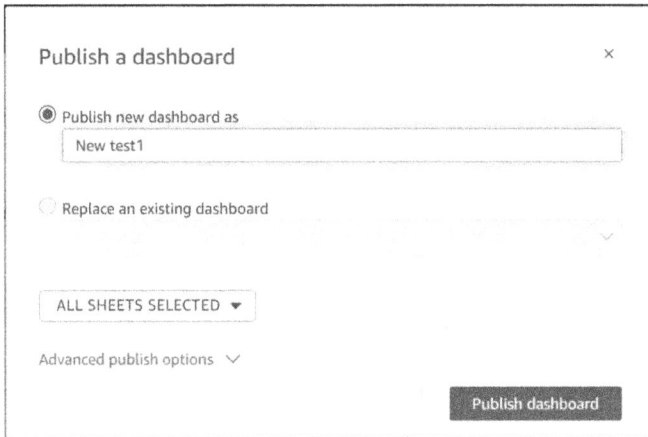

Figure 16.27 Dashboard posted.

On the open dashboard board posting page, select "Post a new dashboard as" and enter the name.

If you select to post on the dashboard, the dashboard will have been posted. Close it by selecting the "X" icon on the open shared dashboard page. You can use the sharing option on the dashboard page to share the dashboard later.

16.7 Practice Questions

Q1. How does Amazon QuickSight enable collaborative analysis without installing applications?

Q2. What are the steps to combine various data sources into one analysis in QuickSight?

Q3. How does Amazon QuickSight facilitate efficient analysis of service metrics like purchase rates or advertising effectiveness?

Chapter 17

Utilizing QuickSight

17.1 QuickSight Tutorial

In this section, lets learn about the QuickSight service provided by AWS. Figure 17.1 shows the QuickSight start page. On the screen, you can see that AWS provides four sample datasets and various features on the left tab. Prepared data in Amazon QuickSight can be reused for various analyses. Through data preparation, you can add calculated fields, apply filters, rename fields, or change data types. If you start from a data source in an SQL database, you can join tables or enter SQL queries if you want to work with more than one table of data. If you want to transform the data source before using it in Amazon QuickSight, you can prepare it according to your needs and save it as part of the dataset. The data created in this way can be edited later.

17.1.1 *QuickSight dataset preparation*

As shown in Figure 17.1, click "Data sets" on the left tab on the Amazon QuickSight start page. After clicking, the screen shown in Figure 17.2 will appear.

When you click on "New Dataset," shown in Figure 17.2, the screen shown in Figure 17.3 will appear. Scroll down on this screen and select the web and social media analysis Amazon S3 data source in the "From existing data sources" section, then click on "Edit dataset" to open the data preparation page.

If you scroll down the screen shown in Figure 17.3, the screen shown in Figure 17.4 will appear.

Click on the "Web and Social Media Analytics" data in the "From the existing data source" section, as shown in Figure 17.4.

Click on [Add dataset] in Figure 17.5. Then, the window shown in Figure 17.6 will appear.

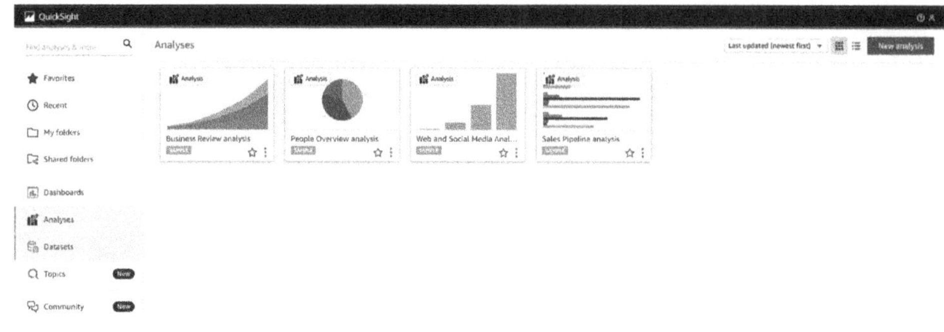

Figure 17.1 QuickSight start page.

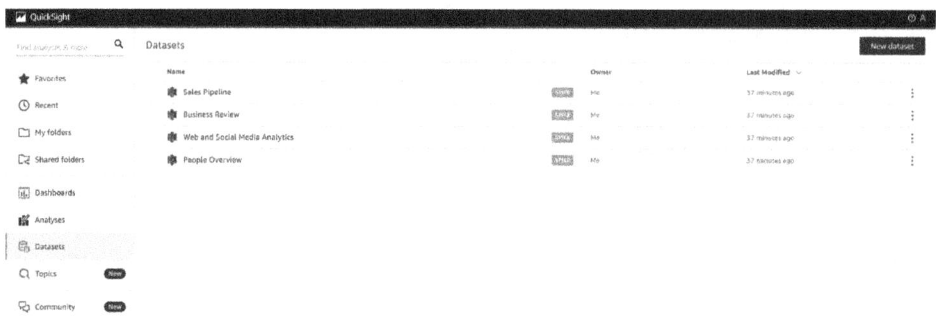

Figure 17.2 QuickSight dataset screen.

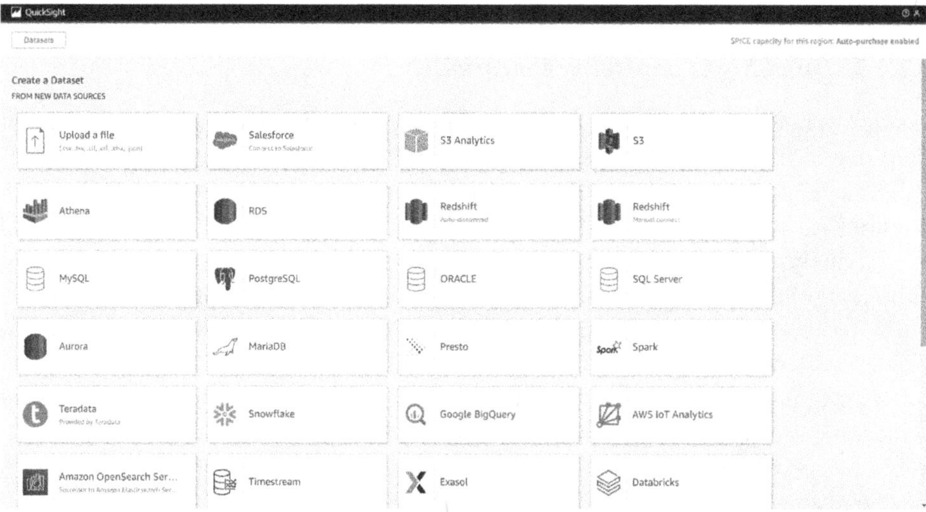

Figure 17.3 QuickSight dataset creation “from a new data source” screen.

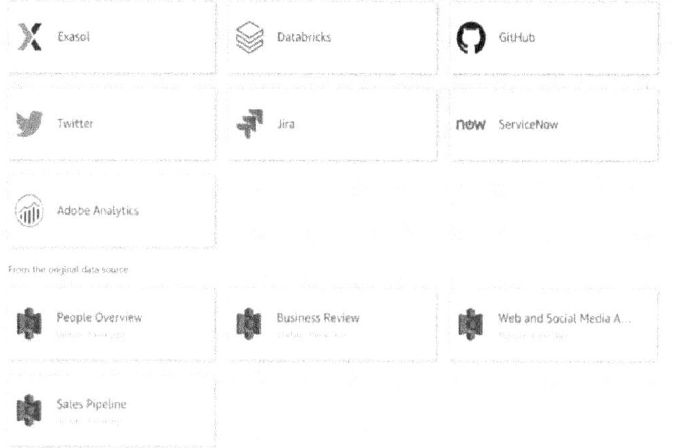

Figure 17.4 QuickSight dataset creation [from existing data source] screen.

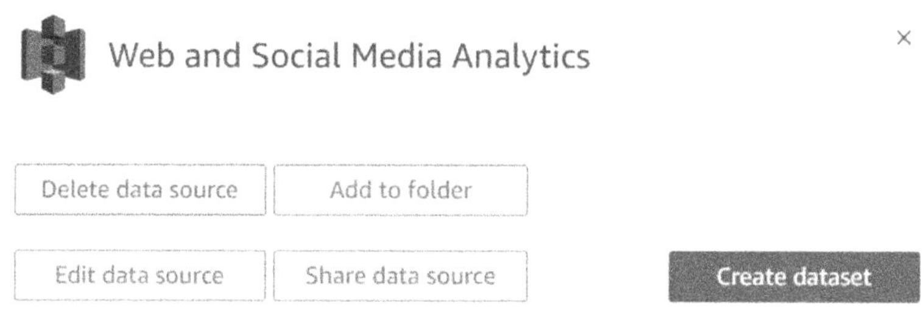

Figure 17.5 Web and social media analytics.

New sheet

Interactive sheet

Single page, interactive content

Layout

Tiled

Optimize for viewing on

1600px

New

Multi-page, highly formatted document

Paper size

Paginated Reports allows you to build highly formatted multi-page reports. Get Paginated Reports

CANCEL CREATE

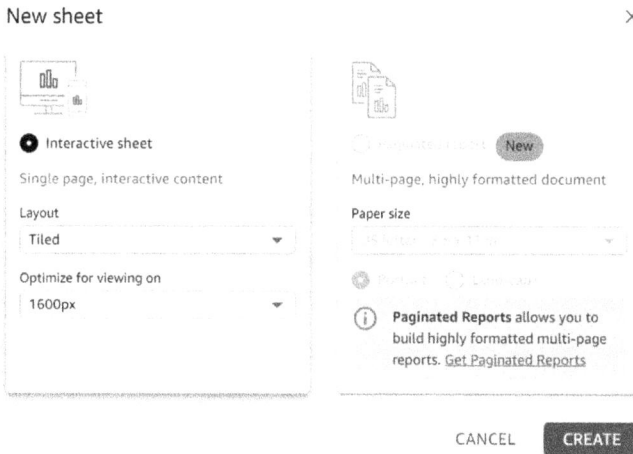

Figure 17.6 Marketing sample data source details.

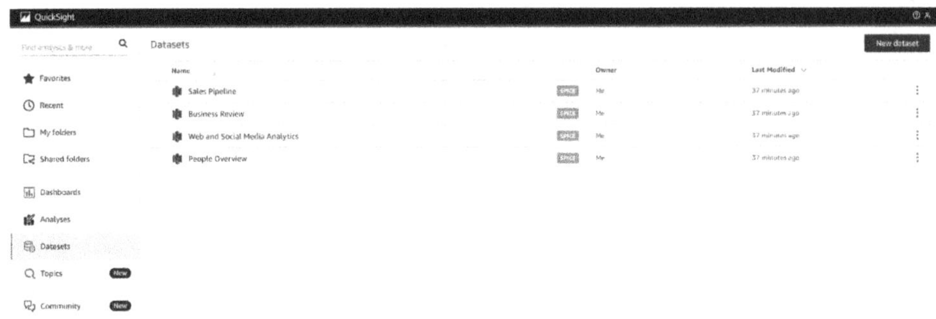

Figure 17.7 QuickSight dataset screen.

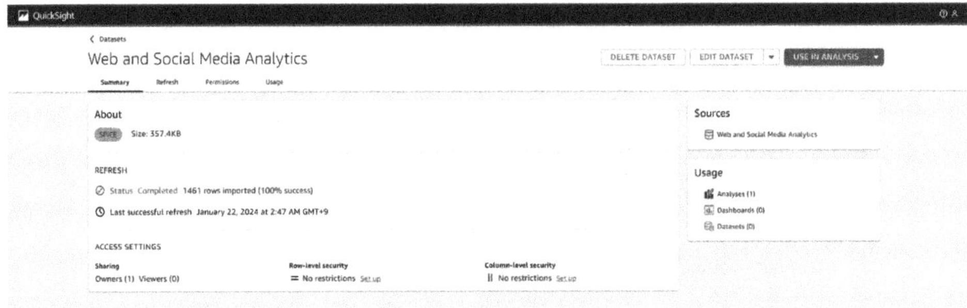

Figure 17.8 Web and social media analytics data screen.

Click the "Create" button at the bottom right here to complete the dataset creation.

17.1.2 *Dataset editing*

In this chapter, let's learn about how to view and edit datasets. Click the "Datasets" tab on the QuickSight home screen, then click the "New dataset" tab in the top right corner.

If you click on the "Web and Social Media Analytics" data shown in Figure 17.7, the screen shown in Figure 17.8 will appear.

Next, click on the "Edit dataset" tab to proceed with editing the dataset.

Figure 17.9 shows an overview of the information regarding the data in Web and Social Media Analytics. Here, in the field panel, select the field menu for the "Twitter followers cumulative" and "Mailing list cumulative" fields, then choose "Exclude field." To select more than one field at once, press the Ctrl key when selecting. This will display the screen shown in Figure 17.10.

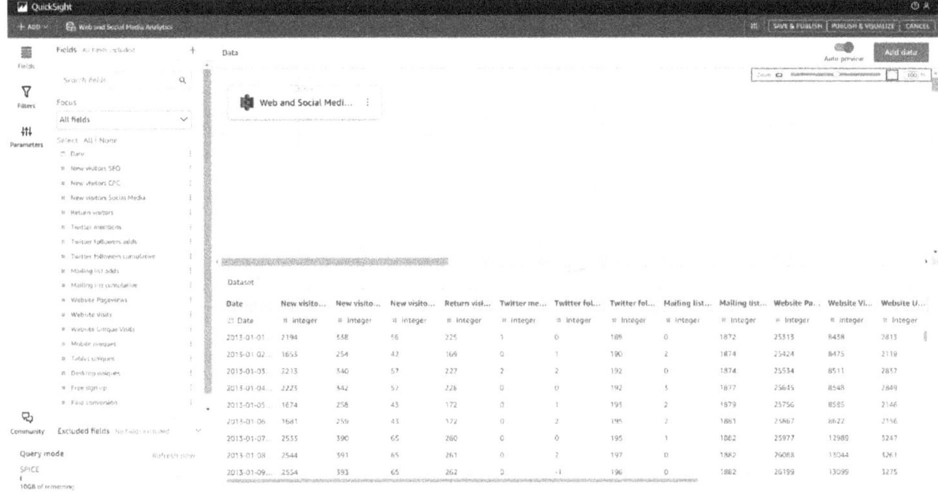

Figure 17.9 Web and social media analytics.

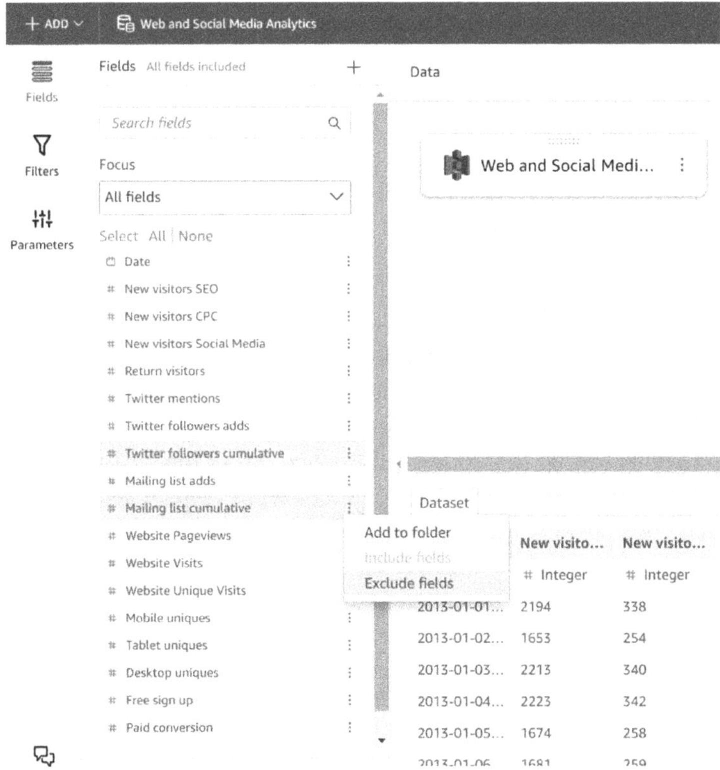

Figure 17.10 Field removal process.

Dataset

w visito...	Return visi...	Twitter me...	Twitter fol...	Mailing list...	Website Pa...	Website Vi...	Website U...	Mobile uni...	Tablet uni...	Desktop u...	Free sign up	Paid conve...
Integer	# Integer	# Integer	# Integer	# Integer	# Integer	# Integer	# Integer	# Integer	# Integer	# Integer	# Integer	# Integer
	225	1	0	0	25313	8438	2813	281	563	1969	1406	84
	169	0	1	2	25424	8475	2119	212	424	1483	847	59
	227	2	2	0	25534	8511	2837	284	567	1986	1135	68
	228	0	0	3	25645	8548	2849	285	570	1995	855	51
	172	0	1	2	25756	8585	2146	215	429	1502	644	45
	172	0	2	2	25867	8622	2156	216	431	1509	862	60
	260	0	0	1	25977	12989	3247	325	649	2273	1299	65
	261	0	2	0	26088	13044	3261	326	652	2283	1304	65
	262	0	-1	0	26199	13099	3275	327	655	2292	982	49

Figure 17.11 Process of removing fields.

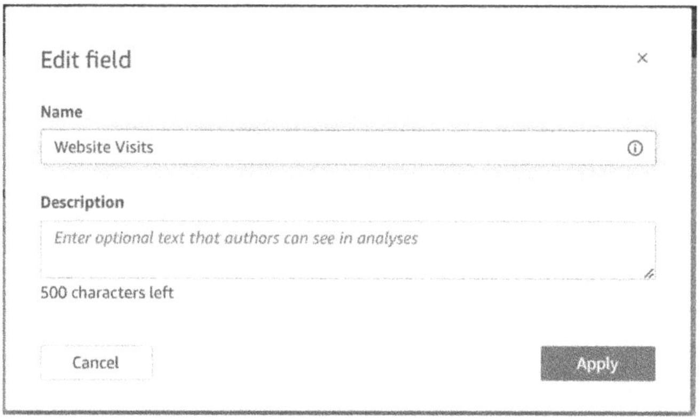

Edit field ×

Name

Website Visits ⓘ

Description

Enter optional text that authors can see in analyses

500 characters left

Cancel Apply

Figure 17.12 Field removal process.

Figure 17.9 shows an overall overview of information about data in Web and Social Media Analytics. Here, in the field panel, select the field menu of Twitter followers accumulated and mailing list accumulated fields, then choose field exclusion. To select more than one field at once, press the Ctrl key and click.

In the open field editing window, enter "website page views" in the name column, then click "Apply" (Figures 17.11 and 17.12).

Once the application is complete, let's now look at adding a calculated field to the Events field that substitutes a text string for any 0-length string value.

As shown in Figure 17.13, when you click on the "Add calculated field" tab on the left side, the "Add calculated field" window appears as shown in Figure 17.14.

Enter "populated_event" as the name, then search for the "if else" unction on the right and click double-click it. This will add the function to the calculated field formula.

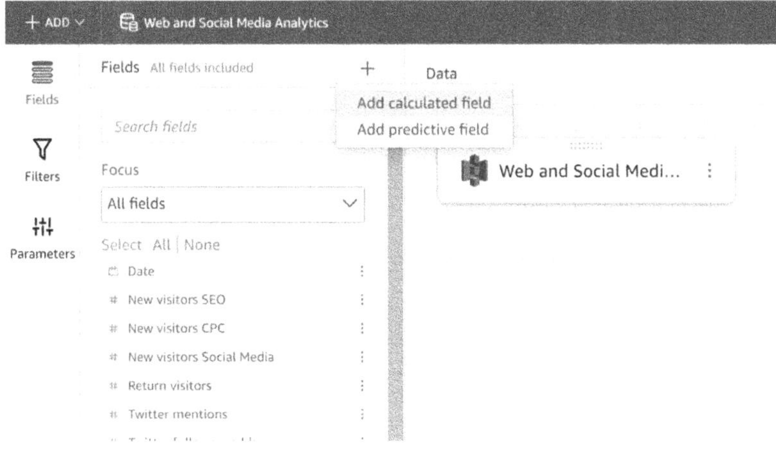

Figure 17.13 Web and social media analytics.

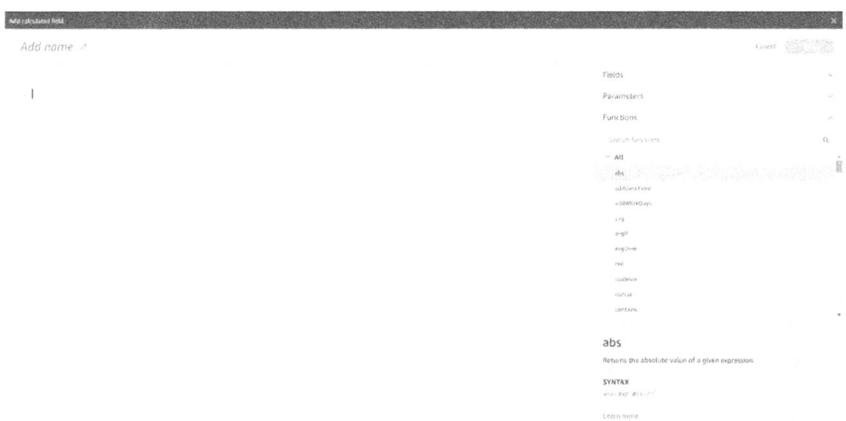

Figure 17.14 Add calculated field page.

Next, in the formula editor, input the additional functions and parameters shown in Figure 17.15, bold them, and then click the "Save" tab in the top right corner to save.

Once you have completed adding the calculated fields shown in Figure 17.15, you can see that the populated_event event has been created in the left part of Figure 17.16. Click the "Save & Publish" tab on the top right to save it.

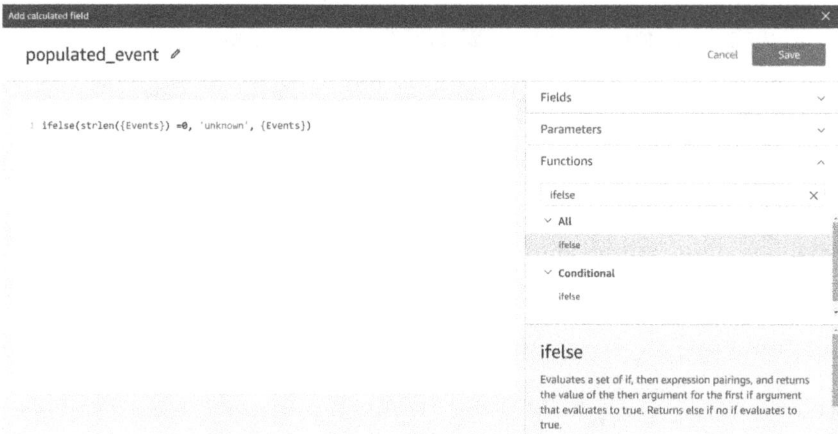

Figure 17.15 Successfully added calculated field.

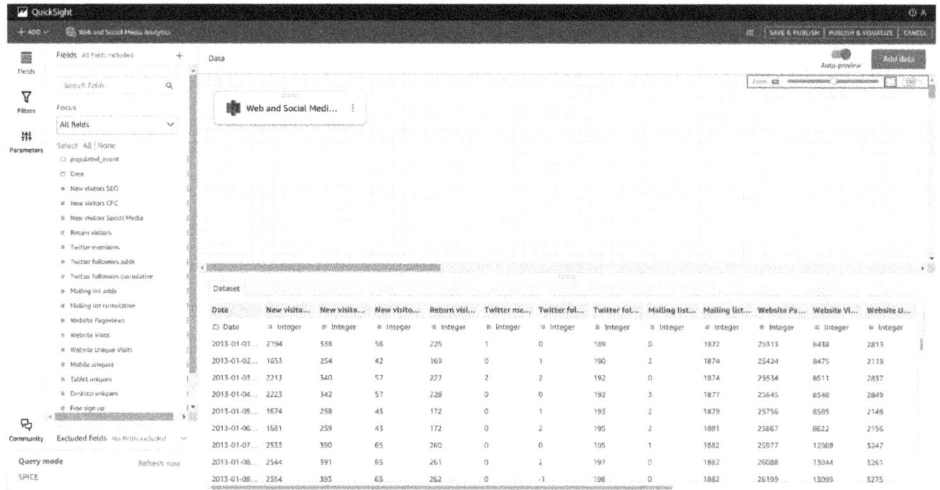

Figure 17.16 Marketing sample screen.

17.2 QuickSight Practice

For this example, we'll explore population data from Seoul, sourced from the Public Data Portal. This portal offers various datasets, including the one we're focusing on, which details the population by each neighborhood (referred to as "dong" in Korean) within Seoul. It includes information on the gender distribution, the number of households, and the average population per household. This

	A	B	C	D	E	F	G	H	I	J
1	Administrative Institutions	Total	Population (Male	Population (Female	Area (Composition Ratio	omposition Ratio (Male	omposition Ratio (Female	Gender Ratio	Number of Households	Population per Household
2	Gwangjin gu	337416	162541	174875	100	48.17	51.83	92.95	169291	1.99
3	Junggok 1dong	15299	7322	7977	4.53	2.17	2.36	91.79	8504	1.8
4	Junggok 2dong	20999	9903	11096	6.22	2.93	3.29	89.25	10656	1.97
5	Junggok 3dong	15945	7897	8048	4.73	2.34	2.39	98.12	8513	1.87
6	Junggok 4dong	28035	13715	14320	8.31	4.06	4.24	95.78	13122	2.14
7	Neung dong	11180	5134	6046	3.31	1.52	1.79	84.92	6406	1.75
8	Guui 1dong	23387	11143	12244	6.93	3.3	3.63	91.01	13077	1.79
9	Guui 2dong	26043	12511	13532	7.72	3.71	4.01	92.45	11517	2.26
10	Guui 3dong	28211	13309	14902	8.36	3.94	4.42	89.31	12079	2.34
11	Gwangjang dong	34168	16524	17644	10.13	4.9	5.23	93.65	11814	2.89
12	Jayang 1dong	21862	10646	11216	6.48	3.16	3.32	94.92	11913	1.84
13	Jayang 2dong	23946	11612	12334	7.1	3.44	3.66	94.15	10887	2.2
14	Jayang 3dong	27553	13166	14387	8.17	3.9	4.26	91.51	11009	2.5
15	Jayang 4dong	17926	8915	9011	5.31	2.64	2.67	98.93	10209	1.76
16	Hwayang dong	23437	11141	12296	6.95	3.3	3.64	90.61	18185	1.29
17	Gunja dong	19425	9603	9822	5.76	2.85	2.91	97.77	11400	1.7

Figure 17.17 Seoul population statistics data.

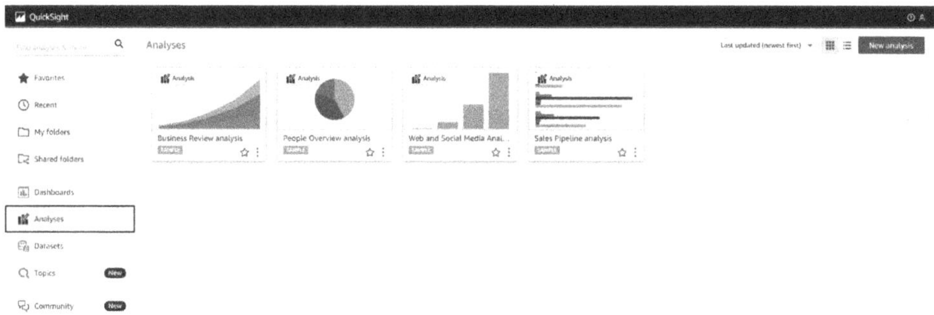

Figure 17.18 QuickSight main screen.

dataset is invaluable for understanding demographic dynamics at a granular level within the district.

If you have saved the population status data file of, Seoul, let's move to the main screen of QuickSight and analyze the population status data (Figure 17.17).

Upload the file you saved from the public data portal to QuickSight as shown in Figure 17.18. To upload the data file, click on the "New analysis" tab in the top right corner, and the screen shown in Figure 17.19 will appear.

Click on the [New Dataset] tab in the top left corner of the screen, as shown in Figure 17.19.

Click "Upload a file," as shown in Figure 17.20, and upload the Seoul population status file downloaded from the Public Data Portal. When you upload the file, a window will appear where you can confirm the settings. Here, you can briefly understand the information about the file. Click on the "Next" tab located in the bottom right corner of the screen, as shown in Figure 17.21, to proceed.

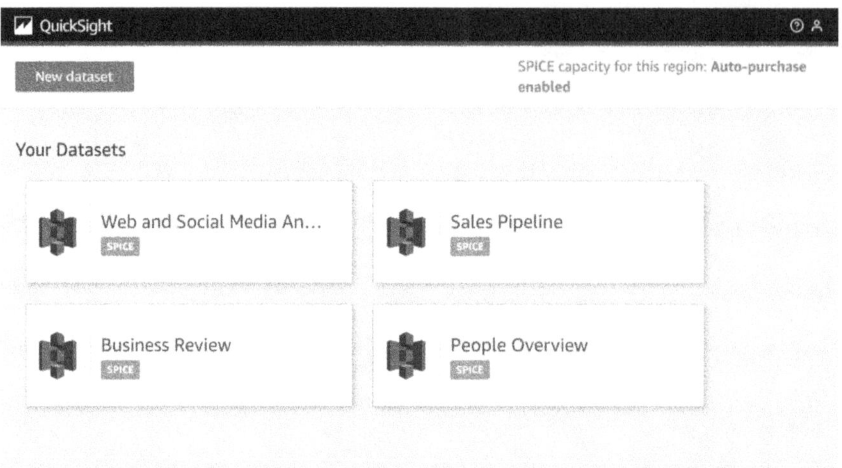

Figure 17.19 QuickSight data.

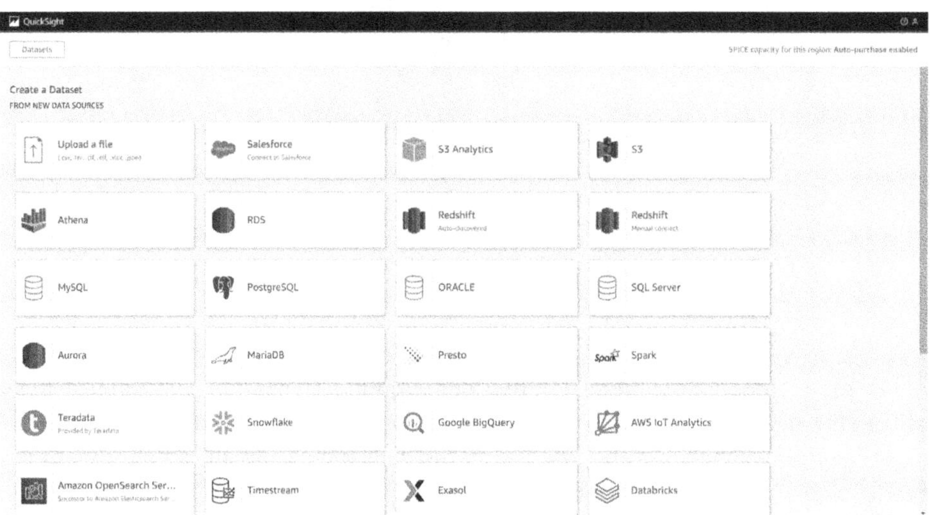

Figure 17.20 QuickSight data upload.

When you click on the "Next" tab shown in Figure 17.21, the window shown in Figure 17.22 will appear. On this screen, you can view detailed information about the data source.

If you have understood the information about the data, as shown in Figure 17.22, click on the "Visualize" tab at the bottom right of the screen to prepare for data analysis. Create a data analysis sheet like the one shown in Figure 17.23. For the

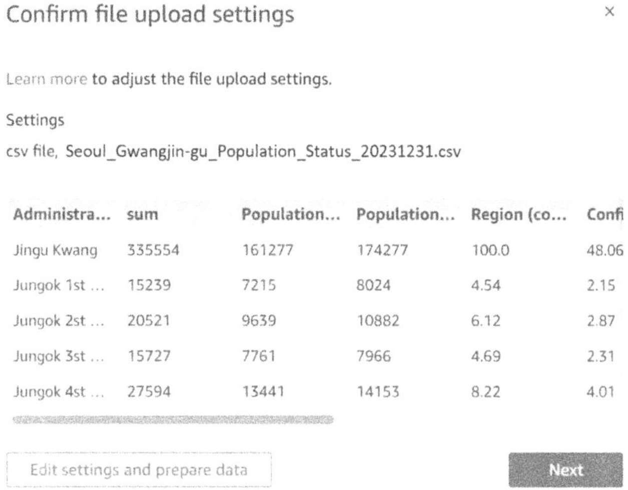

Confirm file upload settings
Learn more to adjust the file upload settings.
Settings
csv file, Seoul_Gwangjin-gu_Population_Status_20231231.csv
Administra... sum Population... Population... Region (co... Confi
Jingu Kwang 335554 161277 174277 100.0 48.06
Jungok 1st ... 15239 7215 8024 4.54 2.15
Jungok 2st ... 20521 9639 10882 6.12 2.87
Jungok 3st ... 15727 7761 7966 4.69 2.31
Jungok 4st ... 27594 13441 14153 8.22 4.01
Edit settings and prepare data
Next

Figure 17.21 Check the file upload settings.

Figure 17.22 Data source details.

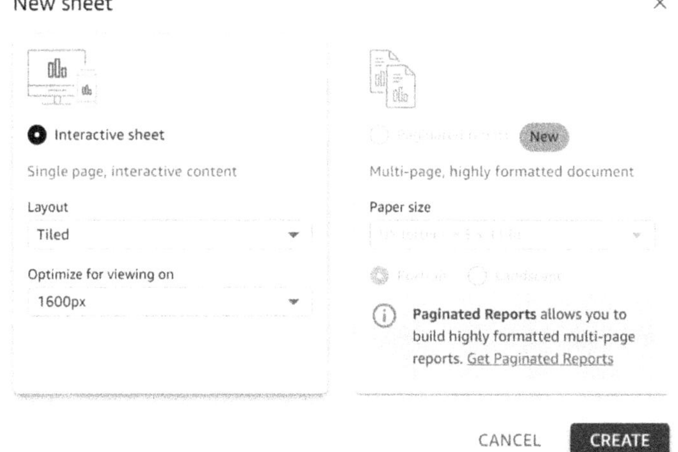

Figure 17.23 Generating data analysis sheet.

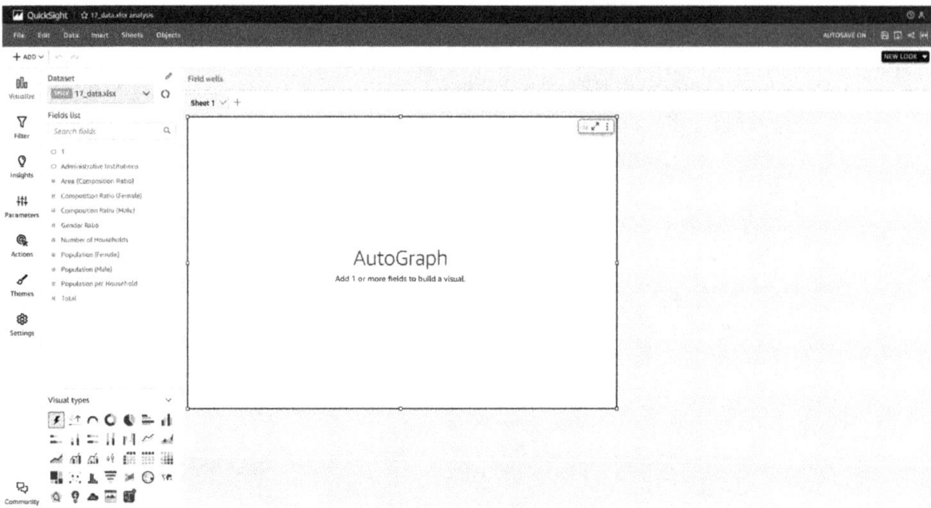

Figure 17.24 Population status data analysis sheet in Seoul.

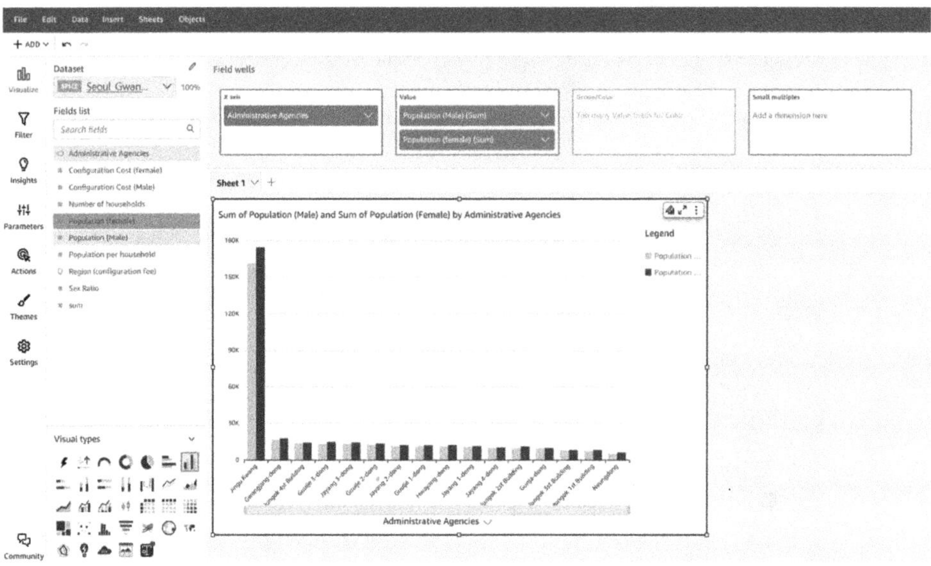

Figure 17.25 Analysis of male and female population by administrative agency.

settings, set the layout as a grid and generate it by setting the optimized pixel value for visualization to 1600px.

Click the [Create] tab in Figure 17.23 to go to the screen shown in Figure 17.24. Let's take a brief look at the population status of Seoul.

As you can see in Figure 17.25, data was analyzed in order to understand the population of males and females by administrative institution. The vertical bar

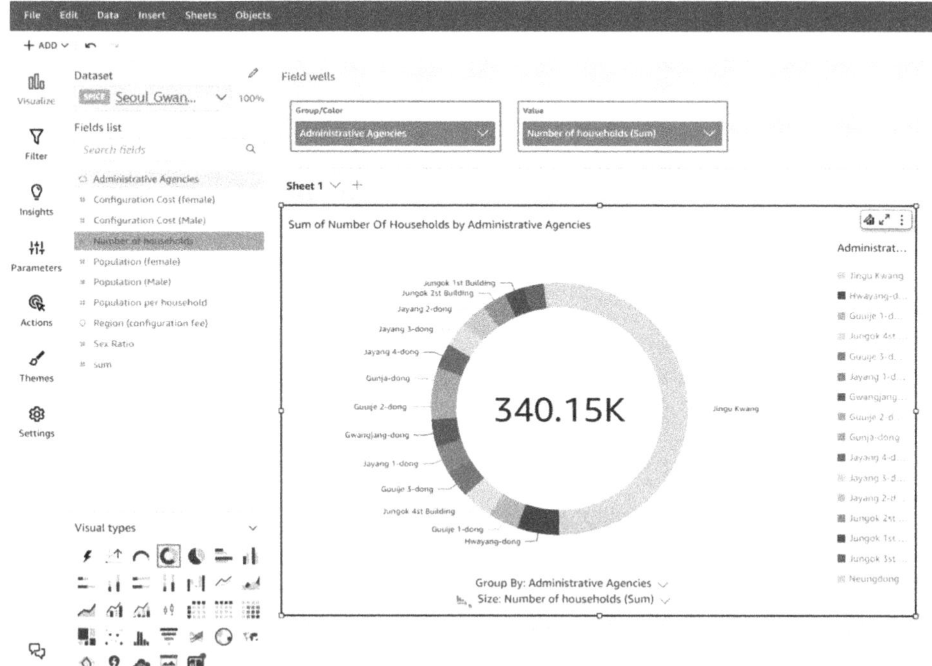

Figure 17.26 Number of households by administrative institution.

chart type has been selected as the method of visualization. Specify the administrative institution field on the X-axis, and the two fields of population (male) and population (female) for the values. After completing all the steps, the screen shown in Figure 17.25 will be displayed.

The following is about finding out the population status of Seoul, specifically the number of households by administrative organization (autonomous district). Create a new sheet and analyze it as shown in Figure 17.25. On the right side of [Sheet 1], click the "+" symbol to create a new sheet. After creating it, set the visual to a donut chart. Then, in the Group/Color tab, specify the [Administrative District] field, and in the Values tab, specify the [Total Number of Households] field.

Next, we will try to delete field values using the filter function. As shown in Figure 17.26, we will delete the field values in Seoul. As shown in Figure 17.26, there should be a "Filter" tab on the left side. Clicking on the "Filter" tab will show the "Add Filter" button. From here, select [Administrative Institution] as the field to be filtered. After selecting, click on the "Edit" button in the following Figure 17.27.

On the filter editing window of the administrative agency filter, select "Custom filter" for the filter type and specify "Exclude" for the filter condition.

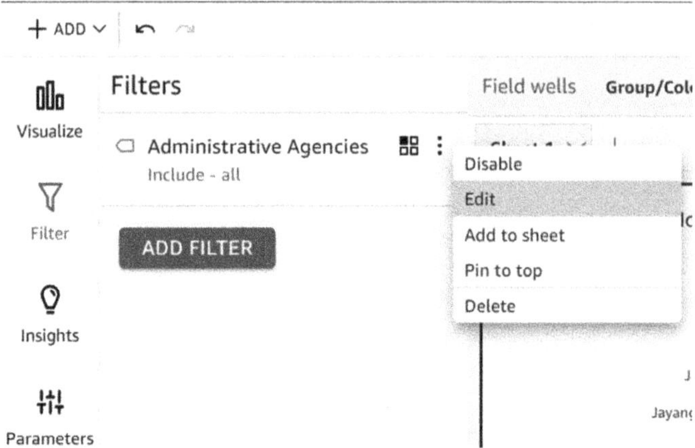

Figure 17.27 Administrative institution filter.

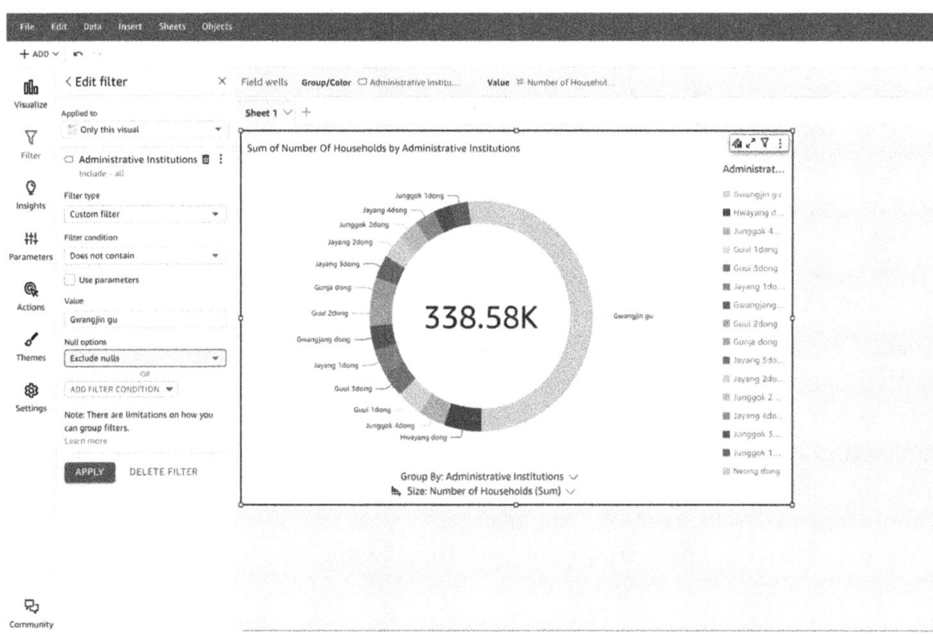

Figure 17.28 Setting the filter value for administrative agencies.

Lastly, if you select Seoul in the value, it will exclude data for the entire field of Seoul. Once all processes are finished, click the "Apply" tab in the bottom left corner to apply the filter settings.

When applying a filter, the screen shown in Figure 17.29 will appear. When you click "Insight" on the left side, you can view organized information about the visualized data.

Figure 17.30 is the proposed insight content for Figure 17.29. Looking at the content, when considering the number of households per neighborhood in Seoul,

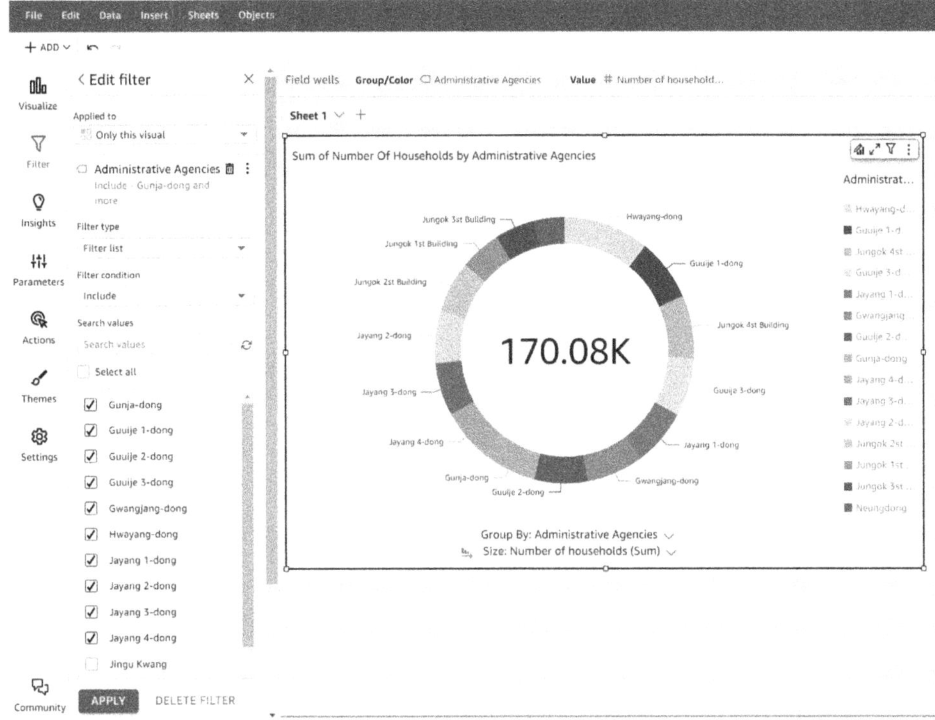

Figure 17.29 Number of households by neighborhood in Seoul.

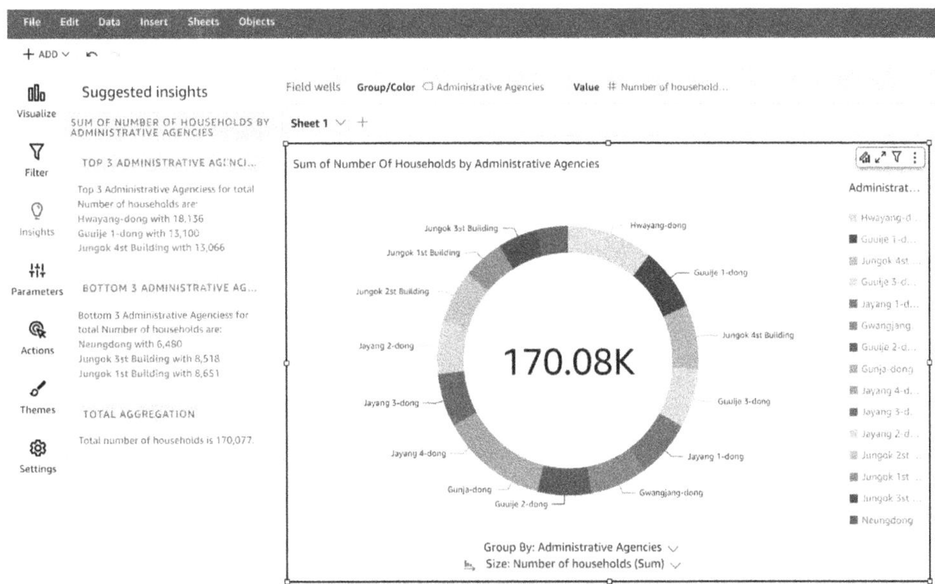

Figure 17.30 Number of households by neighborhood (Dong), Seoul.

the top three neighborhoods with the highest number of households are Hwayang-dong, Junggok 4-dong, and Gwangjin 1-dong in that order.

17.3 Practice Questions

Q1. How does Amazon QuickSight support data preparation for analysis?

Q2. What are the steps for creating and editing datasets in QuickSight for a specific analysis?

Q3. Describe how QuickSight can be used to analyze population data, providing insights into demographic distributions.

Chapter 18

Power BI

Power BI is a business analytics service that utilizes up-to-date information to provide insights for quick decision-making. Anyone can easily connect and model data, as well search for information to create visual reports. Additionally, Power BI allows collaboration and sharing during the creation process and can be used in conjunction with other tools including Microsoft Excel. Once reports are created, they can be shared and distributed within the organization through web and mobile devices. This enables insights about the data to be shared with other team members, and the built-in governance and security features can be expanded throughout the entire enterprise. Therefore, Power BI can be seen as an expandable integrated platform for BI services (Figure 18.1).

18.1 Power BI Components

Power BI consists of Power BI Desktop, Power BI Service, and Power BI Mobile. Power BI Desktop is a Windows desktop application that can be used for free. Power BI Service is an online SaaS, and Power BI Mobile is a mobile app that can be used on Windows phones, tablets, iOS, or Android devices. These three elements allow users to effectively generate or share data in various ways (Figure 18.2).

18.2 Power BI Features

18.2.1 *Utilizing self-service analysis in analyzing business size*

Using an analysis platform that is determined by individuals to organizations as a whole, you can reduce the added costs, complexity, and security risks that occur when using multiple solutions.

Figure 18.1 Power BI.

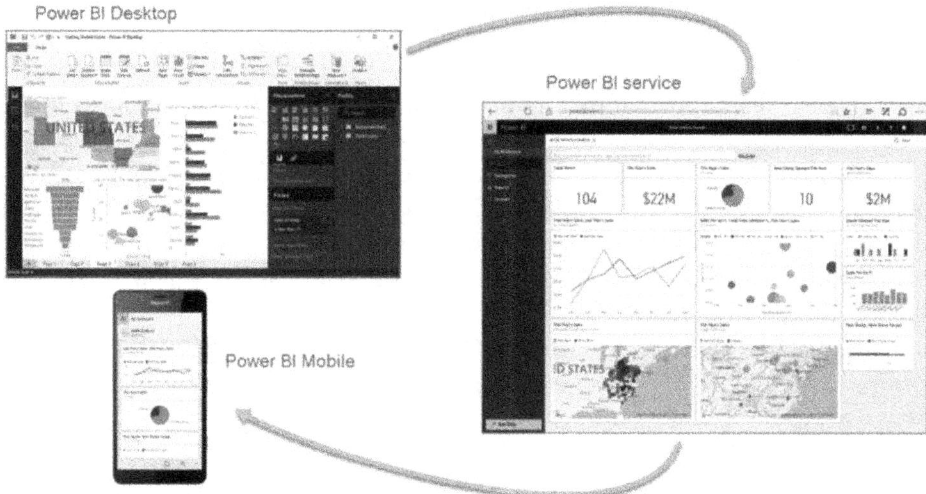

Figure 18.2 Power BI elements.

18.2.2 *Accurate results due to the use of smart tools*

You can secure and share meaningful insights through data visualization, built-in AI functions, tight integration with Excel, and user-customized data connectors in advance.

18.2.3 *Analysis data protection*

You can safely protect data through industry-leading data security features, sensitivity labeling, end-to-end encryption, and real-time access monitoring.

18.3 Power BI Contents

18.3.1 *Dataset*

A dataset is a data collection used in Power BI to import or connect data and build reports and dashboards. It can import various data sources such as Excel, CSV, text files, as well as various data sources like databases and online services (Figure 18.3).

18.3.2 *Visualization charts*

Visualization represents data in various charts. In the Power BI visualization pane, you can use various visualizations such as horizontal bar, vertical bar, pie, ribbon, waterfall, treemap, card, scatter, and gauge charts. Fundamentally, an area chart is based on a line chart with the area between the axis and the line filled in. An area chart can emphasize a size change over time and can be used to highlight the total value between trends. For example, data representing earnings over time can be plotted on an area chart to emphasize the total earnings (Figure 18.4).

The standard is a horizontal bar chart that examines specific values across multiple categories (Figure 18.5).

A decomposition tree chart can visually represent data across multiple dimensions using visual elements (Figure 18.6). It automatically aggregates data and allows drilling down into your dimensions in any order. It is also an AI visualization tool that can be requested to find the next dimension to drill down into based on specific criteria. Therefore, it is a useful tool for conducting ad hoc exploration and root cause analysis. Line charts generally emphasize the overall shape of the total values over time.

Date	Day	MonthNo	Month	QuarterNo	Quarter	Year
01/01/2019 00:00:00	1	1	January	1	Qtr 1	2019
01/02/2019 00:00:00	2	1	January	1	Qtr 1	2019
01/03/2019 00:00:00	3	1	January	1	Qtr 1	2019
01/04/2019 00:00:00	4	1	January	1	Qtr 1	2019
01/05/2019 00:00:00	5	1	January	1	Qtr 1	2019
01/06/2019 00:00:00	6	1	January	1	Qtr 1	2019
01/07/2019 00:00:00	7	1	January	1	Qtr 1	2019
01/08/2019 00:00:00	8	1	January	1	Qtr 1	2019
01/09/2019 00:00:00	9	1	January	1	Qtr 1	2019
01/10/2019 00:00:00	10	1	January	1	Qtr 1	2019

Figure 18.3 Power BI sample data.

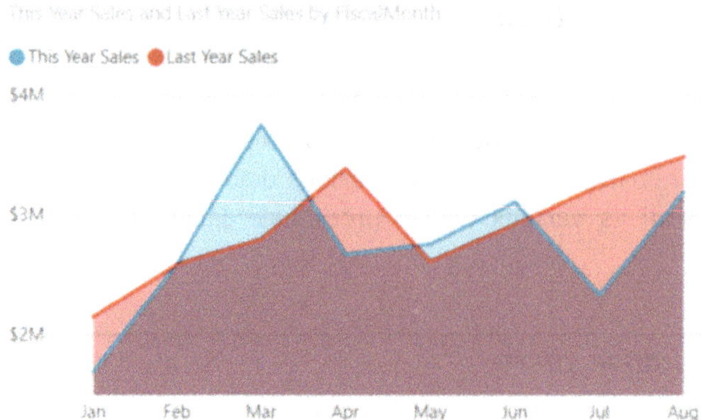

Figure 18.4 Area chart.

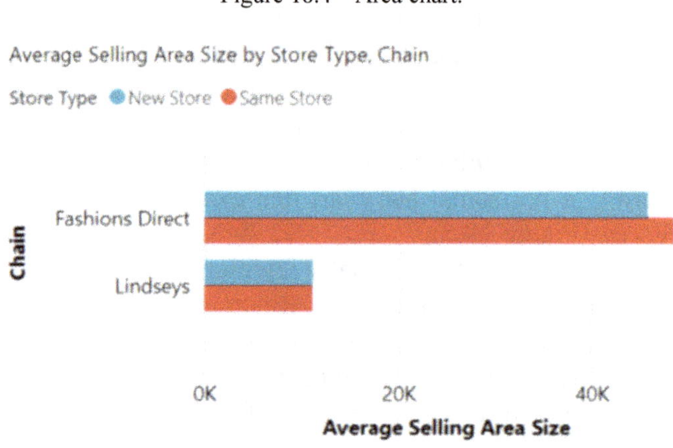

Figure 18.5 Horizontal bar chart.

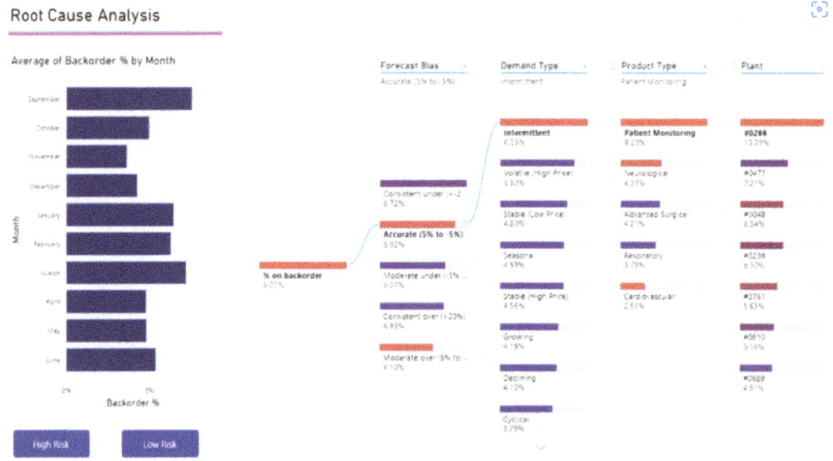

Figure 18.6 Decomposed tree chart.

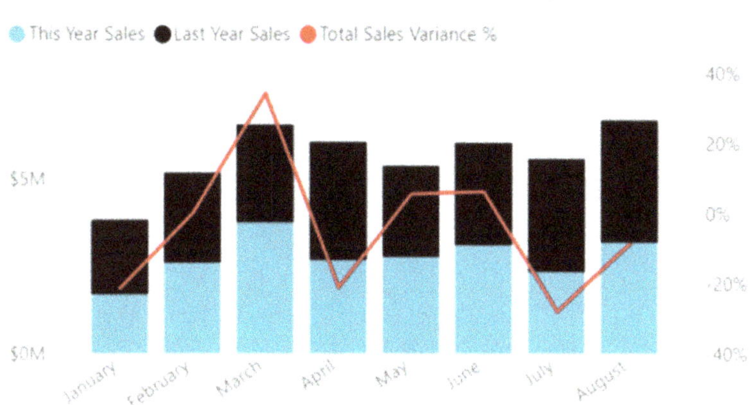

Figure 18.7 Combo chart.

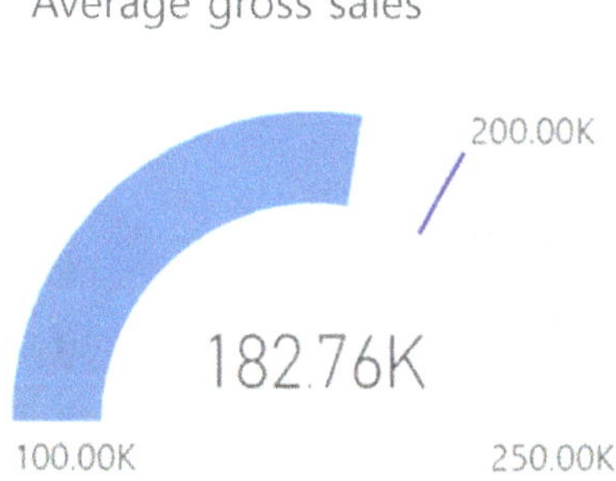

Figure 18.8 Instrument panel chart.

Combo charts combine vertical bar charts and line charts. Combining these two charts allows for faster comparison of data. Combo charts can have one or two Y-axes, so careful consideration is necessary (Figure 18.7).

A gauge chart displays a single value that measures progress towards a goal in the form of a circular shape. The goal or target value is represented by a line, and the progress towards the goal is indicated by shading. The value representing the progress is displayed boldly within the arc. All possible values are evenly distributed along the arc from the minimum value to the maximum value (Figure 18.8).

A funnel-shaped chart helps visualize the process that flows sequentially from one step to the next. The funnel shape shows the status of the process being tracked at a glance, and each funnel stage represents a percentage of the total (Figure 18.9).

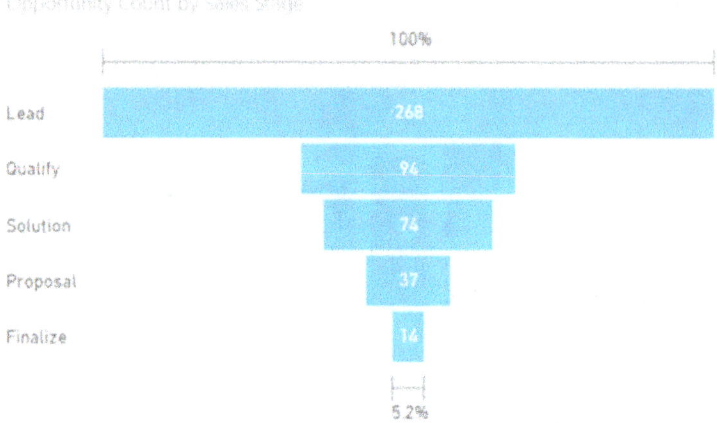

Figure 18.9 Funnel-shaped chart.

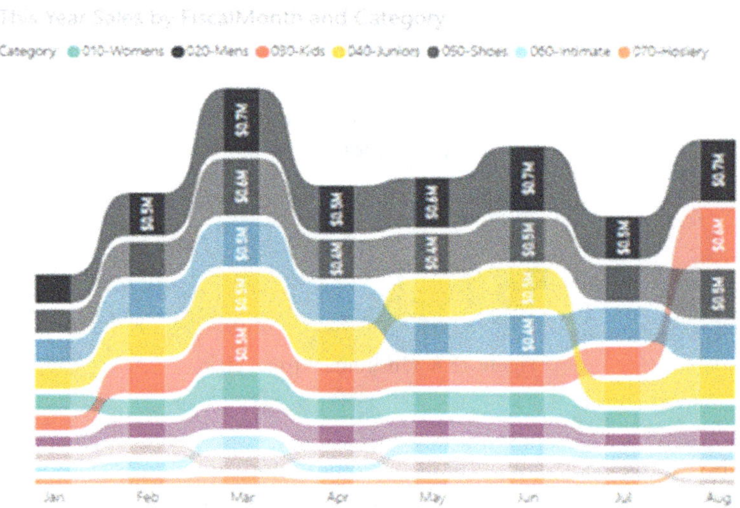

Figure 18.10 Ribbon chart.

The ribbon chart shows the largest value among the data categories with the highest rankings, and always displays the highest range value at the top, effectively showing changes in rankings. The ribbon shows how the values of the data categories change over the visualized period, and by connecting category values across a time continuum, it makes it easy to see increases or decreases. Through visualization, you can determine how a specified category ranks across the entire range of the chart's X-axis (typically the timeline) compared to other categories (Figure 18.10).

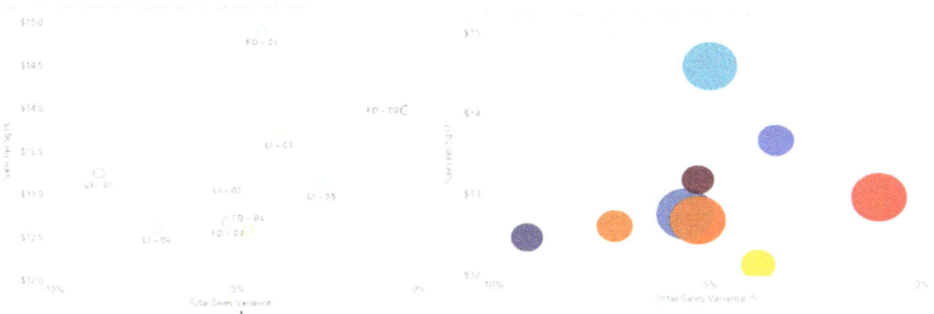

Figure 18.11 Scatter and bubble charts.

Figure 18.12 Treemap chart.

In scatter plots, there are always two value axes — one along the horizontal axis to represent a set of numerical data and another along the vertical axis to represent a different set of numerical values. The intersection of x and y numerical values in the chart is marked with points, and these values are combined into a single data element. Depending on the data, these data elements can be distributed evenly or unevenly along the horizontal axis. Bubble charts replace data elements with bubbles and use bubble size to indicate an additional dimension of the data. Both scatter plots and bubble charts can also include a playback axis to indicate changes over time. Apart from being able to plot numerical or categorical data along the x-axis, dot plot charts are similar to bubble charts and scatter plots (Figure 18.11).

A treemap is a chart consisting of colored squares with sizes that represent values. This chart can have a hierarchical structure format where squares are nested within the main square. The space inside each square is allocated based on the measured values. The squares are arranged in size order, with the largest square in the top left and the smallest square in the bottom right (Figure 18.12).

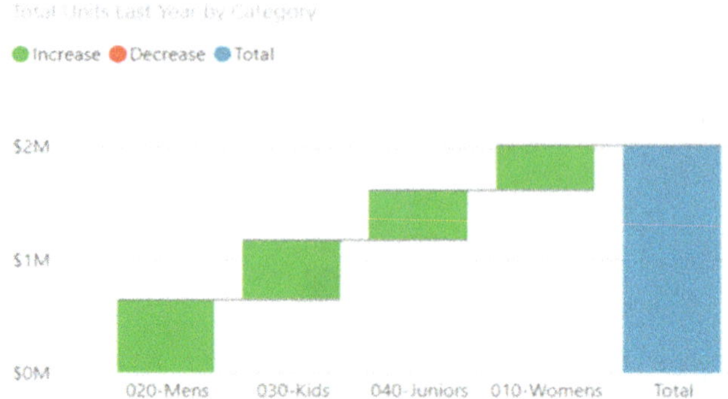

Figure 18.13 Waterfall chart.

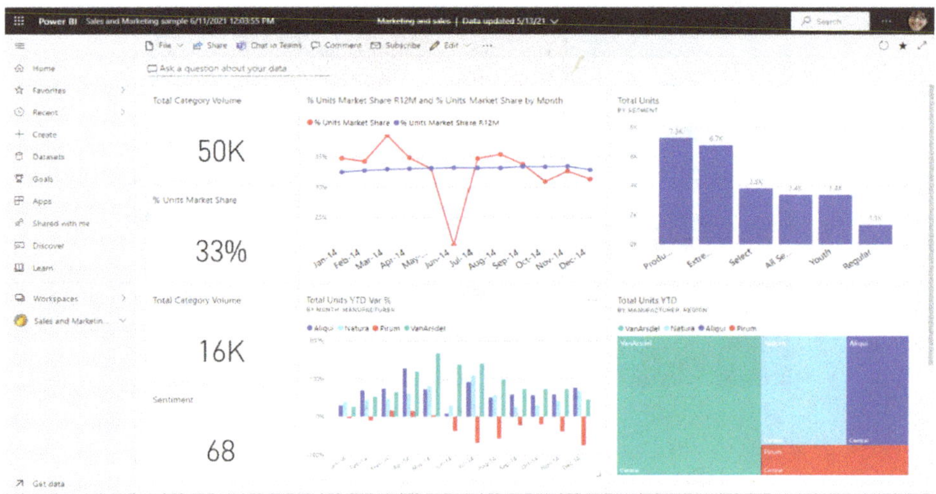

Figure 18.14 Visualization: Power BI report.

A waterfall chart shows the cumulative effect when values are added or subtracted. It is useful for understanding how the initial value is affected by a series of positive and negative changes. Vertical bars are distinguished by color, making it easy to differentiate between increases and decreases. While the intermediate value column is represented by floating vertical bars, the initial and final value columns often start from the horizontal axis (Figure 18.13).

18.3.3 *Report*

A Power BI report is one or more pages consisting of visual objects and graphic text. All visualizations in the report are provided from a single dataset, and one report can be connected to multiple dashboards for use (Figure 18.14).

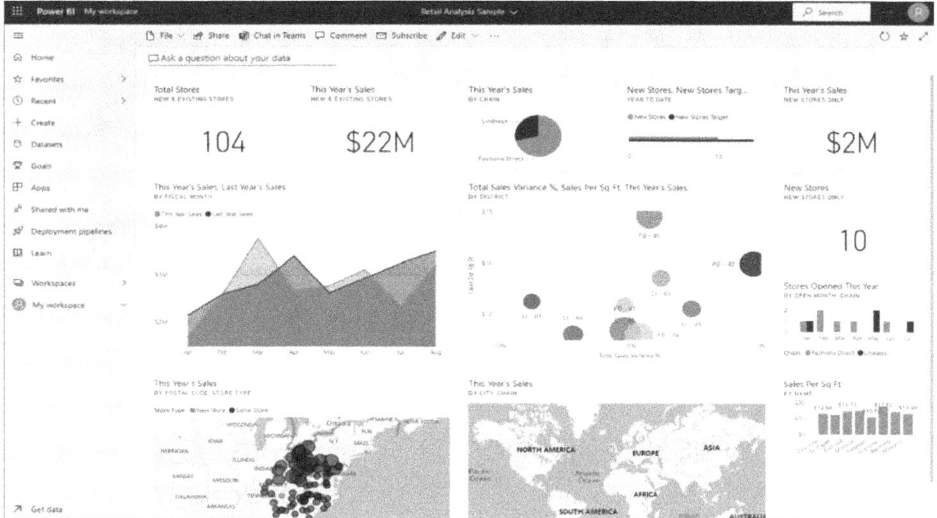

Figure 18.15 Visualization: Power BI dashboard.

18.3.4 *Dashboard*

A dashboard is used to quickly view the necessary information for decision-making or to monitor the most important information in business. A dashboard is a visual object, including tiles, graphics, and text, contained in a single Canvas that can be fixed from reports to dashboards. A dashboard can display visualizations of multiple datasets or multiple reports and can include tiles or entire pages (Figure 18.15).

18.4 Power BI Sign-Up

In this chapter, let's learn about how to sign up for Power BI. As you can see in Figure 18.16, you can sign up by clicking on "Free Trial" in the top right corner of the screen.

First enter your email address to check whether or not you need to create a new Power BI account to enter your email or not.

After entering your email address, as shown in Figure 18.17, the window shown in Figure 18.18 will appear if you need to create a new account.

As shown in Figure 18.19, you need to input the information displayed in the account creation window to create a Power BI account.

After completing all the steps in Figure 18.20, click on the "Start" button.

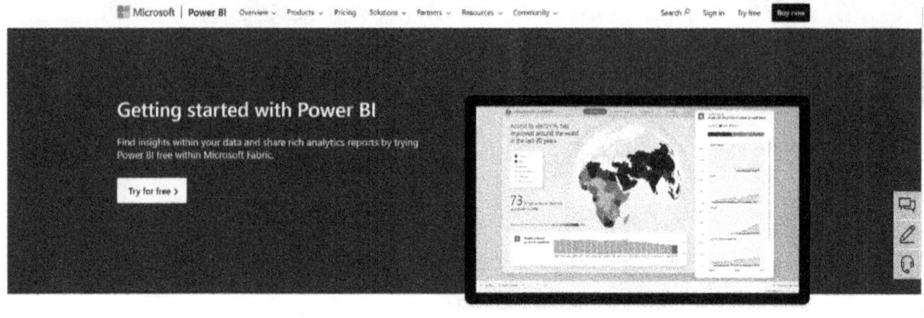

How to try Power BI for free

Figure 18.16 Main screen of Power BI.

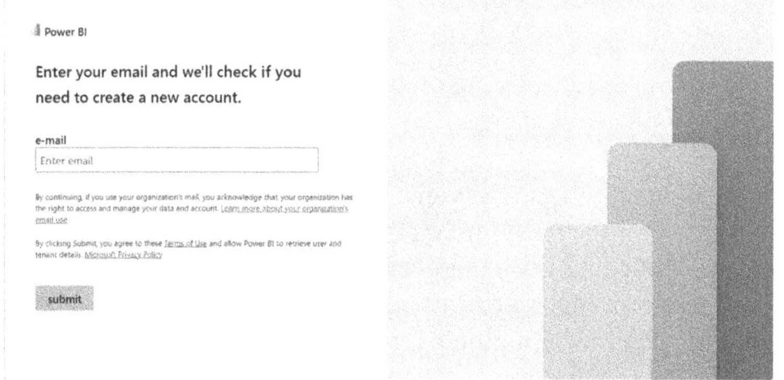

Figure 18.17 Power BI mail entry window.

Microsoft

Microsoft Fabric free selected

1 Go!

If you enter your work or school email address, we will check if you need to create a new account for Microsoft Fabric free.

E-mail

This item is required

If you continue, you acknowledge that your organization has the right to access and manage your data and accounts if you use your organization's mail.

More information

next

2 Create an Account

3 Check Details

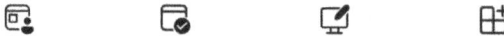

Figure 18.18 Begins membership registration.

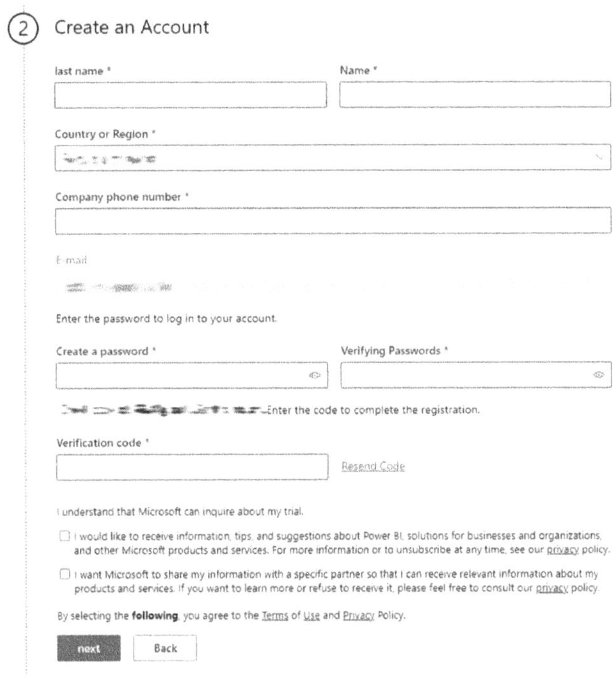

Enter the password to log in to your account.
Create a password *
Verifying Passwords *
Enter the code to complete the registration.

Resend Code
I understand that Microsoft can inquire about my trial.
I would like to receive information, tips, and suggestions about Power BI, solutions for businesses and organizations, and other Microsoft products and services. For more information or to unsubscribe at any time, see our privacy policy.
I want Microsoft to share my information with a specific partner so that I can receive relevant information about my products and services. If you want to learn more or refuse to receive it, please feel free to consult our privacy policy.
By selecting the **following**, you agree to the Terms of Use and Privacy Policy.

Figure 18.19 Creating a membership registration account.

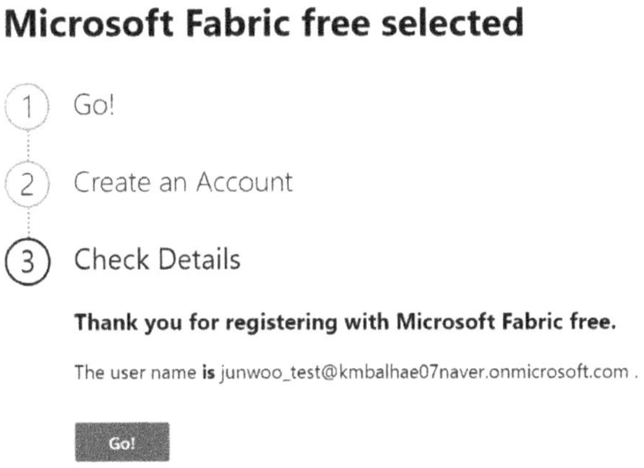

Microsoft Fabric free selected
1 Go!
2 Create an Account
last name *
Company phone number *
3 Check Details
Thank you for registering with Microsoft Fabric free.
The user name **is** junwoo_test@kmbalhae07naver.onmicrosoft.com .
Go!

Figure 18.20 Member registration detailed information confirmation.

When you click on the "Start" button shown in Figure 18.20, you will be able to see Power BI on the web, as shown in Figure 18.21, and you can confirm that the membership registration has been successfully completed.

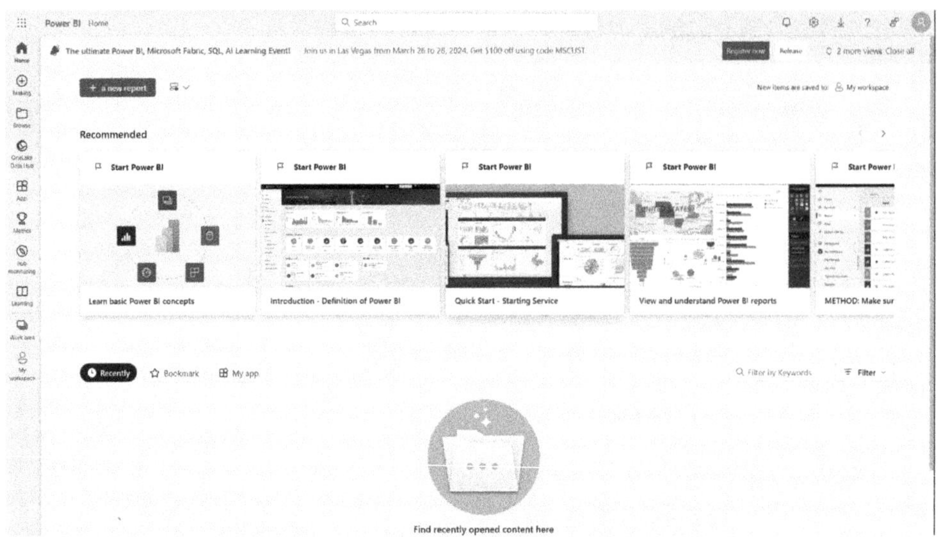

Figure 18.21 Completed membership account creation.

18.5 Power BI Desktop Download

When you log in to the Power BI home screen, you can configure datasets to create reports or dashboards. Additionally, you can easily share and collaborate with users on dashboards and even refresh data. When you go to the official website of Power BI (https://powerbi.microsoft.com), you will see the screen shown in Figure 18.22. Then, at the top of the screen, click on the "Products" tab, and then click on "Power BI," then "Desktop" to download.

The latest version of Power BI Desktop can be obtained from the Microsoft Store or downloaded to your computer as a single executable file that includes support for all languages.

To download Power BI Desktop as shown in Figure 18.23, click "Download free" in the center of the screen to navigate to the download screen.

Once the download is complete, simply click on the "Open" tab shown in Figure 18.24 to start Power BI Desktop.

With Power BI Desktop, you can perform the following tasks.

1. You can connect data by including various data sources.
2. You can share data by using queries that create data models containing insightful information.
3. You can create visualizations and reports using a data model.
4. Files can be shared for other users to utilize, build, and share.

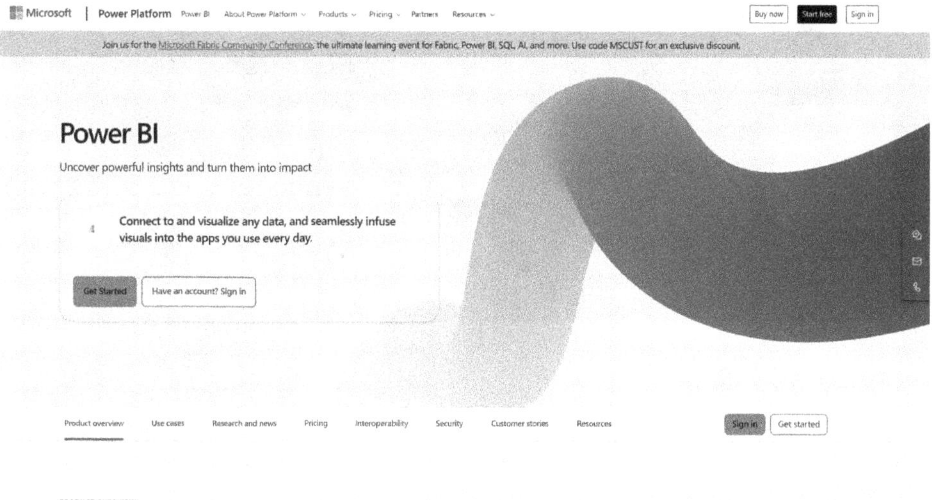

Figure 18.22 Power BI official website.

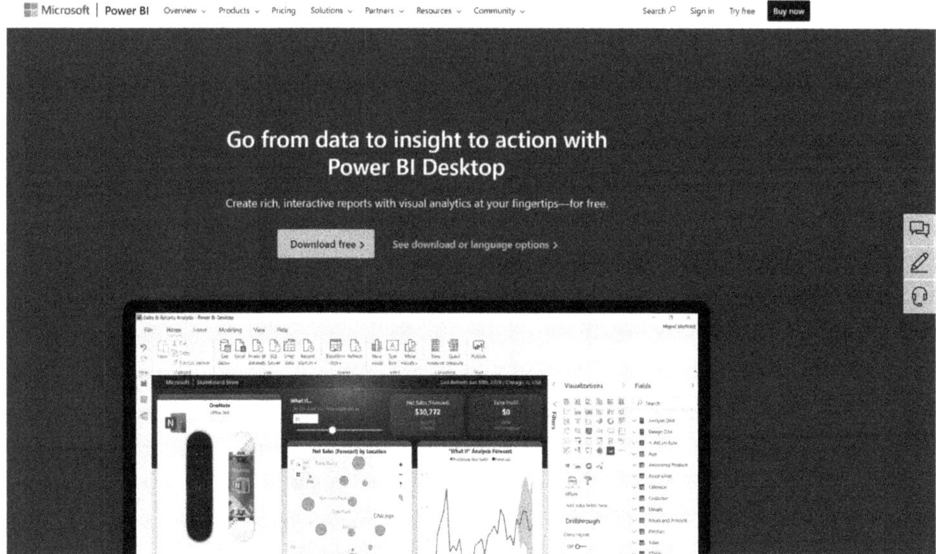

Figure 18.23 Power BI download page.

Power BI Desktop integrates Microsoft Query Engine, data modeling, and visualization technologies. Data analysts and other users can create queries, data connections, models, and collections of reports, and easily share them with other users.

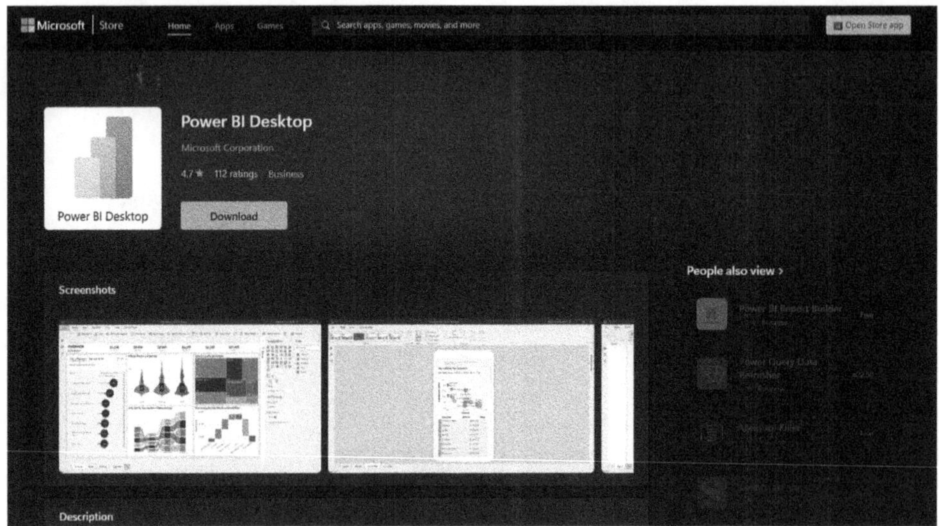

Figure 18.24 Power BI desktop download store.

18.6 Practice Questions

Q1. How does Power BI enable efficient and collaborative report creation and sharing within an organization?

Q2. What are the key components of Power BI and how do they contribute to the analytics process?

Q3. Describe the significance of Power BI's data protection features in maintaining organizational data security.

Chapter 19

Power BI Desktop Tutorial

When you run Power BI Desktop, the start screen shown in Figure 19.1 will appear. From the start screen, you can choose to import data, open another report, or select other links.

After clicking the start button shown in Figure 19.1, input your email address, as shown in Figure 19.2, to proceed with the login process.

After entering your email address, click on your company or school account and press continue, as shown in Figure 19.3. Enter the email and password you previously registered with to log in. If you have logged in, you can confirm the successful login by looking at the name of the user account displayed in the upper right corner of the screen, as shown in Figure 19.3.

19.1 Data Connection

After installing Power BI Desktop, you'll have the capability to access a broad spectrum of data sources. To explore the available data source types, go to the Home tab and click on "Get Data > More." In the "Get Data" window, choose the "All" option to see the full list of data sources. For a practical example, let's learn how to import and manage web data. Simply navigate to the Home tab, click on the "Get Data" button as illustrated in Figure 19.4, and then select "Web" to begin importing your data.

When the dialogue box appears, paste the URL in the URL text field, and click "OK" (https://www.bankrate.com/retirement/best-and-worst-states-for-retirement/) (Figures 19.5 and 19.6).

When a message is displayed on the access web content screen use anonymous authentication and select "Connect." Power BI Desktop runs the query function and connects to web resources. The explorer window returns the contents found on the web page. If you want to understand information about an HTML table, select the table to display a preview. At this point, you can choose to upload

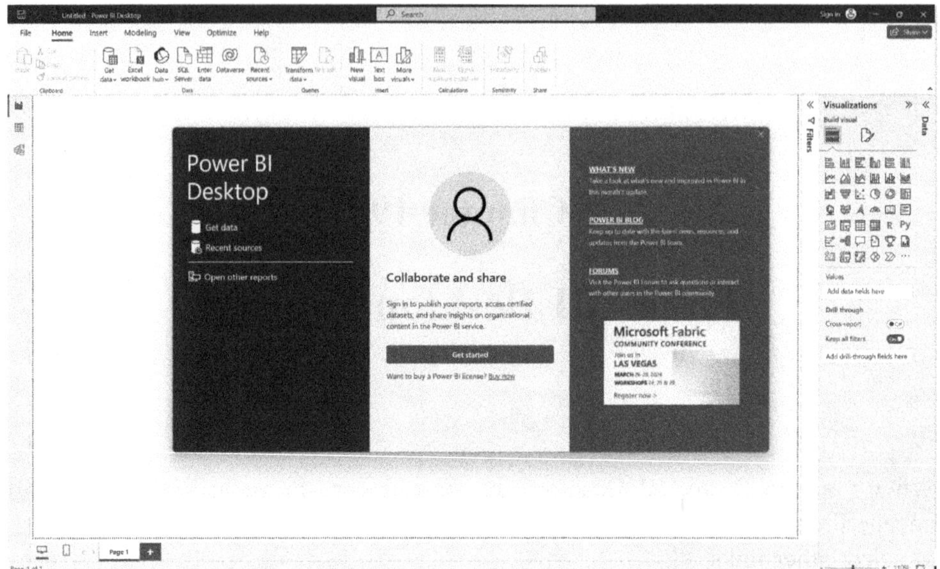

Figure 19.1 Main screen of Power BI Desktop.

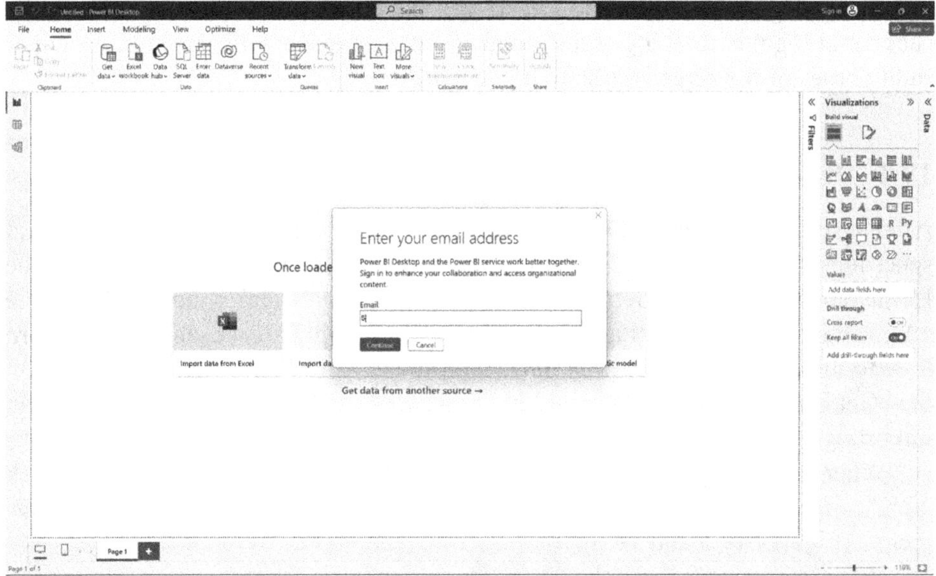

Figure 19.2 Power BI Desktop login.

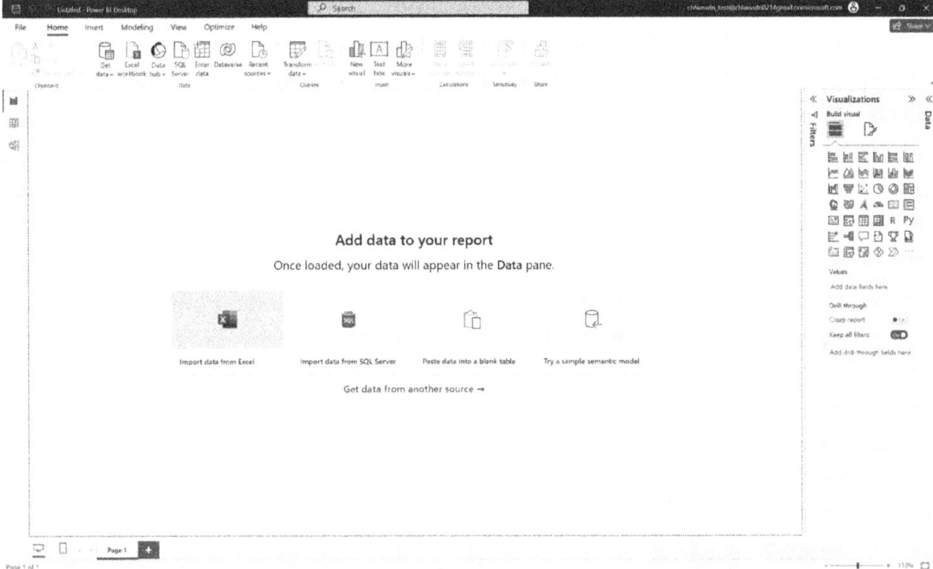

Figure 19.3 Power BI Desktop login complete.

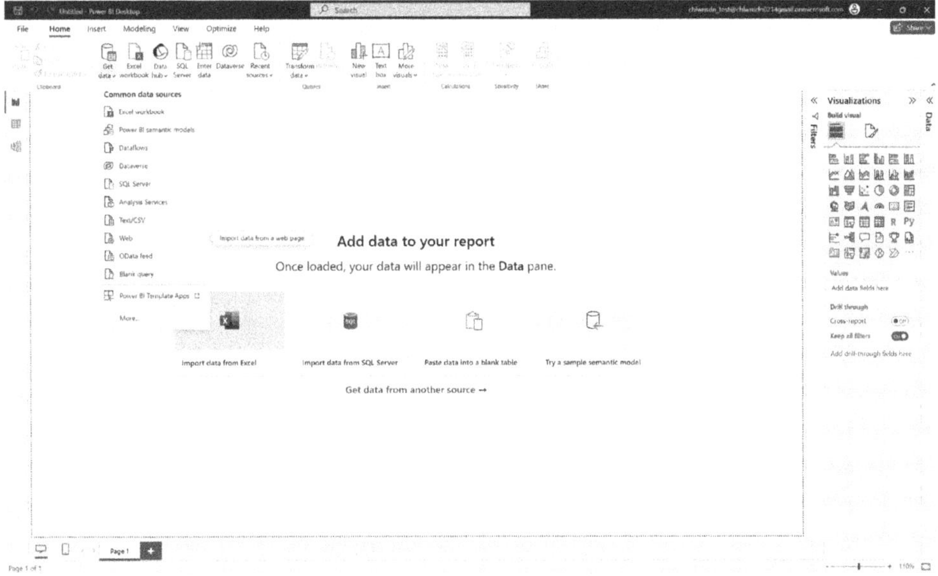

Figure 19.4 Importing data in Power BI.

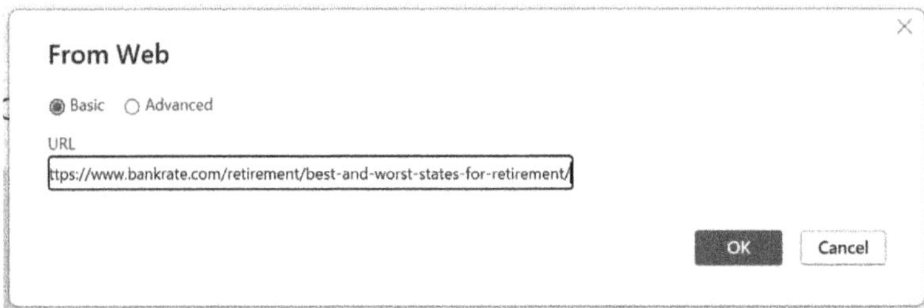

Figure 19.5 Importing web data in Power BI.

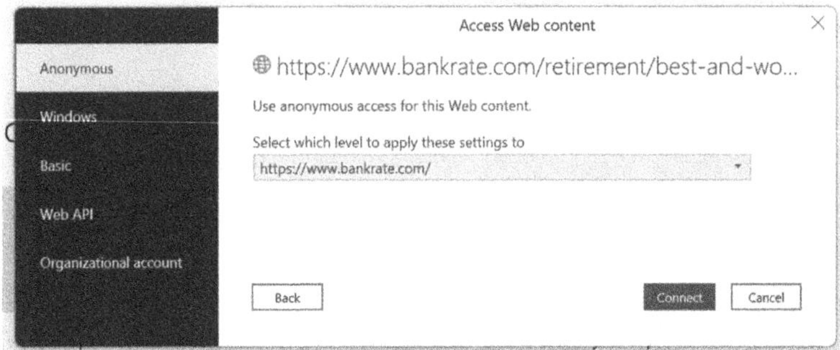

Figure 19.6 Power BI web content access.

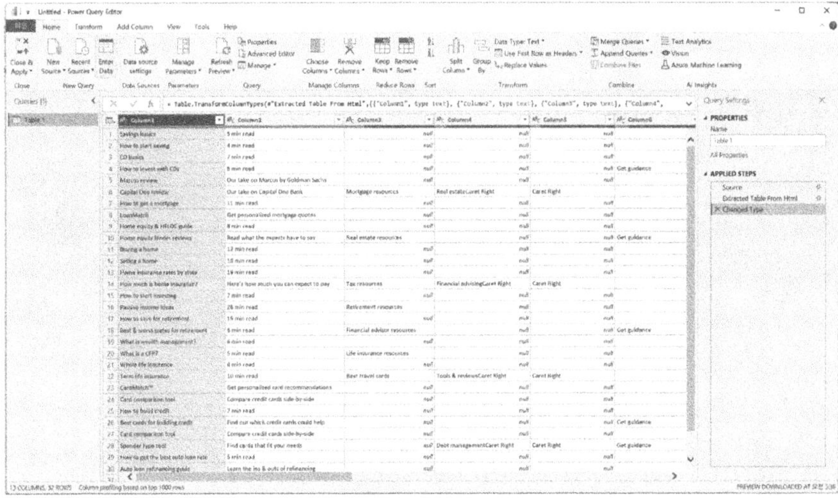

Figure 19.7 Power BI web data.

the table or select table transformation to make changes before uploading it. If you select data transformation, the Power Query Editor starts and the default table view is displayed. The query settings window can be on the right side or you can always select query settings on the "View" tab of the Power Query Editor to display it (Figure 19.7).

19.2 Data Formatting

Since it is connected to the data source, you can specify the data format according to the requirements. To adjust the data format, first upload the data and adjust it while displaying it. Then utilize the functions available in Power Query Editor. In Power BI, the adjustments made in Shaping only apply to specific data views without affecting the original data source. The Power Query Editor captures the steps in sequential order below the Applied Steps in the Query Settings window. As these steps are performed each time the query connects to the data source, the data is always adjusted in the specified manner. This process occurs when using queries in Power BI Desktop or when using shared queries in Power BI services, among others. The Applied Steps in the query settings already include several steps. Each step can be selected to see the effects in the Power Query Editor. First, specify the web source, then execute the table preview in the Navigator pane. In the third step, which is “Changed Type,” Power BI recognized the integer data when importing data and automatically changed the original web text data format to an integer. If you need to change the data format, select the column(s) to be changed. You can select multiple adjacent columns while holding the Shift key or select non-adjacent columns while holding the Ctrl key. Right-click on the column header, select “Change Type,” then choose the new data format from the menu or drop down the list next to the data format in the Transform group on the Home tab and select the new data format (Figure 19.8).

Now, you can apply unique changes and transformations to the data and see the applied steps. Select Row Decrease > Row Removal > Sub row Removal on

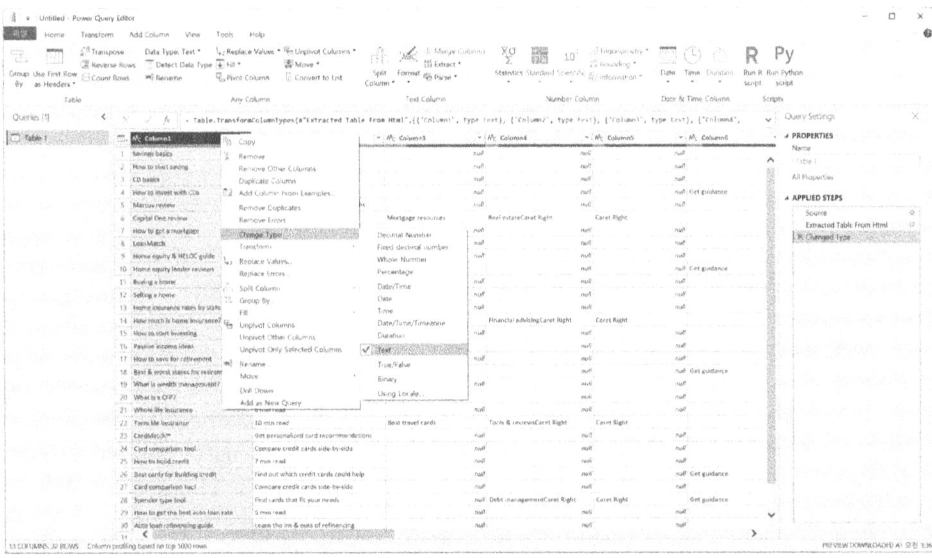

Figure 19.8 Change web data format in Power BI.

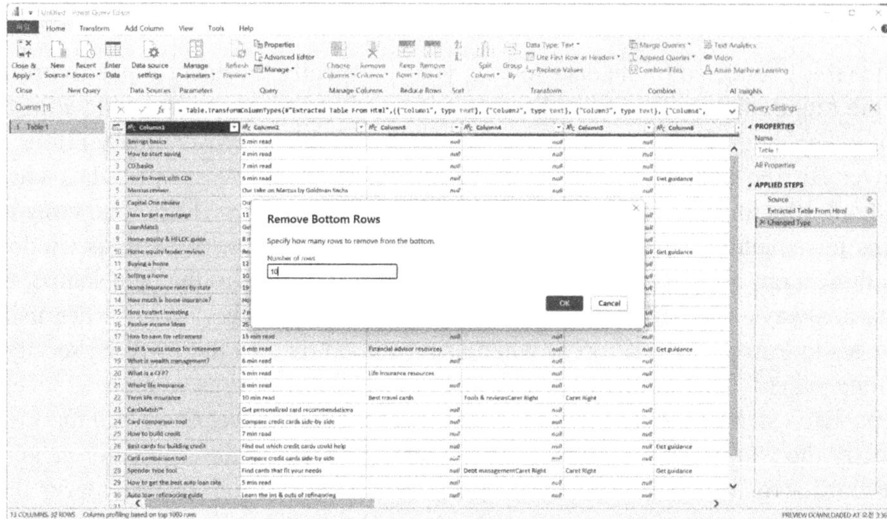

Figure 19.9 Removing sub-rows of Power BI web data.

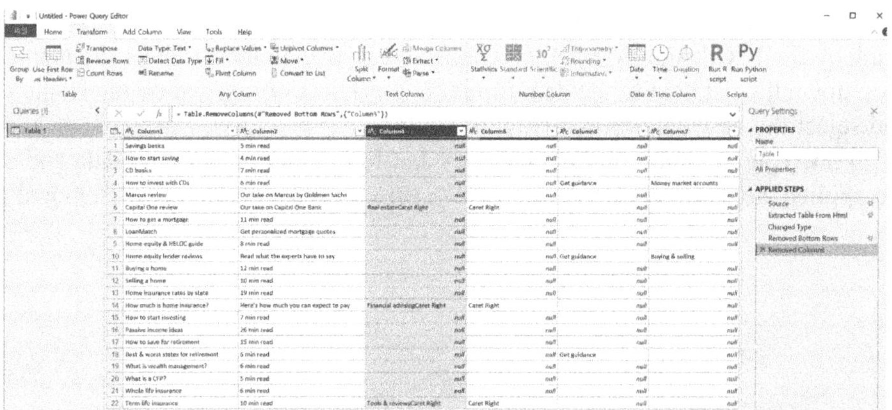

Figure 19.10 After remove sub-rows in Power BI web data.

the Home tab. In the Sub row Removal dialog box, enter 10 and select OK. The bottom 10 rows will be removed from the table, and the applied step will be displayed in the removed sub row step (Figure 19.9).

There is more information on the table than required, so remove the column for Annal Average Temperature. To do this, select the header of the column you want to remove, then hold down the Shift key and select adjacent columns or hold down the Ctrl key and select non-adjacent columns. Then, choose "Remove Column" in the Column Management group on the Home tab. Alternatively, right-click on one of the selected column headers and choose "Remove Column" from the menu. The selected columns will be removed, and the column removal steps will be applied (Figure 19.10).

If you want to bring back the corresponding step, you can select the delete icon next to the step to undo the last step applied in the Applied Steps pane. Now, you can reapply the step by selecting only the column you want to delete, and for better flexibility, you can delete each column as individual steps. From the Applied Steps pane, you can right-click on a step to delete it, rename it, move it up or down in the sequence, or add or remove steps after it. In the middle steps, if the changes affect subsequent steps and if there is a possibility that the query will be interrupted, Power BI Desktop will display a warning.

19.3 Power BI Practice

This practice exercise contains an introduction to some of the features of the Power BI service. Through this exercise, let's learn how to connect data in a dashboard and create reports and dashboards (Figure 19.11).

1. The process of connecting data and generating reports and dashboards is as follows:
2. Import Power BI sample data.
3. Take some data and open it in the report view.
4. Create visualizations using the data and save them as a report.
5. Fix the tiles in the report to create a dashboard.
6. Add different visualizations to the dashboard using the Q&A tool.

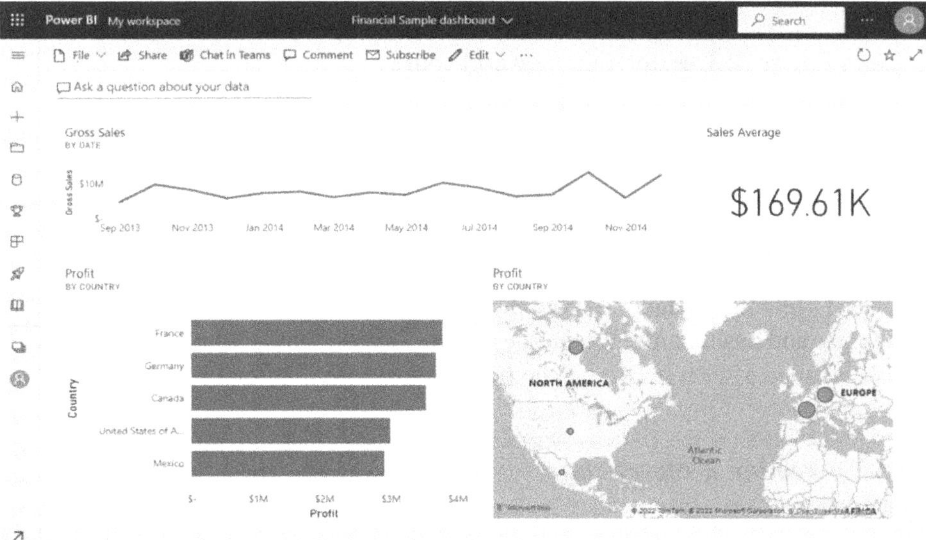

Figure 19.11 Power BI sample data.

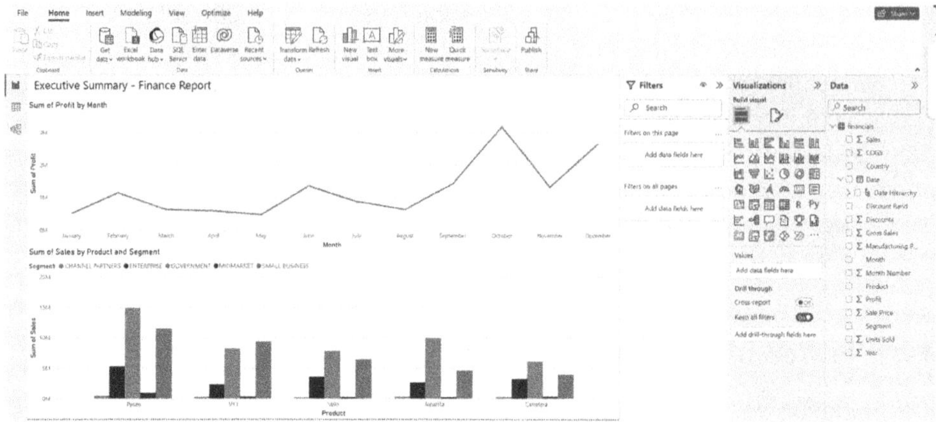

Figure 19.12 Financial sample report.

7. Adjust the size of the tiles on the dashboard, reorganize the tiles, and edit detailed information.
8. Clean up resources by deleting datasets, reports, and dashboards.

19.3.1 *Importing data*

If you want to create a Power BI report, it is common to start with Power BI Desktop. Power BI Desktop provides various features. For example, before starting the report design, you can transform and shape the data. The process of importing data to start the practice is as follows (Figure 19.12):

1. Start Power BI Desktop (app.powerbi.com).
2. Select my workspace in the exploration window.
3. In your workspace, select "New" > "Upload File" and the "Import Data" page will open.
4. In the "Create New Content" section, select File > Local File, and then choose the location where the Excel file is saved.
5. After finding the file on the computer, select "Open."
6. Select "Import" to add the Excel file as a dataset. Then, it can be used to create reports and dashboards. When you select "Upload," the entire Excel workbook is uploaded to Power BI and can be edited by opening it in Excel Online.

19.3.2 *Creating a report chart*

Since it is connected to the data, start by searching. If the user finds the desired content, they can save it to the report Canvas. Then, it can be pinned to the dashboard to monitor and see how it changes over time.

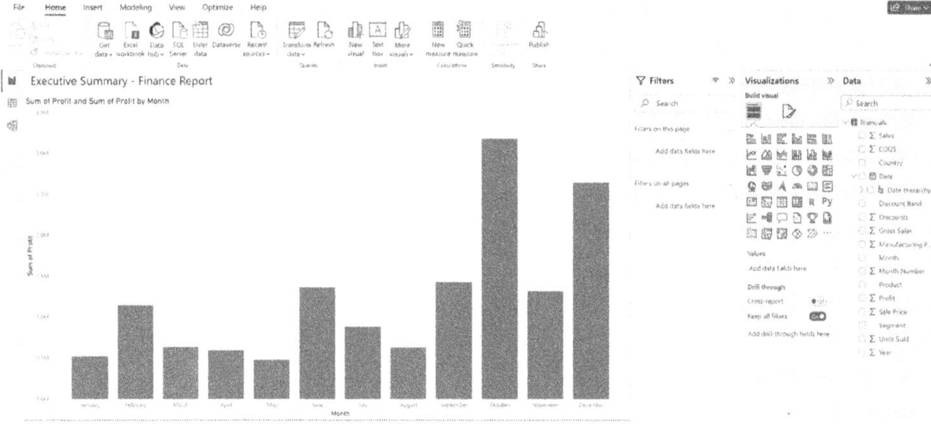

Figure 19.13 Vertical bar chart.

Figure 19.14 Broken line chart.

1. Construct a visualization using the field pane on the right side of the report editor. Select the "Gross Sales" field, then select the "Data" field. Analyze the data in Power BI and create a new bar chart visualization. If you selected the "Data" field instead of the "Gross Sales" field first, a table will be displayed (Figure 19.13).
2. Let's switch to a different way of representing data. Line graphs are useful visualizations for displaying values over time. Select the line graph icon in the visualization window (Figure 19.14).
3. Let's fix the visualization chart to the dashboard. Point your cursor to the visualization and select the pin icon appearing above or below to fix it. Alternatively, click "..." next to the "Show Data Table" tab in the top right corner of the screen and click "Fix Dashboard" (Figure 19.15).

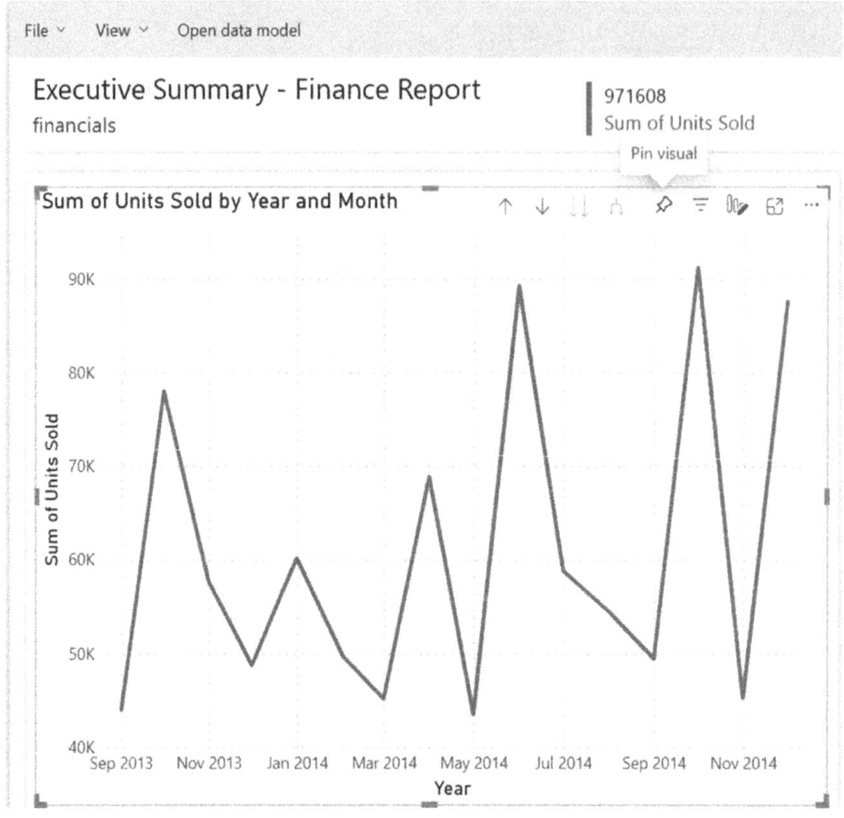

Figure 19.15 Dashboard fix settings.

4. This report is a new item, so a message will be displayed first indicating that it needs to be saved in order to be pinned to a dashboard for visualization. Name the report as "Financial Sample Report" or something similar, and select save. The report will be displayed in the reading view (Figure 19.16).
5. Select the fixed icon again, choose a new dashboard, and name it "Financial Sample Dashboard" (Figure 19.17).
6. Select "Move to Dashboard" to view the line chart, which is fixed as a tile on a new dashboard (Figure 19.18).

Since the visual object has been fixed, the visual object is now stored on the dashboard. As the data is maintained in its latest state, it is possible to track the latest values at a glance. However, if the visualization format of the report is changed, the visualization on the dashboard will not be altered (Figure 19.19).

7. Select a new tile on the dashboard. Then, Power BI returns from the reading view to the report view.

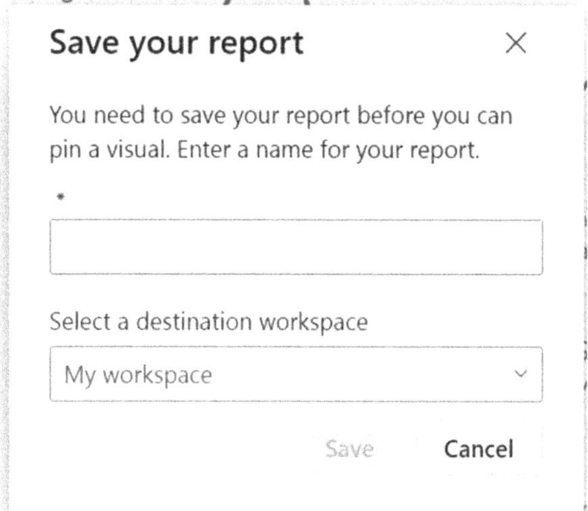

Figure 19.16 Save financial sample report=.

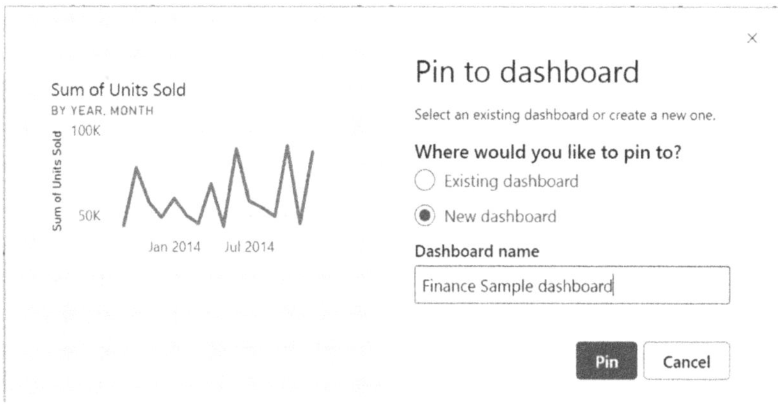

Figure 19.17 Setting the dashboard name=.

8. To switch back to the editing view, select "Edit" from the additional options (...) in the menu bar. By returning to the editing view, you can continue exploring and fixing the tiles.

19.3.3 *Examining using Q&A*

To quickly examine the data, ask questions in the Q&A question box. With Q&A, you can create natural language queries for the data. In the dashboard, the Q&A

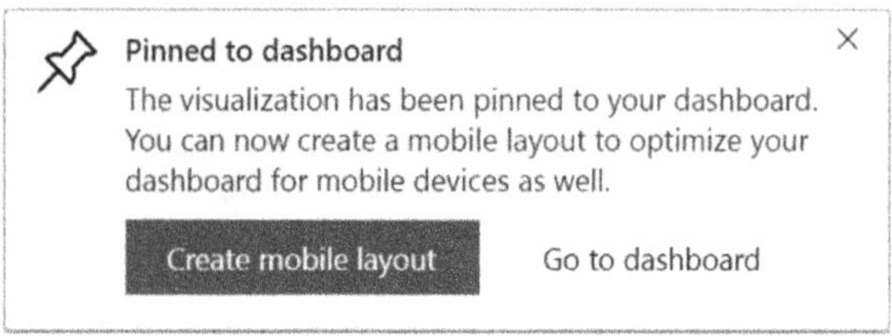

Figure 19.18 Dashboard movement screen.

Figure 19.19 Financial sample graph.

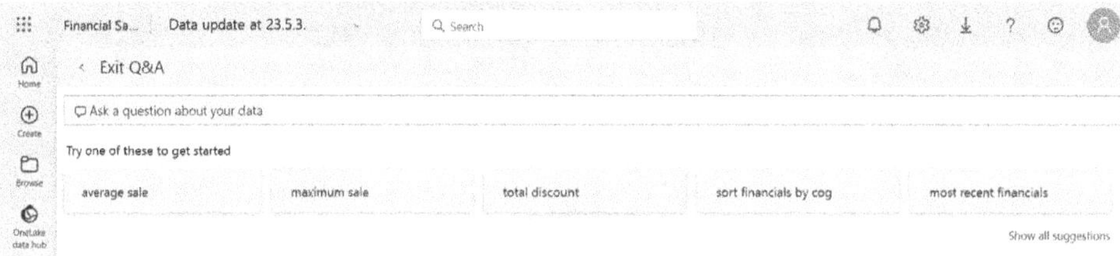

Figure 19.20 Q&A text field.

box is located under the top menu bar, and in the report, it is located in the top menu bar. The steps for using Q&A to explore the data are as follows:

1. To return to the dashboard, select the header "My Workspace" a on the black Power BI bar.
2. Select the dashboard in My Workspace.
3. Choose to ask questions about the data. Q&A automatically provides several suggestions (Figure 19.20).

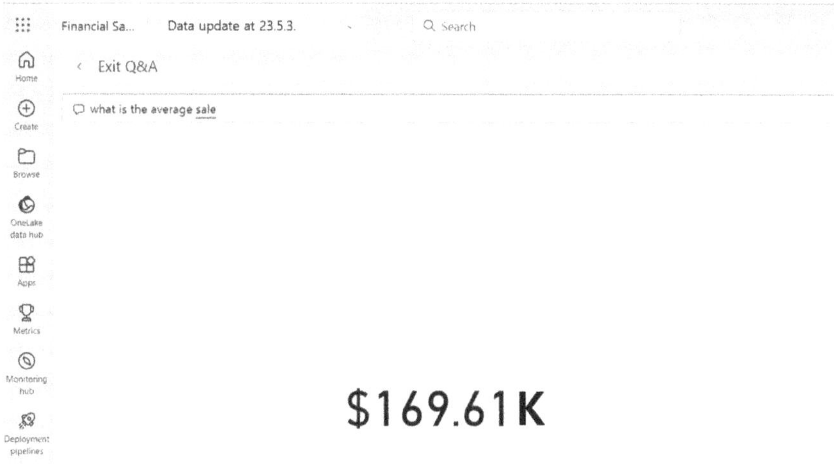

Figure 19.21 Average sales amount.

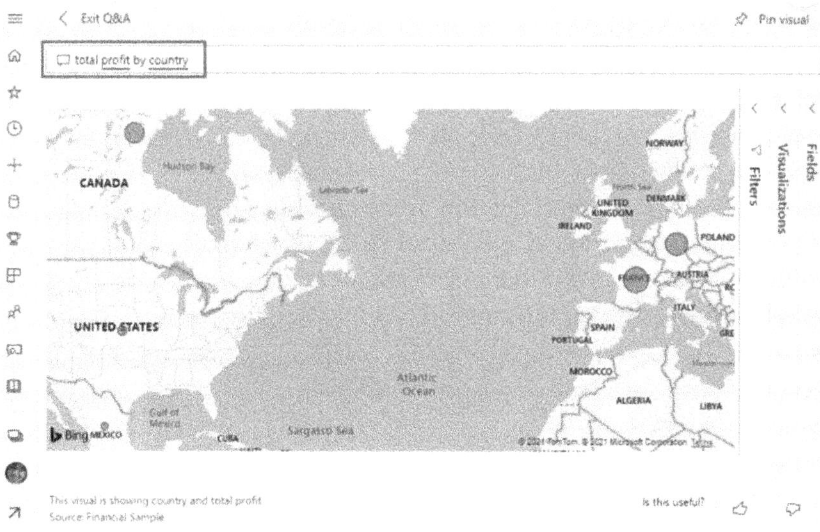

Figure 19.22 Total profits by country.

4. Some suggestions return a single value. For example, if you select “what is the average sale” it will search for the answer in Q&A and provide it in the form of card visualization.
5. Selecting visual objects to fix, lock this visualization on the financial sample dashboard (Figure 19.21).
6. Go back to Q&A and enter “total profit by country” (Figure 19.22).

Figure 19.23 Bar graph of total profits by country.

7. The map is also fixed on the financial sample dashboard.
8. Select the recently pinned map from the dashboard. You can see how to reopen Q&A.
9. Place the cursor behind "by country" in the Q&A box and type "as bar." A bar chart with the results will be created in Power BI (Figure 19.23).
10. Fix bar charts in the financial sample dashboard.
11. If you select to end Q&A and return to the dashboard, a newly created tile will be displayed. Even if you changed the map to a bar chart in Q&A, you can still see that the tile remained as a map. This is because the map was fixed (Figure 19.24).

19.3.4 *Tile position change*

In Power BI, you can rearrange tiles to efficiently utilize dashboard space.

1. Pull up the bottom right corner of the total sales line chart tile until it aligns with the height of the average sales tile, and then release it (Figure 19.25).
2. Select "Edit Details" for additional options about average sales by tile (Figure 19.26).
3. Enter "average sales" in the title box, then click "Apply."
4. Realign the other visual objects together (Figure 19.27).

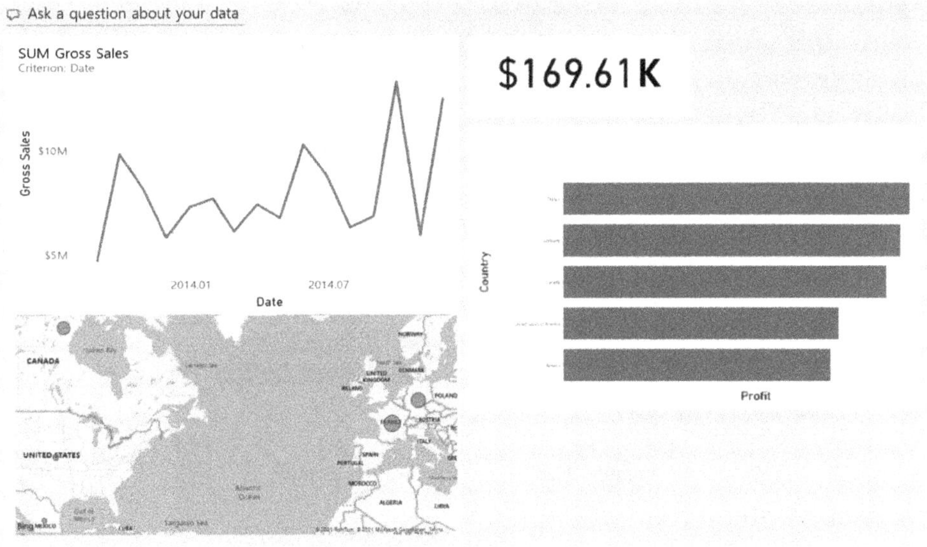

Figure 19.24 Final financial sample dashboard.

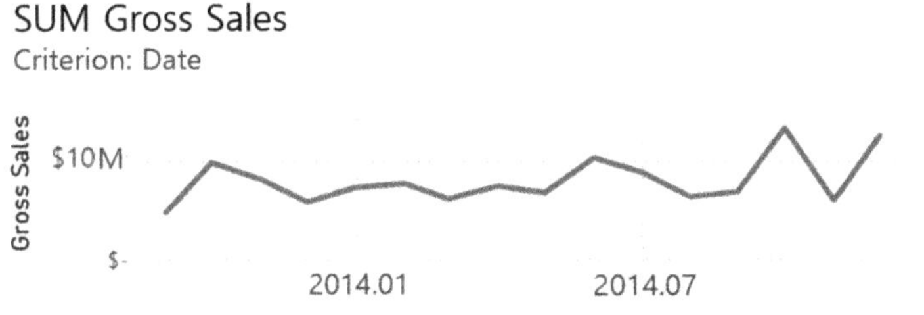

Figure 19.25 Gross sales.

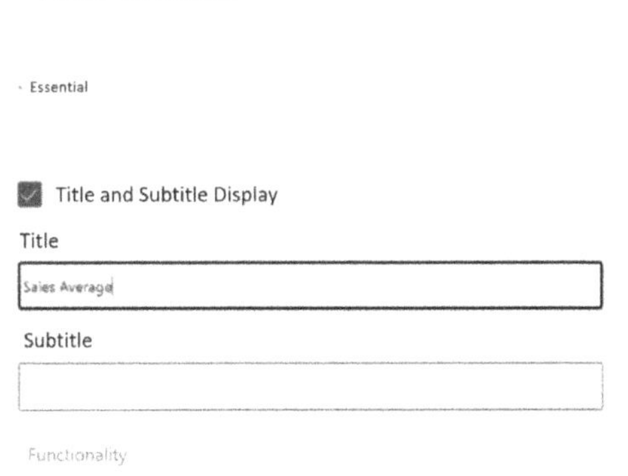

Figure 19.26 Tile detail editing.

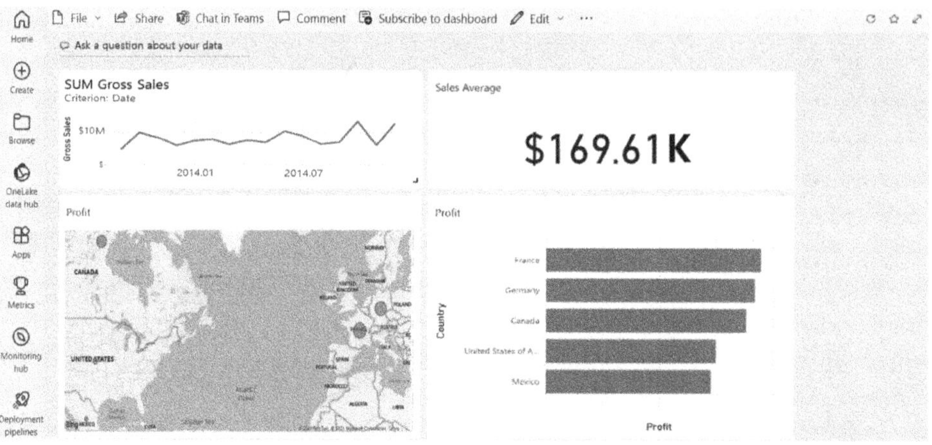

Figure 19.27 Final tile position change.

19.4 Practice Questions

Q1. How does Power BI Desktop facilitate the creation of visual reports and dashboards?

Q2. What steps are involved in importing and preparing data for analysis in Power BI Desktop?

Q3. How can the Q&A feature in Power BI Desktop be used to enhance data analysis?

Bibliography

ACM SIGKDD. (2006, April 30). Data Mining Curriculum. Retrieved from https://www.kdd.org/curriculum/view/data-mining-curriculum. Last modified January 27, 2014.

Altman, N. S. 1992. An Introduction to Kernel and Nearest-Neighbor Nonparametric Regression. *The American Statistician*, 46(3), 175–185.

Appier. (n.d.). A Simple Guide to Unsupervised Learning. Retrieved from https://www.appier.com/ko-kr/blog/a-simple-guide-to-unsupervised-learning. Last modified February 4, 2023.

BBC News. (n.d.). Artificial Intelligence: Google's AlphaGo Beats Go Master Lee Se-dol. Retrieved from https://www.bbc.com/news/technology-35785875. Last modified March 12, 2016.

Ben-Hur, A., Horn, D., Siegelmann, H., & Vapnik, V. N. (2001). Support Vector Clustering. *Journal of Machine Learning Research*, 2, 125–137.

BEOMSU KIM. AlphaGo: Dissecting the AlphaGo Pipeline. Retrieved from https://shuuki4.wordpress.com/2016/03/11/alphago-alphago-pipeline-%ED%97%A4%EC%A7% 91%EA%B8% B0/. Last modified March 11, 2016.

Cortes, C. & Vapnik, V. N. (1995). Support-Vector Networks. *Machine Learning*, 20(3), 273–297.

DeepMind. (n.d.). AlphaGo. Retrieved from https://deepmind.com/research/case-studies/alphago-the-story-so-far.

DeepMind. (n.d.). AlphaGo. Retrieved from https://deepmind.google/technologies/alphago/. Last modified December 23, 2020.

Didu-story. (n.d.). Types and Brief Summary of Loss Functions (feat. keras & pytorch). Retrieved from https://didu-story.tistory.com/27. Last modified March 4, 2021.

DoorBW. (n.d.). Basics of numpy. Retrieved from https://doorbw.tistory.com/171. Last modified June 7, 2018.

Educative. (n.d.). Overfitting and Underfitting. Retrieved from https://www.educative.io/answers/overfitting-and-underfitting. Last modified April 30, 2020.

Excelsior-cjh. (n.d.). [Learning TensorFlow] Chap01 — What Is TensorFlow?. Retrieved from https://excelsior-cjh.tistory.com/148. Last modified May 31, 2018.

FiscalNote. (n.d.). Retrieved from https://fiscalnote.com/press-room/fiscalnote-enters-ai-data-annotation-market-with-new-product-datahunt. Last modified February 25, 2020.

Google Research Blog. (n.d.). Research Blog: AlphaGo: Mastering the Ancient Game of Go with Machine Learning. Retrieved from https://research.googleblog.com/2016/01/alphago-mastering-ancient-game-of-go.html. Last modified January 27, 2016.

happy-obok. (n.d.). Obok Learning Artificial Intelligence. Retrieved from https://happy-obok.tistory.com/55. Last modified February 25, 2022.

Hoonst. (n.d.) Regression Metric. Retrieved from https://hoonst.github.io/2020/09/22/Regression-Metric/. Last modified September 22, 2020.

Jo, Y. I. (2012). *Artificial Intelligence Systems*. Honglong Science Publishing, Korea.

Kaggle. (n.d.). (2018). Fashion MNIST. Retrieved from https://www.kaggle.com/datasets/zalando-research/fashionmnist. Last modified December 7, 2017.

Kim, E. J. (2017). *Introduction to Artificial Intelligence, Machine Learning, and Deep Learning with Algorithms*. WikiBooks.

Kim, S. P. (2016). *First Steps in Deep Learning*. Hanbit Media, Korea.

K-MOOC. (2020). K-MOOC for General Public: Principal Component Analysis and Artificial Neural Networks. Retrieved from http://matrix.skku.ac.kr/math4ai-intro/W13/. Last modified December 18, 2020.

Laboputer. (2020-04-25). Understanding How to Use Numpy. Retrieved from https://laboputer.github.io/machine-learning/2020/04/25/numpy-quickstart/. Last modified April 25, 2020.

Marsland, S. (2014). *Machine Learning: An Algorithm Perspective*, Second Edition. CRC Press, Florida, USA.

McCarthy, J., Minsky, M. L., Rochester, N., & Shannon, C. E. (2006). A Proposal for the Dartmouth Summer Research Project on Artificial Intelligence. *AI Magazine*, 27(4), 12–12.

McCorduck, P. (2004). *Machines Who Think*, Second Edition. A.K. Peters, Ltd., USA.

Medium. (n.d.). [Deep Learning/Machine Learning] Understanding CNN (Convolutional Neural Networks) Easily. Retrieved from https://ireneban.medium.com/%EB%94%A5%EB%9F%AC%EB%8B%9D-%EB%A8%B8%EC%8B%A0%EB%9F%AC%EB%8B%9D-cnn-convolutional-neural-networks-%EC%89%BD%EA%B2%8C-%EC%9D%B4%ED%95%B4%ED%95%98%EA%B8%B0-836869f88375. Last modified October 27, 2020.

Metz, C. (n.d.). In Major AI Breakthrough, Google System Secretly Beats Top Player at the Ancient Game of Go. Wired. Retrieved from https://www.wired.com/2016/01/in-a-huge-breakthrough-googles-ai-beats-a-top-player-at-the-game-of-go/. Last modified January 27, 2016.

Michalski, R. S., Carbonell, J. G., & Mitchell, T. M. (Eds.). (2013). *Machine Learning: An Artificial Intelligence Approach*. Springer Science & Business Media, Germany.

Microsoft. (n.d.). Choosing Algorithms for Azure Machine Learning. Retrieved from https://docs.microsoft.com/ko-kr/azure/machine-learning/studio/algorithm-choice. Last modified March 4, 2024.

Naver Blog. (n.d.). Learning Machine Learning/Deep Learning through Tutorials. Retrieved from https://m.blog.naver.com/ckdgus1433/221599517834. Last modified July 30, 2019.

Naver Blog. (n.d.). What Is TensorFlow? Retrieved from https://m.blog.naver.com/sincc0715/221809045450. Last modified February 14, 2020.

Naver Blog. (n.d.). Miscellaneous Exploration — Cham Study GodGo. Retrieved from https://blog.naver.com/samsjang/220948258166. Last modified February 25, 2022.

NeedJarvis. (2022, January 31). Explanation and Types of Metrics. Retrieved from https://needjarvis.tistory.com/568. Last modified August 3, 2020.

NeedJarvis. (n.d.). Need Jarvis. Retrieved from https://needjarvis.tistory.com/715. Last modified January 31, 2022.

Nillson, N. J. (1998). *Artificial Intelligence*: A New Synthesis. Elsevier, Netherlands.

NVIDIA Korea. (n.d.). Understanding the Differences between Artificial Intelligence, Machine Learning, and Deep Learning. Retrieved from http://blogs.nvidia.co.kr/2016/08/03/difference_ai_ learning_machinelearning/. Last modified August 3, 2016.

oooops.log. (n.d.). Neurons and Perceptrons. Retrieved from https://velog.io/@oooops/%EC%8B%A0%EA%B2%BD%EC%84%B8%ED%8F%AC%EC%99%80-%ED%8D%BC%EC%85%89%ED%8A%B8%EB%A1%A0. Last modified February 14, 2022.

OpenAI. (n.d.). Text Generation Models. Retrieved from https://platform.openai.com/docs/guides/gpt. Last modified January 20, 2024.

Park, Haesun (Translator). (n.d.). Neural Networks. Retrieved from https://ml4a.github.io/ml4a/ko/neural_networks/. Last modified April 17, 2023.

Park, J. H. (n.d.). Starting Data Science by Park Junghyun: Performance Evaluation. Retrieved from https://www.aitimes.com/news/articleView.html?idxno=137087. Last modified June 21, 2021.

Ratsgo. (2017-04-02). Ratsgo's Blog for Textmining: Logistic Regression. Retrieved from https://ratsgo.github.io/machine%20learning/2017/04/02/logistic/. Last modified April 2, 2017.

rubber-tree. (n.d.). Cat Maze. Retrieved from https://rubber-tree.tistory.com/entry/%EB%94%A5%EB%9F%AC%EB%8B%9D-%EB%AA%A8%EB%8D%B8-CNN-Convolutional-Neural-Network-%EC%84%A4%EB%AA%85. Last modified July 11, 2021.

Russell, S. J. & Norvig, P. (n.d.). Artificial Intelligence: A Modern Approach. Retrieved from https://people.eecs.berkeley.edu/~russell/intro.html. Last modified August 22, 2022.

Saito, G. (2017). *Deep Learning from Scratch*. Hanbit Media, Korea.

Samatas, M. (2022). *Actionable Insights with Amazon QuickSight: Develop Stunning Data Visualizations and Machine Learning-Driven Insights with Amazon QuickSight*. Packt Publishing, UK.

Shalev-Shwartz, S. & Ben-David, S. (2014). *Understanding Machine Learning: From Theory to Algorithms*. Cambridge University Press, UK.

SK C&C Blog. (n.d.). Who Is Watson? Retrieved from https://blog.skcc.com/2808. Last modified March 25, 2022.

SlideShare. (n.d.). Microsoft Build: Choosing Algorithms for Azure Machine Learning. Retrieved from https://www.slideshare.net/medit74/ss-74123546?qid=5a76f350-f606-4cb0-aa56-7527fc1d7a67&v=&b=&from_search=2. Last modified April 1, 2017.

SNUAC. (n.d.). Special Feature: Future of Super AI (6) Utilizing GPT Book for Smart Administration. Retrieved from https://snuac.snu.ac.kr/?p=41791. Last modified August 7, 2023.

Steph Su Reads. (2011). Jeopardy, IBM Challenge, and What It Means for AI. Retrieved from http://stephsureads.blogspot.com/2011/02/jeopardy-ibm-challenge-and-what-it.html. Last modified February 25, 2011.

Stock, Data, Dev. (n.d.). Retrieved from http://pubdata.tistory.com/134. Last modified June 9, 2017.

Team AI Korea. (n.d.). Recurrent Neural Network (RNN) Tutorial — Part 1. Retrieved from https://aikorea.org/blog/rnn-tutorial-1/. Last modified September 17, 2015.

Towards Data Science. (n.d.). Understanding the ROC Curve Iin Three Visual Steps. Retrieved from https://towardsdatascience.com/understanding-the-roc-curve-in-three-visual-steps-795b1399481c. Last modified June 12, 2020.

Wikipedia. (n.d.). Cross-Validation. Retrieved from https://en.wikipedia.org/wiki/Cross-validation_(statistics). Last modified March 8, 2024.

Wikipedia. (n.d.) IBM Watson. Retrieved from https://en.wikipedia.org/wiki/Watson_(computer). Last modified January 19, 2024.

Wikipedia. (n.d.). Monte Carlo Tree Search. Retrieved from https://en.wikipedia.org/wiki/Monte_Carlo_tree_search. Last modified January 29, 2024.

woogong80. (n.d.). Stories of Non-Developers Working in IT. Retrieved from https://woogong80.tistory.com/67. Last modified September 23, 2022.

Yang, K. C. (2014). *Theory and Practice of Artificial Intelligence.* Hongreung Science Publishing, Korea.

Yngie-C. (n.d.). Data Science. Retrieved from https://yngie-c.github.io/machine%20learning/2020/04/30/training_test_reg/. Last modified April 30, 2020.

www.ingramcontent.com/pod-product-compliance
Lightning Source LLC
Chambersburg PA
CBHW081041220326
41598CB00038B/6954